DATE DUE

OC 28 '04			
DE 17 04			

DEMCO 38-296

D1601277

FUNDAMENTAL STRUCTURES OF ALGEBRA AND DISCRETE MATHEMATICS

FUNDAMENTAL STRUCTURES OF ALGEBRA AND DISCRETE MATHEMATICS

STEPHAN FOLDES

A Wiley-Interscience Publication
JOHN WILEY & SONS, INC.
New York   /   Chichester   /   Brisbane   /   Toronto   /   Singapore

QA 155 .F65 1994

Foldes, Stephan, 1951-

Fundamental structures of
  algebra and discrete

This text is printed on acid-free paper.

Copyright ©1994 by John Wiley & Sons, Inc.

All rights reserved. Published simultaneously in Canada.

Reproduction or translation of any part of this work
beyond that permitted by Section 107 or 108 of the
1976 United States Copyright Act without the permission
of the copyright owner is unlawful. Requests for
permission or further information should be addressed to
the Permissions Department, John Wiley & Sons, Inc.

Library of Congress Cataloging in Publication Data:
Foldes, Stephan, 1951–
    Fundamental structures of algebra and discrete mathematics/by
  Stephan Foldes.
        p.   cm.
    "A Wiley-Interscience publication."
    Includes bibliographical references and index.
    ISBN 0-471-57180-6 (cloth)
    1. Algebra. I. Title.
QA155.F65   1993
512'.02–dc20   93-8787

Printed in the United States of America

10 9 8 7 6 5 4 3 2 1

In token of my respect and gratitude
this book is dedicated to
Ilonka and Sándor Szász

CONTENTS

ILLUSTRATIONS xi

PREFACE xiii

I. SETS 1

1. Elementary Constructions and Axioms, 1
2. Cardinal and Ordinal Numbers, 11
3. Intersections, 21

   Bibliography, 28

II. ORDERED SETS 31

1. Relations, Orders, and Zorn's Lemma, 31
2. Lattices and Closures, 43
3. Covering Relations, 49
4. Intersecting Convex Sets, 54

   Bibliography, 56

III. GROUPS   59

1. Binary Operations, Homomorphisms, and Congruences, 59
2. Permutation Groups, 68
3. Integers and Cyclic Groups, 75
4. Alternating Groups, 82

   Bibliography, 90

IV. RINGS   93

1. Ideals, 93
2. Polynomials, 104
3. Factorization and the Euclidean Algorithm, 111

   Bibliography, 123

V. FIELDS   125

1. Rational and Real Numbers, 125
2. Galois Groups and Imaginary Roots, 137

   Bibliography, 154

VI. VECTOR SPACES   157

1. Bases, 157
2. Linear Maps and Equations, 167
3. Affine and Projective Geometry, 176
4. Hyperplanes in Linear Programming, 183
5. Time and Speed in Special Relativity, 187

   Bibliography, 194

VII. GRAPHS   197

1. Trees and Median Graphs, 197
2. Games, 202

3. Chromatic Polynomials, 207
Bibliography, 210

VIII. LATTICES — 213
1. Complements and Distributivity, 213
2. Boolean Algebra, 230
3. Modular and Geometric Lattices, 237
Bibliography, 242

IX. MATROIDS — 245
1. Linear and Abstract Independence, 245
2. Minors and Tutte Polynomials, 252
3. Greedy Optimization Procedures, 262
Bibliography, 268

X. TOPOLOGICAL SPACES — 269
1. Filters, 269
2. Closure, Convergence, and Continuity, 272
3. Distances and Entourages, 281
Bibliography, 288

XI. UNIVERSAL ALGEBRAS — 291
1. Homomorphisms and Congruences, 291
2. Algebra of Syntax, 295
3. Truth and Formal Proof, 300
Bibliography, 308

XII. CATEGORIES — 311
Bibliography, 324

INDEX OF DEFINITIONS — 327
INDEX OF NOTATION — 339
INDEX OF THEOREMS — 343

ILLUSTRATIONS

2.1 Symmetric, antisymmetric, and transitive relations, 34
2.2 A relation and its dual, 36
2.3 An order and some of its linear extensions, 51
2.4 Activity precedence order and project schedules, 53
3.1 Integers, 76
3.2 Integers modulo 3, 4, and 5, 80
4.1 Divisibility order of positive integers, 112
4.2 Divisibility order in a power set ring, 112
6.1 Orthogonality, 171
6.2 The Fano plane, 181
6.3 Photons, causalities, and material motion, 191
7.1 A graph on four vertices with four edges, 198
7.2 A graph with two connected components, 199
7.3 A median graph, 201
7.4 A strategic equilibrium, 204
8.1 A bounded lattice in which some elements have no complement, 215
8.2 Nondistributive lattices, 219
8.3 A distributive lattice, 228
9.1 Computing the Tutte polynomial, 258

9.2 A matching with three edges, 265
10.1 Continuity and discontinuity, 276
11.1 A term structure tree, 299
12.1 Product and sum, 323

PREFACE

This book is about the algebraic notion of "structure". In mathematical thinking, a structure crystallizes whenever attention is focused either on combining elementary objects of some kind to form other objects of a similar kind, such as adding numbers to form new numbers, or on relating objects to each other, such as comparing numbers by magnitude. Instead of numbers, two points in space may be combined to define a line, a point and a line may be combined to define a plane, and these geometric objects may also be linked by relations such as inclusion and parallelism. Numbers represent a significant abstraction from whatever is being weighed or enumerated, and straight lines miss much of the reality of land surveying. However, algebra involves a second shift in interest, from the things combined to the ways of combination.

Numbers with addition constitute the historical archetype of algebraic structure. If negatives are included, we have a group; if not, we have a semigroup. If multiplication is taken into consideration as well as addition, then a more complex structure called a ring arises. There is the ring of integer numbers, and the ring of rational numbers, and so on. Most important for the algebraist is the realization that there are rings consisting of objects that are not numbers at all; objects that can be added and multiplied and that obey rules such as $a(b + c) = ab + ac$.

This volume presents a basic theory of groups and rings and other algebraic structures. Like most algebra texts, it has a chapter on fields and one on vector spaces. Like some more recent texts, it includes lattices and universal algebras. The classical number systems $\mathbb{Z}$, $\mathbb{Q}$, $\mathbb{R}$ and $\mathbb{C}$ find their *raison d'être* in abstract algebra, and not the other way round.

The machinery is based on sets, order relations, and closure operators. All mathematical objects are defined in terms of sets. The entire theory is derived from nine set-theoretical axioms. Sets can also be viewed as the simplest kind of all structures. They are the subject of Chapter I.

There are two reasons to study order relations in algebra. First, more or less obvious order relations present in various algebraic structures provide simple explanations of what is going on. Second, the study of ordered sets, as a kind of structure, can be undertaken in the same spirit as the study of structures with a law of composition. To a lesser extent the same is true for graphs.

Binary operations more general than those of groups are needed to discuss ring multiplication, lattices, and word concatenation. Partial binary operations are needed for categories. However, in the last chapter on categories we do not enter into any generalization that is not directly relevant to the material in the preceding chapters, despite (or rather because of) all the new algebra that could be thus presented.

The fundamental role that closure operators play in algebra led us to view matroids and topological spaces as structures of an algebraic nature. This is why we devoted a separate chapter to each, rather than confining them to subsidiary treatment under "geometric lattices" and "filters in Boolean lattices."

Students of algebra and researchers in other areas will find in this book an introduction to, or a clarification of, the basic theories of the twelve kinds of structures. A comprehensive exposition of each particular theory is not the aim of this text. Rather, we seek to identify essentials and to describe interrelationships between particular theories. We hope that the specialist of commutative algebra will find matroids worth the reading and that the student of discrete mathematics will find special relativity close to his or her own field.

The material is self-contained. The reader need not know any mathematical definitions, results, or methods. However, the pace and density of exposition corresponds to those of graduate texts.

Selected advanced results are derived from weak rather than strong hypotheses, whenever this is compatible with the objective of simplicity. Also, several classical concepts are introduced relatively late, in order to demonstrate the simplicity of certain results established without the use of these concepts. Thus, Zermelo's Theorem is proved before set intersections are introduced, elementary group theory (including Lagrange's Theorem) is developed before the theory of integers, and a simple Galois theory is presented without calling on vector space dimensions. And if some major algebraic concept fails to appear altogether, the reader may conclude that it is not required for any of the theorems included in this short volume.

Throughout the text the student is frequently prompted to perform exercises of verification and to explore examples. These integrated exercises are indispensable for any reader not yet familiar with the theory. At the end of each section, there are additional numbered exercises from which to select. Many of these are open-ended questions in the sense that while a satisfactory answer can be given without much difficulty, there is ample room for better and more complete answers. (An exact science mathematics may well be, but mathematical research is not more deterministic than any other intellectual endeavor.)

Each chapter builds on key notions introduced in previous chapters. However, if you are already conversant with some of the structures, then you may go directly to selected chapters or even sections and use the index whenever you suspect a divergence between your definitions and ours.

In a course or seminar, the book should be used as reading material before or after verbal presentations. If the syllabus is limited to certain chapters only, the instructor should summarize for the students the definitions needed from the excluded chapters. In a seminar, we recommend the use of a Socratic approach, with analysis of examples and blackboard exercises, to probe the students' understanding of how constructs relate to theorems and structures to each other. The ultimate object of probing, however, is not progress in learning a science, but the tools and concepts of science itself. It is in the spirit of such questioning that the book was written.

CHAPTER I

SETS

1. ELEMENTARY CONSTRUCTIONS AND AXIOMS

The ability to think about collections of objects with precision and without ambiguity is indispensable in mathematics. Indeed, this is so in any exact science. Students of the physical world care about and count the collection of atoms in a given portion of matter, and chemists concern themselves with the collection of atoms forming a molecule, distinguishing various compounds according to what kind of molecules they contain and in what proportions. Biology in turn makes use of physics and chemistry and describes aggregates of millions of cells forming a tissue. Living and extinct creatures are classified into collections called species and their subcollections called varieties, and resembling species are grouped into families and kinds. Economists conceptualize and measure physical quantities of edible or otherwise useful goods, taking stock of grain, cattle, and money supply, distinguishing raw materials from work-in-process inventories, and discussing such issues as whether home-baked cakes should be included in or excluded from the gross national product. Linguists divide the collection of all words into subcollections such as nouns, pronouns, verbs, and adverbs, and they study small groups of words called sentences as they relate to the former subcollections.

2 SETS

No other science, however, relies as much on the conceptual manipulation of collections as mathematics.

It was already realized by the reflective Greeks of pre-Christian times that some restraint must be exercised in talking about collections. The liar's paradox of Crete goes as follows. Let collection T be the collection of all true sentences uttered on the island of Crete, and let F be the collection of untrue (false) sentences. Then let a mathematician take a boat to Crete and upon disembarking declare: "This very sentence I am pronouncing at this moment belongs to F." Despite the apparent rigor employed in defining T and F, the rules of logic seem to break down. If the mathematician's declaration is true, then the sentence being pronounced does indeed belong to F, and therefore it is false by definition of F. But if the declaration is false, then it must belong to F by definition, and it therefore becomes a true statement. Modern mathematics deals with the paradox by imposing very stringent rules on how collections should be defined. Requiring a precise definition of any mathematical object before making statements about that object is what lends mathematics its reliability. (Look at the contractual text of an insurance policy for an analogy.) Although very restrictive, the rules of definition still allow all usual mathematical objects to be defined in terms of collections. Indeed, the space of three-dimensional geometry will be a collection of vectors, of which points, lines, and planes will be subcollections. Moreover, the vectors themselves will be defined as collections, and the numbers 0, 1, 2, etc., will be formally defined as collections of a most particular kind.

The entire body of mathematical science can be viewed as a theory about collections called *sets*. By using the technical word "set," mathematicians simply indicate that they are talking about a collection and that they strongly believe that they know what they are talking about. Accordingly, the mathematician may wish to avoid the unregulated word "collection." To what extent this suffices to exempt mathematics from the fundamental uncertainty that affects human knowledge is open to debate. First, every mathematical discourse has a small number of primitive concepts that are not defined rigorously but are used in the formal definition of more elaborate concepts. Second, a few simple mathematical propositions are based on belief and observation rather than proof, yet they serve as the very foundation of all further theory. Having disclosed these risk factors, let us proceed.

An object x belonging to a set S is termed an *element*, or *member*, of S, in symbols $x \in S$. If the object x does not belong to S, we write $x \notin S$. Sets are completely determined by their elements, i.e., if two sets A and B have the same elements, then the two sets are the same. In this case we write $A = B$. If A and B are not the same, then we write $A \neq B$. If all elements of A are also elements of B, then A is called a *subset* of B, in symbols $A \subseteq B$. We also say that B is a *superset* of A and write $B \supseteq A$. The negation of $A \subseteq B$ is written $A \nsubseteq B$. Trivially, every set B has at least one subset, for $B \subseteq B$. A subset A of B is a *proper subset* if $A \neq B$, and we then write $A \subset B$. The following axiomatic propositions are adopted, without proof, entirely on the basis of their intuitive plausibility.

(A1) Empty Set Axiom. *There is a set $\emptyset$ which has no element.*

The set $\emptyset$ is called the *empty set* (or *null* or *void* set). Clearly $\emptyset \subseteq A$ for every set A.

(A2) Subset Axiom. *If A is any set, then those elements x of A that satisfy some given condition or possess a given property form a set.*

To designate "the set of those elements x of A that satisfy a certain condition," we usually write $\{x \in A : x$ satisfies a certain condition$\}$. This is of course a subset of A. In practice, the original superset specification "$x \in A$" is often indicated in other, less explicit ways.

(A3) Power Set Axiom. *For any set A, there is a set $\mathcal{P}(A)$, whose elements are all the subsets of A.*

The set $\mathcal{P}(A)$ is called the *power set* of A. Occasionally we write simply $\mathcal{P}A$ instead of $\mathcal{P}(A)$.

Since $A \subseteq A$, we have $A \in \mathcal{P}(A)$, and it follows from (A2) that $\mathcal{P}(A)$ has a subset containing A as unique element. This subset of $\mathcal{P}(A)$ will be denoted by $\{A\}$ and called a *singleton* set. We have $\{A\} = \{x \in \mathcal{P}(A) : x = A\}$.

(A4) Pair Axiom. *If A and B are sets, then there is a set $\{A, B\}$ that has both A and B as elements but has no other elements.*

The set $\{A, B\}$ is called a *pair*. Note that $\{A, B\} = \{B, A\}$. If $A = B$, then the pair is a singleton, $\{A, B\} = \{A\} = \{B\}$.

(A5) Union Axiom. *Given any set A whose members are sets, there is a set $\cup A$ whose elements are the elements of the members of A.*

The set $\cup A$ is called the *union* of A (or of the members of A), and it is also denoted by $\bigcup_{X \in A} X$. The union of a pair $\{A, B\}$ is usually denoted by $A \cup B$ and called the *union* of A and B. Thus if we let $I = \{A, B\}$, then
$$A \cup B = \cup I = \bigcup_{X \in I} X$$

These few axioms immediately allow us to ask and answer a meaningful question that is quintessentially algebraic. Suppose we have sets A, B, and C. Taking the union of $A \cup B$ and C we get some set $(A \cup B) \cup C$. Proceeding differently, taking the union of A with $B \cup C$, we get some set $A \cup (B \cup C)$. But are not $(A \cup B) \cup C$ and $A \cup (B \cup C)$ the same? Yes indeed, because it is easy to verify that they have the same elements. It is time to state the first theorem of algebra:

Proposition 1 *Let A, B, and C be sets.*

(i) $(A \cup B) \cup C = A \cup (B \cup C)$ (*associative law*)
(ii) $A \cup B = B \cup A$ (*commutative law*)
(iii) $A \cup A = A$ (*idempotent law*)

Proof. Associativity has just been observed. Commutativity follows from the earlier made observation that $\{A, B\} = \{B, A\}$, which is often referred to by saying that the pair $\{A, B\}$ is "not ordered." Finally, idempotence is obvious by definition of the union. □

Using axioms (A1) to (A5), we can define a great variety of sets. The reader should verify which particular axioms need to be called upon to construct the following specific examples:

The empty set ∅.
The singleton $\{\emptyset\}$.
The pair $\{\emptyset, \{\emptyset\}\}$.

The power set $\mathcal{P}(\{\emptyset, \{\emptyset\}\})$.
The power set of the above, $\mathcal{PP}(\{\emptyset, \{\emptyset\}\})$.
The power set of the above.

And so forth without end. What is remarkable here is that each of these examples is a *set of sets*, a set whose elements are themselves sets. Indeed we shall only consider sets of sets in this volume. It is the author's view, adopted in this book at least, that in mathematics we need not and should not speak about sets of atoms, molecules, animals, or true or false sentences unless these various objects can be precisely defined as sets themselves. In some cases this may be done meaningfully, such as in theoretical physics, or in mathematical logic where sentences can be defined as proper mathematical objects themselves, i.e., as sets. Neither should we speak about the set of even or odd numbers until these numbers have been defined as sets: this will be done in a while.

It was pointed out that the elements of a pair $\{a,b\}$ are "not ordered," $\{a,b\} = \{b,a\}$. The *ordered pair* $\langle a,b \rangle$ is defined by

$$\langle a,b \rangle = \{\{a\}, \{a,b\}\}$$

and it has the desired property that

$$\langle a,b \rangle = \langle c,d \rangle \quad \text{if and only if both} \quad a = c \quad \text{and} \quad b = d$$

Thus $\langle a,b \rangle = \langle b,a \rangle$ if and only if $a = b$. The reader should verify that $\langle a,b \rangle$ is a subset of $\mathcal{PP}(a \cup b)$. If a and b are elements of sets A and B, respectively, then

$$a \cup b \subseteq (\cup A) \cup (\cup B)$$

and thus $\langle a,b \rangle$ is also a subset of

$$\mathcal{PP}((\cup A) \cup (\cup B))$$

Therefore those ordered pairs $\langle x,y \rangle$ for which $x \in A$ and $y \in B$ form a subset $A \times B$ of

$$\mathcal{PPP}((\cup A) \cup (\cup B))$$

called a *Cartesian product*. A *function, map,* or *mapping from A to B* is then any subset f of $A \times B$ such that for every $x \in A$ there is a unique $y \in B$ with

$$\langle x,y \rangle \in f$$

The set A is called *the domain*, B a *codomain* of f. (A function can have many different codomains, for if $B \subseteq B'$, then $A \times B \subseteq A \times B'$, and thus every function from A to B is also a function from A to B'.) We use the shorthand $f : A \to B$ for "a function f from A to B." For $x \in A$, the unique element y of B such that $\langle x, y \rangle \in f$ is called the *image of x by f*, or the *value of f on x*, and it is denoted by $f(x)$. It is also said that f *associates* $f(x)$ with x, f *maps* x to $f(x)$, or $f(x)$ is obtained by *applying* f to x. The set of all functions from A to B is denoted by B^A; it is a subset of $\mathcal{P}(A \times B)$. For reasons to be seen later, we say that B^A is obtained from B and A by *set exponentiation*. For $f \in B^A$ and $S \subseteq A$ the function

$$\{\langle x, f(x) \rangle : x \in S\}$$

is called the *restriction* of f to S. It is a function from S to B, and it is denoted by $f|S$. We also say that f is an *extension* of $g = f|S$ to A, or that f *extends* g.

Observe that two functions $f, g \in B^A$ are identical if

$$f(x) = g(x) \qquad \text{for every} \quad x \in A$$

A function is thus completely determined by its values on the various elements of its domain, and usually that is how functions are specified.

Informally, a function $A \to B$ is often thought of as a "rule," "procedure," or "machine" that, given any input $x \in A$, "allows us to find" or "produces" an element $f(x) \in B$. Many functions seen in mathematics appear in fact to fit this notion. However, many other functions, perhaps not "seen" but existing nevertheless, have nothing to do with computational procedures. (This issue is of great importance in mathematical philosophy and logic and of practical relevance in computer science. The interested reader is referred to the theory of recursive functions and to the theory of computational complexity.)

For every set A the function $i : A \to A$ defined by

$$i = \{\langle x, x \rangle \in A \times A : x \in A\}$$

[or equivalently by $i(x) = x$ for $x \in A$] is called the *identity function* on A, often denoted by id_A. In a nonempty set B let $b \in B$. Assume also that $A \neq \emptyset$. The function $c = A \times \{b\}$ from A to B is said to be *constant* because

$$c(x) = b \quad \text{for all} \quad x \in A$$

An arbitrary map $f : A \to B$ is said to be *constant on a subset S* of its domain A if the restriction $f|S$ is constant.

The *image of a set* $S \subseteq A$ by a function $f : A \to B$ is the set

$$\{y \in B : y = f(x) \text{ for some } x \in S\}$$

It is denoted by $f[S]$. The *inverse image of a set* $T \subseteq B$ is

$$\{x \in A : f(x) \in T\}$$

It is denoted by $f^{\text{inv}}[T]$. The *image*, or *range*, of the function $f : A \to B$ is $f[A]$; it is denoted by $\text{Im} f$. If $\text{Im} f = B$, then f is said to be *surjective to B* (or a function *onto B*). All identity functions are surjective onto their own domains. On the other hand, the image of a constant function is a singleton, and therefore a constant function $c : A \to B$ is not surjective onto B unless the codomain B is a singleton.

A function $f : A \to B$ is *injective* (or an *injection*) if there are no distinct elements $x \neq x'$ of A with $f(x) = f(x')$. All identity functions are injective. On the other hand, a constant function is not injective unless its domain is a singleton. (The reader should verify this.)

An injective function surjective onto a codomain B is called *bijective* (or a *bijection*) to B. All identity functions are bijective to their own domains. For a nontrivial example, let S be any set and let the *complementation function* $f : \mathcal{P}(S) \to \mathcal{P}(S)$ be defined by

$$f(A) = \{x \in S : x \notin A\} \quad \text{for every} \quad A \in \mathcal{P}(S)$$

This complementation function is bijective from $\mathcal{P}(S)$ to $\mathcal{P}(S)$.

There are two facts that we should take note of at this juncture. First, for any set S there is an injection $f : S \to \mathcal{P}(S)$. Indeed, f can be defined by $f(x) = \{x\}$ for all $x \in S$, i.e.,

$$f = \{\langle x, \{x\}\rangle : x \in S\}$$

Second, let us prove that there is no injection $g : \mathcal{P}(S) \to S$ from the power set of S into S. Suppose that such a g exists: we shall derive a contradiction. Consider those subsets A of S for which $g(A) \notin A$. Let F be the set of the corresponding elements $g(A)$,

$$F = \{x \in S : x = g(A) \text{ for some } A \subseteq S \text{ such that } g(A) \notin A\}$$

If we had $g(F) \notin F$, then by letting $A = F$, it follows from the definition of F that $g(A)$ belongs to F, i.e., $g(F) \in F$. And if $g(F) \in F$, then $g(F) = g(A)$ for some $A \subseteq S$ such that $g(A) \notin A$, again referring to the definition of F verbatim. Since g is supposed to be injective, $g(F) = g(A)$ implies $F = A$. But then $g(F) \in F$ and $g(A) \notin A$ are contradictory, proving the absurdity of the alleged existence of an injective g. This argument is inspired by the liar's paradox. However, what is reduced to absurdity here is not the universal dichotomy of truth and falsehood but merely the possibility of injecting $\mathcal{P}(S)$ into S. The argument is indeed a domesticated variety of the liar's paradox of Crete.

If $f : A \to B$ and $g : B \to C$ are functions, then the *composition* $g \circ f$ is the function from A to C defined by

$$(g \circ f)(x) = g(f(x)) \qquad \text{for all} \quad x \in A$$

i.e., $g \circ f$ is the set of all ordered pairs $\langle x, z \rangle$ such that for some $y \in B$,

$$f(x) = y \qquad \text{and} \qquad g(y) = z$$

Occasionally we shall write simply gf instead of $g \circ f$. The reader can see that the composition of two injective functions is injective and the composition of surjective functions is surjective. Hence, the composition of bijections is bijective. Observe further that a function $f : A \to B$ is bijective to B if and only if the set

$$\{\langle y, x \rangle \in B \times A : \langle x, y \rangle \in f\}$$

is itself a function from B to A. Denoting this new function by f^*, we have $f^* \circ f = id_A$ and $f \circ f^* = id_B$, and f^* is called the *inverse* of f. The reader should verify that f^* itself is a bijection from B to A, having in turn f as its inverse. Moreover, a function $f : A \to B$ is bijective if and only if there is a function $g : B \to A$ such that

$$g \circ f = id_A \qquad \text{and} \qquad f \circ g = id_B$$

and in that case g must coincide with f^*. For all $T \subseteq B$ we have

$$f^*[T] = f^{\text{inv}}[T]$$

Note, however, that $f^*[T]$ is only defined for bijective f, while $f^{\text{inv}}[T]$ is always defined.

The following proposition, of constant use in mathematics, anticipates the subject of the last chapter on category theory:

Proposition 2 *Let $f : A \to B$, $g : B \to C$, and $h : C \to D$ be functions.*

(i) $h \circ (g \circ f) = (h \circ g) \circ f$ *(associative law)*
(ii) $id_B \circ f = f$ *and* $f \circ id_A = f$ *(neutrality of the identities)*

Proof. Let $x \in A$. The image of x by $h \circ (g \circ f)$ is given by h applied to
$$(g \circ f)(x) = g(f(x))$$
i.e., it is $h[g(f(x))]$. But the image of x by $(h \circ g) \circ f$ is $h \circ g$ applied to $f(x)$, i.e., $h[g(f(x))]$ again. Thus $(h \circ g) \circ f$ and $h \circ (g \circ f)$ take the same value on every $x \in A$, and therefore they are identical functions. The neutrality of the identities can be verified by the reader. □

Here is now an early result in equational algebra:

Proposition 3 *Let f be a bijection of a set A into itself. Then for any function $g : A \to A$ there is a unique function $x : A \to A$ satisfying the equation*
$$g = f \circ x \qquad (1)$$
There is also a unique function y satisfying
$$g = y \circ f \qquad (2)$$

Proof. Since f^* is the inverse of f, $x = f^* \circ g$ is a solution of (1), because
$$f \circ (f^* \circ g) = (f \circ f^*) \circ g = id_A \circ g = g$$
Also, if x is any function satisfying $g = f \circ x$, then
$$f^* \circ g = f^* \circ (f \circ x) = (f^* \circ f) \circ x = id_A \circ x = x$$
This proves that x must be equal to $f^* \circ g$ and cannot be any other function. The unique solvability of (2) is shown similarly. □

A set A is said to be *equipotent* to B if there is a bijection from A to B. We shall then write $A \simeq B$.

Proposition 4 *Let A, B, and C be sets.*

(i) $A \simeq A$ (*reflexivity*)
(ii) if $A \simeq B$, then $B \simeq A$ (*symmetry*)
(iii) if $A \simeq B$ and $B \simeq C$, then $A \simeq C$ (*transitivity*)

Proof. Reflexivity results from the fact that the identity id_A is a bijection. Symmetry follows from the observation, made earlier, that the inverse of any bijection is a bijection. Finally, since the composition of two bijections is again a bijection, we have transitivity. □

A bijection f from a set A to a set B establishes what is often called a one-to-one correspondence between the elements of A and those of B. The elements of A and B are matched into ordered pairs $\langle a,b \rangle \in f$, with each element a of A being matched to a unique $b \in B$, and each $b \in B$ corresponding to a unique element of A. It is then tempting to say that A and B have the same number of elements. While we must refrain from using the word "number" until it is defined, this is actually what the term "equipotent" is meant to convey. The following result on set exponentiation may then be thought of as the first theorem of arithmetic:

Proposition 5 *For any sets A, B, and C we have*

$$(C^B)^A \simeq C^{B \times A}$$

Proof. Define a function F from $(C^B)^A$ to $C^{B \times A}$ as follows. If $f \in (C^B)^A$, then for every element $a \in A$, $f(a)$ is a function from B to C. Let then $\overline{f}$ be a function from $B \times A$ to C defined by

$$\overline{f}(\langle b,a \rangle) = f(a)(b)$$

Let $F(f)$ be defined as $\overline{f}$. The function F is injective for if $F(f) = F(g)$, then

$$f(a)(b) = g(a)(b) \qquad \text{for all} \quad \langle b,a \rangle \in B \times A$$

i.e., for $a \in A$ fixed the functions $f(a)$ and $g(a)$ from B to C are the same function, $f(a) = g(a)$. This being true for all $a \in A$, f and g are one and the same function from A to C^B.

To prove surjectivity, let $h \in C^{B \times A}$. Let $h_A : A \to C^B$ be defined as follows. For $a \in A$, $h_A(a)$ is the function from B to C specified by

$$h_A(a)(b) = h(\langle b,a \rangle) \qquad \text{for all} \quad b \in B$$

It can now be verified that $F(h_A) = h$. Thus every $h \in C^{B \times A}$ belongs to Im F, and F is surjective onto $C^{B \times A}$. Since it was also shown to be injective, it must be bijective to $C^{B \times A}$, establishing the equipotence of its domain $(C^B)^A$ and the codomain $C^{B \times A}$. □

EXERCISES

1. For any sets A, B, C verify that
   (a) $A \subseteq B$ is equivalent to $A \cup B = B$,
   (b) $A \times (B \cup C) = (A \times B) \cup (A \times C)$,
   (c) $A \times \emptyset = \emptyset$,
   (d) if I is a singleton, then $A \times I \simeq A$,
   (e) $A \times B \simeq B \times A$,
   (f) $A \times (B \times C) \simeq (A \times B) \times C$,
   (g) $(A \times B)^C \simeq (A^C) \times (B^C)$,
   (h) $A^\emptyset$ is a singleton,
   (i) $\emptyset^A = \emptyset$ unless $A = \emptyset$,
   (j) if I is a singleton, then $A^I \simeq A$ and I^A is a singleton,
   (k) if $A \neq \emptyset$, then the set of constant functions $A \to B$ is equipotent to B,
   (l) for $A \neq \emptyset$, a function $f : A \to B$ is injective if and only if there is a function $g : B \to A$ with $g \circ f = id_A$.

2. Write and run a computer program that produces a complete list of members of the set $\mathcal{PPPPPP}(\emptyset)$. Introduce any notation you wish to make the printout readable.

2. CARDINAL AND ORDINAL NUMBERS

An *ordinal* (or *ordinal number*) is a set α satisfying the following conditions:

(i) every element of α is a subset of α, $\alpha \subseteq \mathcal{P}(\alpha)$,
(ii) for $b, c \in \alpha$, $c \subset b$ if and only if $c \in b$,

(iii) every nonvoid subset S of α, $\emptyset \subset S \subseteq \alpha$, has a member $p \in S$ that is a subset of all members $s \in S$, $p \subseteq s$; p is then called the *first element* of S.

Examples. The null set $\emptyset$, the singleton $\{\emptyset\}$, and the pair $\{\emptyset, \{\emptyset\}\}$ are ordinals. There will be many more.

Condition (i) tells us that for an ordinal α, if $b \in \alpha$ and $c \in b$, then $c \in \alpha$. The reader can see that every element b of an ordinal α is again an ordinal.

Condition (ii) implies that an ordinal never belongs to itself, because no set can be a proper subset of itself. Similarly, an ordinal never belongs to any of its own elements.

From these facts it easily follows that for every ordinal α, the set $\alpha' = \alpha \cup \{\alpha\}$ is again an ordinal, called the *successor* of α. This allows us to construct some very important ordinals. We define:

$0 = \emptyset$ "*zero*"
$1 = 0' = \{\emptyset\}$ "*one*"
$2 = 1' = \{\emptyset, \{\emptyset\}\}$ "*two*"
$3 = 2'$ "*three*"
$4 = 3'$ "*four*"
$5 = 4'$ "*five*"
$6 = 5'$ "*six*"
$7 = 6'$ "*seven*"
$8 = 7'$ "*eight*"
$9 = 8'$ "*nine*"

Proposition 6 *For any ordinals α and β, we have $\alpha \subset \beta$ if and only if $\alpha \in \beta$.*

Proof. If $\alpha \in \beta$, then $\alpha \subseteq \beta$ by the definition of an ordinal applied to β. Also $\alpha \neq \beta$ because $\beta \notin \beta$. Thus $\alpha \subset \beta$.

If $\alpha \subset \beta$, then let φ be the first element of

$$S = \{x \in \beta : x \notin \alpha\}$$

For any $a \in \alpha$, the first element of $\{a, \varphi\}$ cannot be φ, for that would imply $\varphi \in \alpha$, so it is a, and thus $a \in \varphi$. Hence $\alpha \subseteq \varphi$. If we had the strict inclusion $\alpha \subset \varphi$, then φ being a subset of the ordinal β, it would have as a member some element x of S, $x \in \varphi$. But by definition of S and φ we would also have $\varphi \in x$, which is impossible because an ordinal never belongs to any of it own elements. Thus $\alpha = \varphi$ and therefore $\alpha \in \beta$, which proves the proposition. $\square$

In view of this proposition, instead of writing $\alpha \subset \beta$ or $\alpha \in \beta$ when such is the case, it is customary to write $\alpha < \beta$ and to say that α is *less than* β, or that β is *greater than* α. We write $\alpha \leq \beta$ to mean that "$\alpha < \beta$ or $\alpha = \beta$" and we say that α is *less than or equal to* β, or β is *greater than or equal to* α. For example, every ordinal α is less than its successor α', but neither $\alpha' < \alpha$ nor $\alpha' \leq \alpha$ is true. The inequality $\alpha \leq \beta$ is equivalent to $\alpha \subseteq \beta$.

A key property of the relation $\leq$ is that it permits the comparison of any two ordinals. For assume that neither $\alpha \leq \beta$ nor $\beta \leq \alpha$ holds. The set σ of common elements of α and β is easily seen to be an ordinal. By assumption, σ is distinct both from α and β. But then, by Proposition 6, σ belongs to both α and β, i.e., σ belongs to σ, which is impossible. This proves that at least one of $\alpha \leq \beta$ or $\beta \leq \alpha$ must hold. Combining with earlier remarks, we obtain:

Proposition 7 *Let α, β, and γ be ordinals.*

 (i) $\alpha \leq \alpha$ *(reflexivity)*
 (ii) *if $\alpha \leq \beta$ and $\beta \leq \alpha$, then $\alpha = \beta$* *(antisymmetry)*
 (iii) *if $\alpha \leq \beta$ and $\beta \leq \gamma$, then $\alpha \leq \gamma$* *(transitivity)*
 (iv) *at least one of $\alpha \leq \beta$ or $\beta \leq \alpha$ holds* *(total comparability)*

Indeed a property stronger than total comparability holds:

Proposition 8 *Every nonempty set S of ordinals has an element φ that is less than any other element of S.*

Proof. Take any $\alpha \in S$. Let S_α be the set of those elements of S that are less than α. If $S_\alpha = \emptyset$, then let $\varphi = \alpha$. Otherwise S_α is a nonvoid subset of α, and we let φ be the first element of S_α. $\square$

Corollary. *The union of any set of ordinals is an ordinal.*

The ordinal $\varphi \in S$ whose existence was established by Proposition 8 is called the *first element* of S, which is in accordance with the earlier use of this term.

Every ordinal number α is a set, namely the set of ordinals less than α. The union $\cup\alpha$ is always an ordinal and $\cup\alpha \leq \alpha$. Now, either $\cup\alpha < \alpha$ or $\cup\alpha = \alpha$. If $\cup\alpha < \alpha$, then let ρ be an element of α that does not belong to $\cup\alpha$. We have $\cup\alpha \leq \rho$. But since $\rho \in \alpha$, also $\rho \subseteq \cup\alpha$. Thus $\rho = \cup\alpha$, and the only element of α not in $\cup\alpha$ is $\rho = \cup\alpha$, i.e.,

$$\alpha = (\cup\alpha) \cup \{\cup\alpha\}$$

which means that α is the successor of $\cup\alpha$. Can α be at the same time the successor of some other ordinal β? The answer is no, because from $\alpha = \beta \cup \{\beta\}$ it follows that

$$\cup\alpha = (\cup\beta) \cup \beta = \beta$$

This argument also shows that α is not the successor of any ordinal β if $\cup\alpha = \alpha$. There are two kinds of nonzero ordinals α. On the one hand, there are those for which $\cup\alpha < \alpha$. Then α is the successor of $\cup\alpha$, and $\cup\alpha$ is termed the *predecessor* of α. On the other hand, there are those ordinals α for which $\cup\alpha = \alpha$. These have no predecessor, and they are called *limit ordinals*. Every limit ordinal is the union of lesser ordinals. An ordinal α that is either 0 or such that $\cup\alpha < \alpha$ is said to be of the *first kind*, while nonzero limit ordinals are sometimes said to be of the *second kind*. We now arrive at a most important concept: an ordinal is called *finite* if it is of the first kind and all its elements are also of the first kind. An ordinal that is not finite is called *infinite*. Finite ordinals are also called *natural numbers*.

Examples. The ordinals $0, 1, 2, \ldots, 9$ defined earlier are natural numbers. The successor of any natural number is again a natural number.

But does there exist any infinite ordinal? We are unable to prove it. We have seen how to make "one" out of "zero," "two" out of "one," and so on to trillions. However, even a megaquadrillion is just finite dust. It is time to postulate three new axioms.

(A6) Axiom of Existential Infinity. *There is a set to which all finite ordinals belong.*

(A7) Axiom of Limited Infinity. *There is no set having among its members sets equipotent to every ordinal.*

Otherwise stated, for every set S, there is an ordinal α such that no set belonging to S is equipotent to α.

(A8) Axiom of Choice. *For every set of nonvoid sets S, there exists a function $c : S \to \cup S$ such that $c(A) \in A$ for every $A \in S$.*

The function c is called a *choice function*; for each $A \in S$ it is said to *choose* the element $c(A)$ in the set A.

An immediate consequence of the Axiom of Existential Infinity is that there is a set whose elements are precisely the natural numbers. In view of the paramount importance that it claims in the spiritual life of mathematicians, *the set of all natural numbers* is denoted by the Greek letter ω. The Axiom of Limited Infinity, on the other hand, implies that the "set of all ordinals" is nonexistent.

With the intention of using ordinal numbers for enumeration, we now further develop the theory of equipotence, in particular as regarding ordinals. For any sets S and R the *difference* $S \backslash R$ is defined by

$$S \backslash R = \{x \in S : x \notin R\}$$

Counting Lemma. *If a set S is equipotent to a set T, and if s and t are elements of S and T, respectively, then $S \backslash \{s\}$ and $T \backslash \{t\}$ are equipotent.*

Proof. If $f : S \to T$ is a bijection, and if $f(s) = t$, then

$$g = \{\langle x, y \rangle \in f : x \neq s, \, y \neq t\}$$

is a bijection from $S \backslash \{s\}$ to $T \backslash \{t\}$. If $f(s) \neq t$, then let $r \in S$ such that $f(r) = t$. Now

$$g = \{\langle x, y \rangle \in f : x \neq s, \, y \neq t\} \cup \{\langle r, f(s) \rangle\}$$

is a bijection from $S \backslash \{s\}$ to $T \backslash \{t\}$. □

It is now easy to show that no natural number is equipotent to any other natural number. (If this is not true, let n be the first natural number equipotent to some natural number distinct from itself, say to $m \neq n$. As no bijection can exist between the empty set and a nonempty set, neither of n or m is 0. Then by the Counting Lemma, the predecessors of n and m are equipotent, contradicting the definition of n.) Thus a natural number is not equipotent to any of its own elements. This is not true for ordinal numbers in general. For example, the successor of ω, the ordinal $\omega' = \omega \cup \{\omega\}$, is equipotent to ω. A bijection $f : \omega' \to \omega$ can be defined by

$$f(n) = n' \quad \text{for all} \quad n \in \omega', \quad n \neq \omega, \quad \text{and} \quad f(\omega) = 0$$

This observation motivates the following definition. A *cardinal* (or *cardinal number*) is an ordinal that is not equipotent to any of its own elements (i.e., not equipotent to any ordinal less than itself). Thus natural numbers are cardinal numbers. Let us verify that so is ω. Were this not so, there would be a smallest $n \in \omega$ equipotent to ω. Obviously $n \neq 0$, so let m be the predecessor of n. By the Counting Lemma, the sets

$$m = n \setminus \{m\} \quad \text{and} \quad \bar{\omega} = \omega \setminus \{0\}$$

would be equipotent. But $\bar{\omega}$ is also equipotent to ω, via the bijection

$$f = \{\langle n, n' \rangle : n \in \omega\}$$

so by transitivity (Proposition 4) m would be equipotent to ω, contradicting the minimal choice of n.

Zermelo's Theorem (First Formulation). *Every set is equipotent to a cardinal.*

Proof. Observe first that it will be enough to prove that every set S is equipotent to some ordinal α. For if α is not a cardinal, then let β be the first element of α that is equipotent to α. Obviously β is a cardinal equipotent to S.

To prove that every set S is equipotent to an ordinal, we use the Axiom of Choice, which assures us of the existence of a choice function c from the set $\mathcal{P}^*(S)$ of nonempty subsets of S into $\cup \mathcal{P}^*(S) = S$. With a fixed choice function c in mind, we call *ordinal function* into S any injection $f : \alpha \to S$ from some ordinal α into S, such that for every $\beta \in \alpha$

$$f(\beta) = c(S \setminus f[\beta])$$

In particular, if $\alpha \neq 0$, then $f(0) = c(S)$ for every ordinal function $f : \alpha \to S$. Further, we claim that if $f : \alpha \to S$ and $g : \rho \to S$ are ordinal functions and $\alpha \leq \rho$, then $f(\beta) = g(\beta)$ for every $\beta \in \alpha$. For if β were the first element of α for which $f(\beta) \neq g(\beta)$, then $f(\gamma) = g(\gamma)$ for all $\gamma \in \beta$ and the sets

$$S_f = S \setminus f[\beta]$$
$$S_g = S \setminus g[\beta]$$

would be the same, hence $c(S_f) = c(S_g)$. But $f(\beta) = c(S_f)$ and $g(\beta) = c(S_g)$, and then $f(\beta) = g(\beta)$, proving our claim. This implies in particular that, with respect to a fixed choice function c, there can be at most one ordinal function $\alpha \to S$ for any ordinal α.

Ordinal functions are injective. Therefore, their images are equipotent to their ordinal domains. Hence, by the Axiom of Limited Infinity, there is an ordinal σ without ordinal function $f : \sigma \to S$. Then define the ordinal β as follows. If for every ordinal $\rho < \sigma$ there are ordinal functions from ρ to S, then let $\beta = \sigma$. Otherwise let β be the first element of ρ for which no ordinal function exists from β to S. In either case, β has the following properties: there is no ordinal function $\beta \to S$; and for every $\rho < \beta$ there is an ordinal function $\rho \to S$. Observe that β cannot be a limit ordinal, for in that case we could define an ordinal function $\beta \to S$ as the union of all ordinal functions with domains less than β. Also, β cannot be 0 for the empty set is surely an ordinal function $0 \to S$. Thus β has a predecessor α, and there is an ordinal function f from α to S. Is f surjective onto S? If it were not, then letting

$$a = c(S \setminus \operatorname{Im} f)$$

we could define an ordinal function g by $g = f \cup \{\langle \alpha, a \rangle\}$ on the domain β, which is impossible. Thus f must be surjective, and since ordinal functions are injective, this implies that f is a bijection from α to S. The proof is finished. □

Since by definition no cardinal is equipotent to any other cardinal, it follows from Zermelo's Theorem that every set S is equipotent to a unique cardinal, called the *cardinal* (or *cardinality*) of S and denoted by Card S. We also say that Card S is the *number of elements* of S. Note that $A \subseteq B$ implies Card $S \leq$ Card B. Sets are called *finite* or *infinite* according to whether their cardinal is finite or infinite. The following is now elementary:

Proposition 9 *Two sets are equipotent if and only if they have the same cardinal.*

As, by an earlier remark, there is no injection, and therefore no bijection, from $\mathcal{P}(\omega)$ to ω, $\operatorname{Card}\mathcal{P}(\omega)$ is distinct from ω. Thus there are infinite cardinals other than ω.

We now introduce a terminological redundancy. A *family* is simply a function $f : A \to B$. The domain A is called the *index set* of the family, and f is said to be a *family (of elements of B) indexed by (the elements of)* A. For $i \in A$, the element $f(i)$ of B is denoted by b_i, while the family f itself is often denoted by $(b_i : i \in A)$ or $(b_i)_{i \in A}$. For example, for any set S, the identity function id_S is nothing else but the family $(x : x \in S)$. The image set of a family $(b_i : i \in A)$ is denoted by $\{b_i : i \in A\}$. Families generalize the set concept. For example, we define the *union of a family of sets* $(b_i)_{i \in A}$, in symbols $\bigcup_{i \in A} b_i$, as the set

$$\cup\{b \in B : b = b_i \text{ for some } i \in A\}$$

It is important to notice that in a family $(b_i)_{i \in A}$ we may have $b_i = b_j$ even if the indices i and j are distinct. A family indexed by an ordinal is usually called a *sequence*. A sequence indexed by a natural number n is called an *n-tuple* (*couple, triple, quadruple, quintuple* for $n = 2, 3, 4, 5$) and it is usually written as a string of n elements of B, possibly in brackets and separated by commas, such as (u), $(u\,v)$, $(u\,v\,w)$, $(u\,v\,w\,t)$ for $n = 1, 2, 3, 4$ and

$$(u_0, u_1, \ldots, u_i, \ldots)$$

in general. The position i of u_i, $0 \leq i < n$, in the string indicates that the sequence in question, as a function from n to B, maps i to u_i. The image set of the sequence can be written accordingly as

$$\{u_0, u_1, \ldots, u_i, \ldots\}$$

or explicitly as $\{u, v, w\}$, $\{u, v, w, t\}$ for $n = 3, 4$, etc.

Let $(A_i)_{i \in I}$ be a family of sets indexed by a set I. The *product* $\prod_{i \in I} A_i$ *of the family* is the set of all functions $f : I \to \bigcup_{i \in I} A_i$ such that

$$f(i) \in A_i \quad \text{for every} \quad i \in I$$

For each $j \in I$ the function $\operatorname{pr}_j : \prod A_i \to A_j$ defined on the product set by $\operatorname{pr}_j(f) = f(j)$ is called the *j*th *projection*.

Proposition 10 *The product of a family of nonempty sets is not empty.*

Proof. By the Axiom of Choice there is a choice function c from the image set $\{A_i : i \in I\}$ of the family $(A_i)_{i \in I}$ to the set $\bigcup_{i \in I} A_i$ such that $c(A_i) \in A_i$ for every A_i, $i \in I$. Then $f : I \to \bigcup_{i \in I} A_i$ defined by $f(i) = c(A_i)$ is an element of the product $\prod_{i \in I} A_i$. $\square$

Note that the product $\prod A_i$ is empty as soon as any one of the factors A_i is empty.

Recall that for sets A, B the power A^B was defined as the set of all functions from B to A. Consider now the constant family $(A_i)_{i \in B}$ where all $A_i = A$.

Proposition 11 *If each A_i is identical with A, then the sets $\prod_{i \in B} A_i$ and A^B are identical, $\prod_{i \in B} A_i = A^B$.*

The *cardinal product* of a family $(\alpha_i)_{i \in I}$ of cardinals is defined as the cardinality of the set $\prod_{i \in I} \alpha_i$. If $I = 2$, then the cardinal product is denoted by $\alpha_0 \cdot \alpha_1$. Clearly for any cardinals β, γ there is a family $(\alpha_i)_{i \in 2}$ indexed by $2 = \{0, 1\}$ such that $\beta = \alpha_0$, $\gamma = \alpha_1$, and thus the cardinal product $\beta \cdot \gamma$ of any cardinals β and γ is always well defined. The following important proposition then becomes a matter of simple observation.

Proposition 12 *For any cardinal numbers α, β, and γ we have*

$\alpha \cdot \beta = \beta \cdot \alpha$ *(commutativity)*
$(\alpha \cdot \beta) \cdot \gamma = \alpha \cdot (\beta \cdot \gamma)$ *(associativity)*
$1 \cdot \alpha = \alpha \cdot 1 = \alpha$ *(neutrality of 1)*
$0 \cdot \alpha = \alpha \cdot 0 = 0$ *(absorption by 0)*

The role of cardinal numbers is to count the elements of sets. Cardinal products can be used to count the elements of a product of sets. Indeed, if $(A_i)_{i \in I}$ is a family of sets and each A_i has α_i elements, then the cardinality of $\prod_{i \in I} A_i$ is the cardinal product of the family $(\alpha_i)_{i \in I}$. Further, if $I = 2$, then $\prod_{i \in 2} A_i$ is equipotent to the Cartesian product $A_0 \times A_1$. Thus $A_0 \times A_1$ has $\alpha_0 \cdot \alpha_1$ elements. If the two sets are finite, then $\alpha_0 \cdot \alpha_1$ is just the usual product of the natural numbers α_0 and α_1. The reader may find it instructive to prove rigorously such elementary-school theorems as $2 \cdot 3 = 6$. If a mule needs a bushel of forage every day, two mules on a three-day expedition need six bushels. This is more than pedestrian arithmetic. But we still do not know how to add natural numbers.

Historical Note. The foundations of transfinite arithmetic were laid down toward the end of the last century by Georg Cantor. Mathematical attempts both to negate and to comprehend infinity continue to this day.

EXERCISES

1. Is $\{\{\emptyset\},\{\{\emptyset\}\}\}$ an ordinal? Is $\mathcal{P}(0)$ an ordinal? What about $\mathcal{P}(1)$ and $\mathcal{P}(2)$? And $\mathcal{P}(n)$ for an ordinal n greater than 2?

2. Show that
   (a) β is the successor ordinal of α if and only if $\alpha < \beta$ and there is no ordinal γ with $\alpha < \gamma < \beta$,
   (b) if S is a finite set and $x \in S$, then $\text{Card}\,S$ is the successor ordinal of $\text{Card}(S \setminus \{x\})$,
   (c) $(\omega \setminus n) \simeq \omega$ for all $n \in \omega$,
   (d) every infinite subset of ω is equipotent to ω,
   (e) $A \times A \simeq A^2$ for any set A,
   (f) $\mathcal{P}(A) \simeq 2^A$ for any set A,
   (g) a set A is finite if and only if for every $B \subset A$ we have $\text{Card}\,B < \text{Card}\,A$,
   (h) a function $f : A \to B$ is surjective onto B if and only if there is a function $g : B \to A$ with $f \circ g = id_B$.

3. For any sets A, B, C verify that
   (a) $A \subseteq B$ implies $A \times C \subseteq B \times C$ and $A^C \subseteq B^C$,
   (b) $\text{Card}\,A \leq \text{Card}\,B$ implies $\text{Card}(C^A) \leq \text{Card}(C^B)$,
   (c) $\text{Card}\,A \leq \text{Card}\,B$ if and only if there is an injection $A \to B$, or equivalently, if and only if there is a surjection $B \to A$.

4. For nonempty sets A and B, show that the following conditions are equivalent:

(a) Card A < Card B,
(b) there is no surjection $A \to B$,
(c) there is no injection $B \to A$.

5. Show that for any set A the following conditions are equivalent:
   (a) A is finite,
   (b) every injection $A \to A$ is surjective,
   (c) every surjection $A \to A$ is injective.

6. Let $X = 9 \cup \{9\}$ and let $S_n = \{v \in X^n : v(0) \neq 0 \text{ if } n \neq 1\}$ for each $n \in \omega$, $n \neq 0$. Let

$$N = \bigcup_{\substack{n \in \omega \\ n \neq 0}} S_n$$

Show the existence of a bijection $N \to \omega$.

7. Call a natural number d a *divisor* of $n \in \omega$ if for some $q \in \omega$ we have $n = d \cdot q$. Write a computer program that for any natural number input n finds all the divisors of n.

8. Let $D(n)$ denote the set of divisors of a natural number n. Verify that $D(n)$ is finite. Define $f : \omega \to \omega$ by $f(n) = \text{Card } D(n)$. Write a computer program that for input n calculates $f(n)$. What is the biggest number n you can find with $f(n) = 2$?

3. INTERSECTIONS

Let E be a set of sets. By axiom (A2), $\cup E$ has a subset $\cap E$ consisting of those elements which belong to all members of E. The set $\cap E$ is called the *intersection of* (*the members of*) E and it is also denoted by

$$\bigcap_{X \in E} X$$

If this set is nonvoid, then the members of E are said to *intersect*. The intersection of a pair $\{A, B\}$ is also denoted by $A \cap B$ and called the intersection of A and B. Thus

$$A \cap B = \cap \{A, B\}$$

Similarly to unions (Proposition 1) we have:

Proposition 13 *Let A, B, and C be sets.*

(i) $(A \cap B) \cap C = A \cap (B \cap C)$ *(associativity)*
(ii) $A \cap B = B \cap A$ *(commutativity)*
(iii) $A \cap A = A$ *(idempotence)*

Further, union and intersection are linked by several algebraic identities, some of which involve the concept of complementation. The *Boolean sum*, or *symmetric difference* of two sets A, B, is the set

$$A + B = (A \setminus B) \cup (B \setminus A)$$

If $A \supseteq B$, then the set $A \setminus B$ is called the *complement of B in A*, denoted by $c_A B$.

Proposition 14 *Let A, B, and C be subsets of a set S.*

(i) $(A \cap B) \cup B = B$ and
$(A \cup B) \cap B = B$ *(absorption)*
(ii) $A \cap (B \cup C) = (A \cap B) \cup (A \cap C)$
$A \cup (B \cap C) = (A \cup B) \cap (A \cup C)$ *(distributivity of union and intersection)*
(iii) $A \cap (B + C) = (A \cap B) + (A \cap C)$ *(distributivity of intersection over the sum)*
(iv) $c_S(c_S A) = A$ *(involution of complementation)*
(v) $c_S(A \cap B) = c_S A \cup c_S B$
$c_S(A \cup B) = c_S A \cap c_S B$ *(De Morgan's laws)*

Proof. Quite straightforward, based on the definitions. Let us just carry it out for (iii). $A \cap (B + C)$ is the set of those elements of A that belong to either B or C but not to both. Thus each element of $A \cap (B + C)$ belongs either to $(A \cap B)$ or to $(A \cap C)$ but not to both,

$$A \cap (B + C) \subseteq (A \cap B) + (A \cap C)$$

Conversely, if $x \in (A \cap B) + (A \cap C)$, then x belongs to either $A \cap B$ or $A \cap C$ but not to both. If $x \in A \cap B$, then $x \notin C$ and thus $x \in A \cap (B + C)$. Similarly, if $x \in A \cap C$, then $x \notin B$ and $x \in A \cap (B + C)$. Thus

$$(A \cap B) + (A \cap C) \subseteq A \cap (B + C)$$

This completes the proof of (iii). □

Statement (i) in the above proposition may be viewed as the first result of lattice theory, and distributivity of intersection over the Boolean sum (iii) is a precursor example of ring algebra.

Two sets A and B are called *disjoint* if their intersection $A \cap B$ is empty. Let α and β be two cardinals. Then $A = \{0\} \times \alpha$ and $B = \{1\} \times \beta$ are disjoint, Card $A = \alpha$ and Card $B = \beta$. We define the *cardinal sum*, or *addition*, $\alpha + \beta$ as the cardinality of $A \cup B$. Observe that if A' and B' are any two disjoint sets with α and β as respective cardinalities, then
$$\text{Card}(A' \cup B') = \alpha + \beta$$
The reader should verify this by showing the equipotence of $A' \cup B'$ to the union $A \cup B$ used in the definition of the cardinal sum.

Remark on Abusive Notation. We are using the same symbol "+" for cardinal sum as for Boolean sum. We really should not do this, as these concepts are distinct. However, we are running short of simple symbols. We trust that the context will always indicate what is meant by "+."

Proposition 15 *Let α, β, and γ be cardinals. Cardinal sum and cardinal product obey the following rules:*

(i) $(\alpha + \beta) + \gamma = \alpha + (\beta + \gamma)$ (*associativity*)
(ii) $\alpha + \beta = \beta + \alpha$ (*commutativity*)
(iii) $0 + \alpha = \alpha$ (*neutrality of 0 for addition*)
(iv) $\alpha \cdot (\beta + \gamma) = \alpha \cdot \beta + \alpha \cdot \gamma$ (*distributivity of product over sum*)

Proof. Associativity follows from the associativity of the union (Proposition 1) if we take any three sets A, B, C with respective cardinals α, β, γ and such that
$$A \cap B = B \cap C = A \cap C = \emptyset$$
Commutativity follows trivially from commutativity of set unions. The neutrality of 0 is obvious. For distributivity, let A, B, C be *pairwise disjoint* sets as just described and observe the identity of $A \times (B \cup C)$ and $(A \times B) \cup (A \times C)$.   □

A most important observation of arithmetic is that for each natural number n, $n + 1$ coincides with the successor of n. This fact can be used in the proof of such arithmetic statements as $2 + 2 = 4$ or $5 + 3 = 8$.

We have now seen a fair amount of algebra by example. The concept of intersection leads us to general algebraic tools of great importance: closure systems, closure operators, and generators. They will be recurrently called upon in the study of each particular algebraic structure.

Let U be any set. A set $\mathcal{C}$ of subsets of U is called a *closure system* (*on* U) if $U \in \mathcal{C}$ and the intersection of each nonempty subset of $\mathcal{C}$ belongs to $\mathcal{C}$. The members of $\mathcal{C}$ are then called the *closed sets* of $\mathcal{C}$. Clearly for every subset A of U, the intersection of all closed supersets of A is closed. We denote it by $\overline{A}$ and call it the *closure* of A in $\mathcal{C}$. Obviously A is a subset of $\overline{A}$ and $\overline{A}$ is the smallest closed superset of A (i.e., $\overline{A} \subseteq K$ for any closed superset K of A). Further, if $A \subseteq B$ are both subsets of U, then every closed superset of B is also a superset of A, from which it follows that $\overline{A} \subseteq \overline{B}$. Finally, the closure of a closed set is always itself. Let us summarize:

Proposition 16 *With respect to any closure system on U we have, for any $A, B \subseteq U$:*

(i) $A \subseteq \overline{A}$ (*extensive law*)
(ii) *if* $A \subseteq B$, *then* $\overline{A} \subseteq \overline{B}$ (*isotone law*)
(iii) $\overline{\overline{A}} = \overline{A}$ (*idempotence*)

Otherwise stated, the function $F_{\mathcal{C}} : \mathcal{P}(U) \to \mathcal{P}(U)$ given by $F_{\mathcal{C}}(A) = \overline{A}$ for $A \subseteq U$ is extensive, isotone, and idempotent. Any function $F : \mathcal{P}(U) \to \mathcal{P}(U)$ having these three properties, i.e., such that

$$A \subseteq F(A), \quad A \subseteq B \text{ implies } F(A) \subseteq F(B), \quad \text{and} \quad F(F(A)) = F(A)$$

is called a *closure operator* on U. It is easily verified that the image of a closure operator F is always a closure system $\mathcal{C}$ in which the closure $\overline{A}$ of any subset A of U coincides with $F(A)$. There is a bijective map from the set of closure operators on U to the set of closure systems on U, namely the map associating with each closure operator F its image set $\text{Im}\, F$. The members of $\text{Im}\, F$ are also called the *closed sets of the operator* F. For a given closure system, a closed set K is said to be *generated by* a subset $G \subseteq K$ if $\overline{G} = K$. In this case, G is also called a *set of generators* for K. Every closed set K has at least one set of generators, namely K itself. However, a typical task in algebra is to find proper subsets that generate a closed set K, proper subsets that contain as few elements as possible. Sometimes a closed set K is generated by a singleton $\{g\}$. In this important case we also say that K is *generated by the element* g, or g is a *generator of* K. We urge the reader familiar with other concepts of generation to adopt our nonalgorithmic, syntax-free concept as a working tool throughout this volume.

In the subsequent chapters we shall see many examples of closure systems and generators. However, the theory presented so far already allows the following examples.

Let f be any function from a set U to itself. Then

$$\mathcal{C} = \{A \subseteq U : f(x) \in A \text{ for all } x \in A\}$$

is a closure system. The members of $\mathcal{C}$ are said to be *closed under the function* f. For instance, let $U = \omega$, and let f be the function mapping each $n \in \omega$ to its successor $n + 1$. Then the closed sets are those of the form $\omega \setminus m$ with natural numbers m. Every such closed set $\omega \setminus m$ is generated by its first element m.

Let n be any natural number, U any set. A function f from U^n to U is called an *n-ary operation on* U (in particular *nullary, unary, binary, ternary, quaternary,* for $n = 0, 1, 2, 3, 4$). The set

$$\mathcal{C} = \{A \subseteq U : f(x) \in A \text{ for all } x \in A^n\}$$

is a closure system; its members are said to be *closed under the operation* f. For instance, a binary operation f is defined on $U = \omega$ by the cardinal sum of natural numbers,

$$f(a\,b) = a + b$$

We then have a large variety of sets closed under the cardinal sum. Sets closed under multiplication are defined similarly.

Given any set S, those sets of subsets of S that are closed under the binary operation "intersection,"

$$f(A\,B) = A \cap B$$

form a closure system on $U = \mathcal{P}(S)$. Another closure system on $\mathcal{P}(S)$ is defined by the binary operation "union." It is worthy of notice that the set of finite subsets of S is closed under both intersection and union. The set of infinite subsets of S is closed under union but need not be closed under intersection.

Observe that the set of all closure systems on a set U is itself a closure system on $\mathcal{P}(U)$.

Let S be any set. A nonempty set P of bijections from S to itself, such that for every $f, g \in P$ their composition $f \circ g$ belongs to P, and also the inverse of f belongs to P whenever f does, is called a *permutation group* on S. The set of permutation groups is a closure system on the set of bijections $S \to S$.

A nonempty set R of subsets of a set S that is closed under both binary operations $A + B$ (Boolean sum) and $A \cap B$ (intersection) is called a *ring of sets on S*. The set of rings of sets is a closure system on $U = \mathcal{P}(S)$.

The next two technical lemmas are often used implicitly in algebraic constructions.

Disjoint Copy Lemma. *If A and B are two sets, then there is a set A' that is equipotent to A and disjoint from B.*

Proof. Let $O = \{b \in B : b \text{ is an ordinal}\}$. The union $\cup O$ is an ordinal β. Following the idea of Zermelo's Theorem, show the existence of an injective function f defined on a set of ordinals greater than β and surjective onto A. Let A' be the domain of f. □

Injection–Extension Lemma. *If $f : A \to B$ is an injection, then there is a set $E \supseteq A$ and a bijection $g : E \to B$ such that $g|A = f$.*

Proof. Let C be a set equipotent to $B \setminus \operatorname{Im} f$ and disjoint from A. Let $h : C \to (B \setminus \operatorname{Im} f)$ be a bijection. Let $E = A \cup C$ and define $g : E \to B$ by

$$g(x) = \begin{cases} f(x) & \text{for } x \in A \\ h(x) & \text{for } x \in C \end{cases}$$

□

Warning. The expression "A is contained in B" is ambiguous. It is widely used in mathematical literature to mean both $A \in B$ and $A \subseteq B$. The expression "A is included in B" is equivalent, in mathematical usage, to "A is contained in B." Note that we have so far refrained from using any of these two equally ambiguous expressions.

In the sequel we shall use them only if the context clearly indicates their meaning and often in the active voice, e.g., "B contains A."

EXERCISES

1. Verify that for any sets A, B, C
   (a) $(A \cup B) \cap (A \cup C) \cap (B \cup C) = (A \cap B) \cup (A \cap C) \cup (B \cap C)$,
   (b) $A \times (B \cap C) = (A \times B) \cap (A \times C)$,
   (c) $A^{B \cup C} \simeq (A^B) \times (A^C)$ if $B \cap C = \emptyset$.

2. Show that if α, β are cardinal numbers, then $\alpha \leq \beta$ if and only if there is a cardinal δ with $\alpha + \delta = \beta$.

3. Show that for every cardinal α
   (a) $\alpha + \alpha = 2 \cdot \alpha$,
   (b) α is infinite if and only if $\alpha + 1 = \alpha$.

4. Verify that for any nonempty set S of ordinals, the first element of S is $\cap S$.

5. Let A and B be subsets of a set S on which a closure operator is defined. Does $\overline{A \cap B} = \overline{A} \cap \overline{B}$ or $\overline{A \cup B} = \overline{A} \cup \overline{B}$ hold?

6. Suppose a unary operation is defined on a set. Consider the set of closed subsets under this operation. Is it true that any union of closed sets is closed?

7. Let $U = 5$, $A = \{3, 5 \setminus 2, 1 \cup \{4\}\}$. What are the elements of the closure system $\mathcal{C}$ on U generated by A? Determine $\text{Card}\,\mathcal{C}$.

8. For any natural number n greater than 2, show that
   (a) n is not closed under cardinal sum but $\omega \setminus n$ is,
   (b) n is not closed under cardinal product but $\omega \setminus n$ is.

9. Call a set S of natural numbers closed if $0 \in S$ and the successor of each member of S also belongs to S. What are the closed subsets of ω?

10. Find all the permutation groups on the set $3 = \{0, 1, 2\}$. How would you go about finding all the permutation groups on some larger $n \in \omega$?

11. Show that for any finite set S there is a permutation group G on S that is generated by a single $g \in G$ and such that $\operatorname{Card} G = \operatorname{Card} S$. Find examples of permutation groups G not generated by any single $g \in G$.

12. Show the existence of an infinite permutation group G that is not generated by any finite subset $F \subseteq G$.

13. Show that if a permutation group G is generated by a single $g \in G$, then $f \circ h = h \circ f$ for all $f, h \in G$.

BIBLIOGRAPHY

"I have no list to submit," wrote Sheldon Glashow in the Harvard Guide to Influential Books. What list to submit indeed? Over 7000 bibliographical items are catalogued yearly by *Mathematical Reviews* in the area of algebra. Who am I to tell you what is important? In Glashow's words, "To read is the thing, voraciously and eclectically. No guide is needed." On the other hand, by writing this book, I already renounced my claim to neutrality in the great promotional war of mathematical ideas. Since you take the trouble to read what I write, I may as well tell you what I read.

> Heinz BACHMANN, *Transfinite Zahlen*. Springer 1967. A most thorough and detailed book on cardinal and ordinal numbers.
>
> Andrée BASTIANI, *Théorie des ensembles*. Centre de Documentation Universitaire, Paris 1970. A limpid and rigorous introduction to sets, ordered sets, cardinals, ordinals, the basic number systems, and very large sets of sets—the latter in view of the requirements of category theory.
>
> Garrett BIRKHOFF and Thomas C. BARTEE, *Modern Applied Algebra*. McGraw-Hill 1970. The reader who wishes to reflect upon the concepts of function, finiteness, and infinity will find most instructive the elementary discussion of finite state machines and how they relate to more general Turing machines.
>
> Nicolas BOURBAKI, *Theory of Sets*. Addison-Wesley 1968. This formalistic approach to set theory starts with 60 pages of rigorous syntax. Integers are defined only after another hundred

pages, but there is much good reading on functions, equivalence relations, and ordered sets in between.

Paul R. HALMOS, *Naive Set Theory*. Springer 1974. A concise introduction to axiomatic set theory, unencumbered by formalistic notation.

Thomas J. JECH, *The Axiom of Choice*. North-Holland 1973. Provides an in-depth appreciation of just how important it is to include the Axiom of Choice among the basic postulates of a mathematical theory. Uses advanced tools of mathematical logic.

P. T. JOHNSTONE, *Notes on Logic and Set Theory*. Cambridge University Press 1987. A one hundred page presentation of axiomatic set theory embedded in formal, but very readable, logic. Designed for advanced undergraduates.

Irving KAPLANSKY, *Set Theory and Metric Spaces*. Chelsea Publishing Co. 1977. Concise, rich, and reasonably self-contained. A short appendix also describes selected applications of set theory to advanced algebra.

Waclaw SIERPINSKI, *Cardinal and Ordinal Numbers*. Hafner Publishing Co. 1958. This introduction to classical set theory raises many questions of logical and combinatorial interest.

Marcel VAN DE VEL, *Theory of Convex Structures*. To appear, Elsevier 1993. A contemporary research monograph on closure systems. A "convex structure" is what we call, in Chapter II, an algebraic closure system.

CHAPTER II

ORDERED SETS

1. RELATIONS, ORDERS, AND ZORN'S LEMMA

Historical Note. It should already be clear that the notion of order is unavoidable, even in an elementary exposition of set theory. Some properties of the order of natural numbers were observed long before Cantor. The impossibility of infinite descent in ω was explicitly noted, and exploited, by Pierre Fermat in the seventeenth century. However, the first broad study of order relations, for their own sake, was published only in 1908, by Felix Hausdorff. The historical context included nascent set theory, Boolean algebra, ideals in rings of numbers, and of course Hausdorff's own preoccupation with topology. Many of the concepts investigated by Hausdorff apply to relations other than order relations.

For any set A, consider the set A^2 of couples of elements of A. Any subset R of A^2 is called a *(binary) relation on* A. The couple (A, R) is a *relational structure*. If $B \subseteq A$, then the *restriction* $R|B$ is the relation $R \cap B^2$ on B. The reader should verify that two relations R and S on A are the same if and only if their restrictions to each subset of A of cardinality at most 2 are the same. Only binary relations shall be studied in this volume.

On any set A, we have the *empty relation* $R = \emptyset$. We also have the *identity relation*
$$I = \{(a,a) : a \in A\}$$
and the *distinctness relation*
$$D = \{(a,b) \in A^2 : a \neq b\}$$
As relations are subsets of A^2, the union, as well as the intersection, of any set of relations on A is again a relation on A, and so is the difference of two relations. For the above examples, $I \cap D = \emptyset$.

A relational structure (α, R) many times encountered in the sequel is defined on any ordinal α by
$$R = \{(b,c) \in \alpha^2 : b \leq c\}$$
More suggestively we write $(\alpha, \leq)$ for (α, R).

Let (A, R) and (B, S) be two relational structures and let $h : A \to B$ be a function such that $(h(a), h(b)) \in S$ whenever $(a, b) \in R$. Then h is said to be a *relation-preserving function* from (A, R) to (B, S), in shorthand $h : (A, R) \to (B, S)$. If h is bijective to B and the inverse h^* is also relation preserving, from (B, S) to (A, R), then h is an *isomorphism*. The structures (A, R) and (B, S) are *isomorphic* if there is an isomorphism $(A, R) \to (B, S)$. An isomorphism of a relational structure to itself is called an *automorphism*.

Examples. The function $h : \omega \to \omega$ given by $h(x) = x + 1$ is relation preserving $(\omega, \leq) \to (\omega, \leq)$. If D and I are the distinctness and identity relations on a nonempty set A, then a function h from A to A is relation preserving $(A, D) \to (A, I)$ if and only if it is constant.

Proposition 1 *The composition of two relation-preserving functions $h : (A, R) \to (B, S)$ and $g : (B, S) \to (C, T)$ is relation preserving from (A, R) to (C, T). For any relational structure (A, R) the identity map on A is relation preserving from (A, R) to itself.*

Note that if (A, R) is any relational structure, then the set of all automorphisms of (A, R) is a permutation group on A. We shall denote it by $\mathrm{Aut}(A, R)$ or simply $\mathrm{Aut} R$.

For a relation R on a set A we generally write aRb instead of $(a,b) \in R$. The relation R is called

reflexive on A	if aRa for all $a \in A$
irreflexive	if aRa for no $a \in A$
symmetric	if aRb implies bRa
antisymmetric	if aRb and bRa together imply $a = b$
transitive	if aRb and bRc together imply aRc

A reflexive and transitive relation R on a set A is called a *preorder*. The relational structure (A, R) is called a *preordered set*. A symmetric preorder is an *equivalence* relation; an antisymmetric preorder is an *order*. In the latter case the structure (A, R) is also referred to as an *order*, or more properly, an *ordered set* and it is often suggestively denoted by $(A, \leq)$. Often we refer to the "ordered set A" when it is implicitly understood which order relation on A we have in mind. The order $\leq$ is *total* (or *linear*) if $a \leq b$ or $b \leq a$ holds for each $a, b \in A$. A totally ordered set is called a *chain*. Whether the order $\leq$ on A is total or not, the term chain is also used to designate any subset B of A such that the restriction of $\leq$ to B is a total order. If the restriction of $\leq$ to B is the identity relation, then B is called an *antichain*.

Examples. Identity relations are equivalence relations. Distinctness relations are irreflexive and symmetric. For any ordinal α, Proposition 8, Chapter I, implies that $(\alpha, \leq)$ is a chain, called the *natural order* on α. More generally, for any set of sets A,

$$R = \{(a,b) \in A^2 : a \subseteq b\}$$

is an order. We write customarily $(A, \subseteq)$ for (A, R) and we call it the *inclusion order* on A. In view of Proposition 4, Chapter I, for any set of sets A,

$$R = \{(a,b) \in A^2 : a \text{ and } b \text{ are equipotent}\}$$

is an equivalence relation.

A relation on a finite set can be represented by a diagram of points and arrows, an arrow from point a to b signifying the pres-

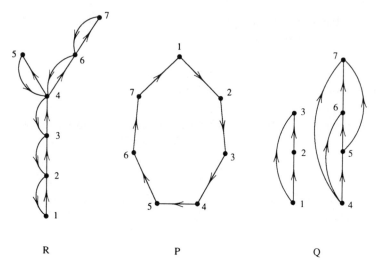

FIGURE 2.1 Symmetric, antisymmetric, and transitive relations.

ence of the couple (a,b) in the relation. In Figure 2.1, R is a symmetric relation, P is antisymmetric, and Q is antisymmetric as well as transitive.

The members of a set $\mathcal{C}$ of sets are *pairwise disjoint* if no two distinct members of $\mathcal{C}$ have a common element. A *partition* of a set A is a set $\mathcal{C}$ of pairwise disjoint nonempty subsets of A such that $\cup \mathcal{C} = A$. Then every $a \in A$ belongs to one and only one *partition class* $C \in \mathcal{C}$, called the *class of a*. Let E be an equivalence relation on a set A. Then there is a unique partition $\mathcal{C}$ of A such that $a, b \in A$ belong to the same class if and only if aEb. (Verify.) This is called the *partition associated with E*. The partition classes are also referred to as *equivalence classes*. Conversely, it is easy to see that for every partition $\mathcal{C}$ of a set A, the relation

$$E = \{(a,b) \in A^2 : a,b \in C \text{ for some } C \in \mathcal{C}\}$$

is an equivalence relation on A. It is called the *equivalence associated with $\mathcal{C}$*.

For every function f defined on a domain A there is an *induced equivalence* E on A, where aEb means $f(a) = f(b)$. Every equivalence E on A arises this way: let $f : A \to \mathcal{P}(A)$ map $a \in A$ to the class of a in the partition associated with E. The set of equivalence classes is also called *quotient set*, denoted by A/E, and $f : A \to A/E$ is called *canonical surjection*. Why the term "quotient set"?

Proposition 2 *If all classes of an equivalence E on A have the same cardinality k, then* $\operatorname{Card} A = k \cdot \operatorname{Card}(A/E)$.

Proof. Let C be any class and show the equipotence of A and the Cartesian product $C \times (A/E)$. □

A preorder R on a set A is usually denoted by the more suggestive symbol $\lesssim$. We write $a < b$ for "$a \lesssim b$ and not $b \lesssim a$," $a \sim b$ for "$a \lesssim b$ and $b \lesssim a$," and $a \parallel b$ for "neither $a \lesssim b$ nor $b \lesssim a$." [Read *a less (smaller) than b, a equivalent to b, a incomparable with b*, respectively.] The relation E defined by

$$E = \{(a,b) \in A^2 : a \sim b\}$$

is an *equivalence relation, said to be associated with* R. The relation O on A/E given by

$$O = \{(C,D) \in A/E : c \lesssim d \text{ for some } c \in C, d \in D\}$$

is an *order, said to be associated with* R.

The *dual* R^* of a binary relation R on A is given by aR^*b if and only if bRa. If R is a preorder (order, linear order), then R^* is also a preorder (order, linear order, respectively). We have $(R^*)^* = R$, and R is symmetric if and only if $R^* = R$. If R is a preorder $\lesssim$, or order $\leq$, then we write $\gtrsim, \geq$ for the duals of $\lesssim, \leq$. Instead of $b < a$ we also write $a > b$ and say "*a greater (larger) than b.*" Figure 2.2 illustrates a relation and its dual.

Example. Let $(\mathcal{P}(S), \subseteq)$ be the set of subsets of any set S, ordered by inclusion. It is isomorphic to its dual: consider $h : \mathcal{P}(S) \to \mathcal{P}(S)$ given by $h(X) = S \setminus X$. On the other hand, $(\omega, \leq)$ is not isomorphic to its dual: what would the image of 0 be in a hypothetical isomorphism?

Let $(A, \lesssim)$ be a preordered set, and let $B \subseteq A$. An element $b \in B$ is called *minimal in B* if no element of B is smaller than b, *maximal in B* if no element of B is larger. Stronger, $b \in B$ is a *minimum (smallest element,* min) of B if $b < x$ for all $x \in B \setminus \{b\}$, and a *max-*

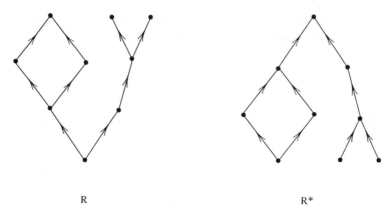

FIGURE 2.2 A relation and its dual.

imum (*largest element*, max) of B if $x < b$ for all $x \in B \setminus \{b\}$. An element $a \in A$ is a *lower bound* of B if $a \lesssim b$ for all $b \in B$, and a *greatest lower bound* (glb) if in addition $a' \lesssim a$ for every other lower bound a' of B. Dually, $a \in A$ is an *upper bound* of B if $b \lesssim a$ for all $b \in B$, and a *least upper bound* (lub) if in addition $a \lesssim a'$ for every other upper bound a'. Any two greatest lower bounds of a set are equivalent, and similarly for lub's. If B has a minimum (maximum), then it is also a unique glb (lub) of B. Every nonempty finite set $B \subseteq A$ has a minimal element as well as a maximal element. (Verify this assertion by assuming its falsehood, taking the first natural number n such that some B with n elements has no minimal element, choose any $b \in B$, then see if $B \setminus \{b\}$ has a minimal element c and compare c to b.) A set B can have several distinct minimal or maximal elements (example?), but any two minima of B must coincide, and the same holds for maxima: we write $\min B$ and $\max B$ for these unique elements of B if they exist. In an ordered set $(A, \leq)$, a set B can have at most one glb and at most one lub: these are denoted by $\text{glb}\, B$ and $\text{lub}\, B$ if they exist.

Examples. For any set S, glb's and lub's always exist in $(\mathcal{P}(S), \subseteq)$. For $X \subseteq \mathcal{P}(S)$

$$\text{glb}\, X = \cap X \quad \text{if} \quad X \neq \emptyset \quad \text{and} \quad \text{lub}\, X = \cup X$$

In $(\omega, \leq)$ the entire set ω does not possess any upper bound.

An ordered set $(A, \leq)$ is *well ordered* if every nonempty $B \subseteq A$ has a minimum. Applying this requirement to two-element subsets, we see that every well-ordered set is a chain. According to Proposition 8, Chapter I, $(\alpha, \leq)$ is a well-ordered set for every ordinal α. Moreover:

Proposition 3 *An ordered set $(A, \leq_A)$ is well ordered if and only if it is isomorphic to the natural order $(\alpha, \leq)$ for some ordinal α.*

Proof. (We denote the order on A by $\leq_A$ to clearly distinguish it from the ordinal inclusion order $\leq$ on α.) If

$$h : (\alpha, \leq) \to (A, \leq_A)$$

is an isomorphism, then for any nonempty $B \subseteq A$, the image by h of the first element of $h^*[B]$ is a minimum of B. The converse is proved just like Zermelo's Theorem (First Formulation, Chapter I). A choice function $c : \mathcal{P}^*(A) \to A$ is defined by

$$c(B) = \min B \quad \text{for all nonempty} \quad B \subseteq A$$

An *ordinal function* into A is defined as an injection $f : \beta \to A$ from some ordinal β into A such that for every $\gamma \in \beta$,

$$f(\gamma) = c(\{x \in A : x \neq f(\delta) \text{ for any } \delta \in \gamma\})$$

Then a particular ordinal β is defined such that there is no ordinal function $\beta \to A$, but for the predecessor α of β there is one, say $f : \alpha \to A$. Now f is not only a bijection but an order isomorphism as well. □

Zermelo's Theorem (Second Formulation). *Every set can be well ordered.*

Proof. Every set A is equipotent to some cardinal α, according to the First Formulation of Zermelo's Theorem given in Chapter I. Let $f : A \to \alpha$ be a bijection. Define $(A, \leq_A)$ by

$$a \leq_A b \quad \text{if and only if} \quad f(a) \leq f(b) \qquad □$$

Zorn's Lemma. *Let $(A, \leq)$ be an ordered set. If every chain $B \subseteq A$ possesses an upper bound in A, then A has a maximal element.*

Proof. Once again we proceed along the lines of the proof of Zermelo's Theorem (First Formulation). Let $c : \mathcal{P}^*(A) \to A$ be any choice function. For a totally ordered $B \subseteq A$ let $u(B)$ denote the set of its upper bounds in A. Observe that $u(B) \subseteq B$ if and only if $u(B)$ is a singleton consisting of a maximum of B. A function c' is defined, from the set $\mathcal{C}(A)$ of totally ordered subsets of A, to A, by letting

$$c'(B) = \begin{cases} \max B & \text{if } u(B) \subseteq B \\ c(u(B) \setminus B) & \text{otherwise} \end{cases}$$

Then *ordinal functions* are defined as order-preserving injections $f : \alpha \to A$ from an ordinal α to A such that for every $\beta \in \alpha$,

$$f(\beta) = c'(\{f(\gamma) : \gamma \in \beta\})$$

Ordinal functions are fully determined by their domains, and their images are well-ordered subsets of A.

The existence of an ordinal ν is ascertained for which no ordinal function $\nu \to A$ exists but such that ordinal functions exist from all ordinals less than ν. These ordinal functions are shown to agree on the common elements of their domains. The ordinal ν must have a predecessor λ, or else an ordinal function could be defined on the whole of ν by calling on the ordinal functions defined on the lesser ordinals. The image B of the ordinal function $\lambda \to A$ must have a maximum, and it can have no other upper bound in A, for else $u(B) \setminus B \neq \emptyset$ and an ordinal function mapping λ to $c'(B)$ could be defined on ν. Then $\max B$ is maximal in A. $\square$

The proofs of Zermelo's Theorem (First Formulation) and of Zorn's Lemma are often referred to as "proofs by transfinite induction." They rely on inference from lesser to greater ordinals. Any argument based on such inference but limited to finite ordinals is usually called "inductive" or "recursive."

The reader will find it instructive to prove the following propositions:

Proposition 4 *For every preordered set $(A, \leq)$ the following are equivalent:*

(i) *every nonempty subset $B \subseteq A$ has a maximal element,*

(ii) *no restriction of $\lesssim$ to a subset B of A is isomorphic to the natural order $(\omega, \leq)$,*
(iii) *if $B \subseteq A$ and the restriction $(B, \lesssim)$ is a chain, then the dual chain $(B, \gtrsim)$ is well ordered.*

Proposition 5 *For every preordered set $(A, \lesssim)$ the following are equivalent:*

(i) *every nonempty subset $B \subseteq A$ has a minimal element,*
(ii) *no restriction of $\lesssim$ to a subset B of A is isomorphic to $(\omega, \geq)$,*
(iii) *if $B \subseteq A$ and the restriction $(B, \lesssim)$ is a chain, then $(B, \lesssim)$ is well ordered.*

A preordered set $(A, \lesssim)$ obeying the conditions of Proposition 4 is said to satisfy the *ascending chain condition*. Dually, if the conditions of Proposition 5 hold, $(A, \lesssim)$ is said to satisfy the *descending chain condition*. Note that a chain is well ordered if and only if it satisfies the descending chain condition.

Let $(A, \leq)$ be any order and let $\mathcal{C}(A)$ be the set of its chains (totally ordered subsets of A). A *maximal chain* is a maximal element of the inclusion-ordered set $(\mathcal{C}(A), \subseteq)$. Observe that the union of the members of any chain in $(\mathcal{C}(A), \subseteq)$ is a chain in $(A, \leq)$. Thus Zorn's Lemma applies to $(\mathcal{C}(A), \subseteq)$:

Maximal Chain Theorem. *Every order has a maximal chain.*

In contrast, the next application of Zorn's Lemma concerns the relationships of implication between different orders (A, R), (A, Q) on the same set A. If $R \subseteq Q$ (i.e., aRb implies aQb), then Q is said to *extend* R or to be an *extension* of R. Indeed the definition applies not just to orders, but to any two binary relations on A. Of special interest are the preorders on A: the set of these preorders is denoted by $\mathcal{P}r(A)$. We have

$$\mathcal{P}r(A) \subseteq \mathcal{P}(A^2)$$

In fact, $\mathcal{P}r(A)$ is a closure system on A^2. It is in describing this closure that the combinatorial concepts of path and cycle will first show their usefulness.

Warning. Restriction to a subset and extension of relations are not "opposite operations" in any reasonable sense.

Let
$$(a_i : i \in n+1) = (a_0, \ldots, a_n)$$
be a sequence of $n+1$ distinct elements (natural $n \geq 1$, $a_i \neq a_j$ if $i \neq j$). The relation P defined by
$$P = \{(a_i, a_{i+1}) : i \in n\}$$
is called a *path* (of *length* n) *from* a_0 *to* a_n. If $n \geq 2$, then
$$P \cup \{(a_n, a_0)\}$$
is called a *cycle* (of *length* $n+1$). A binary relation R is called *acyclic* if no relation $P \subseteq R$ is a cycle.

Examples. $\{(1,5),(5,3),(3,6)\}$ is a path of length 3, from 1 to 6, but the relation $\{(1,5),(3,5),(3,6)\}$ is not a path. The relation $\{(1,5),(5,3),(3,1)\}$ is a cycle of length 3, and $\{(1,5),(3,5),(3,1)\}$ is acyclic.

Let R be any binary relation on a set A. Let $\lesssim$ be the preorder generated by R in the closure system $\mathcal{P}r(A)$. Then $a \lesssim b$ if and only if $a = b$ or there is a path in R from a to b.

A closure system related to $\mathcal{P}r(A)$ consists of the set $\mathcal{T}r(A)$ of all transitive relations on A. Obviously, $\mathcal{P}r(A) \subseteq \mathcal{T}r(A)$, i.e., the *preorder closure* of any relation is an extension of the *transitive closure*. For reflexive relations, the preorder and transitive closures coincide.

Linear Extension Theorem. *Every order relation has a linear order extension.*

Proof. Let (A, R) be an ordered set. Let $(O(R), \subseteq)$ be the set of all order relations on A extending R, ordered by inclusion. For any nonvoid $\mathcal{C} \subseteq O(R)$ that is totally ordered by $\subseteq$, $\cup \mathcal{C}$ is again an order relation on A, and it is an upper bound of $\mathcal{C}$ in $(O(R), \subseteq)$. Thus Zorn's Lemma applies and $(O(R), \subseteq)$ has a maximal element. We need only to show that every maximal member M of $O(R)$ is a total order. Were this not so, for some $a, b \in A$ we would have neither aMb nor bMa. Consider the relation
$$M' = M \cup \{(a,b)\}$$

It is acyclic because M is acyclic and (a,b) cannot belong to a cycle (if it did, bMa would follow from the transitivity of M). The transitive (preorder) closure of M' is then an order on A, contradicting the maximality of M. □

This result can be restated by saying that linear orders are precisely the maximal members of the inclusion-ordered set of all orders on a given set.

Linear Conjunction Theorem. *A relation on a set A is an order relation if and only if it is the intersection of some total orders defined on A.*

Proof. The "if" part can be easily verified by the reader. To prove the converse, let R be an order on A. For incomparable $a, b \in A$ the transitive closure of $R \cup \{(a,b)\}$ is an order R_{ab} on A. (Obviously $R_{ab} \neq R_{ba}$.) Let $\mathcal{L}_{ab}$ be the set of linear extensions of R_{ab}—a fortiori these are also linear extensions of R. Obviously $\mathcal{L}_{ab} \cap \mathcal{L}_{ba} = \emptyset$. Let

$$\mathcal{L} = \cup\{\mathcal{L}_{ab} : a, b \text{ incomparable in } (A, R)\}$$

(Obviously both $\mathcal{L}_{ab}$ and $\mathcal{L}_{ba}$ are included in $\mathcal{L}$ for incomparable a, b.) We have $R = \cap \mathcal{L}$. [Check membership of (a,b) in R and in $\cap \mathcal{L}$ according to whether aRb, bRa, or none of these holds.] □

The set $\mathcal{L}$ of linear extensions of R constructed in the above proof is generally not the only set of linear extensions that produces R by intersection. For example, if R is the identity relation on $3 = \{0, 1, 2\}$, then $\mathcal{L}$ contains all the six linear orders on 3. However, two linear orders will do: the natural order and its dual. The minimum cardinality of a set of total orders the intersection of which yields a given order R is called the *order dimension* of R. All identity relations have order dimension 2. On the other hand, the research-minded reader now has all the tools required to demonstrate that the order dimension of any inclusion-ordered power set $\mathcal{P}(S)$ is $\operatorname{Card} S$.

Zorn's Lemma is more than just a clever tool. Its depth is attested to by its axiomatic potential. If we postulated Zorn's Lemma without proof, calling it the "Maximality Axiom" if you wish, we could deduce from it the statement of the Axiom of Choice as follows. For a set S of nonvoid sets, let $\mathcal{A}$ be the set of all choice functions defined on subsets of S. For example, for $X \in S$ and $p \in X$ the function $c : \{X\} \to X$ given by $c(X) = p$ belongs to $\mathcal{A}$. Then order $\mathcal{A}$ by inclusion (remember that functions are sets), and apply Zorn.

EXERCISES

1. Show that given any preorder $\lesssim$ on a set A, the associated equivalence $\sim$ is the largest symmetric relation contained in $\lesssim$.

2. Determine the order-preserving functions
   (a) $(2, \leq) \to (2, \leq)$,
   (b) from a chain to an antichain,
   (c) from an antichain to a chain.

3. What are the minimal elements of $(\mathcal{P}(A)\setminus\{\emptyset\}, \subseteq)$? Does this ordered set have a minimum?

4. Verify that if A and B are equipotent sets, then $(\mathcal{P}(A), \subseteq)$ and $(\mathcal{P}(B), \subseteq)$ are isomorphic orders, and conversely.

5. Show that every finite chain is order-isomorphic to some natural order $(n, \leq)$, $n \in \omega$. Show that this does not generalize to infinite chains.

6. Let R and S be equivalence relations on sets A, B, respectively. Show that a bijective map $h : A \to B$ is an isomorphism $(A, R) \to (B, S)$ if and only if both of the following conditions hold:
   (a) for every equivalence class C of R, $h[C]$ is a class of S,
   (b) for every class K of S, $h^*[K]$ is a class of R.
   Verify that such an isomorphism h gives rise to a bijection between A/R and B/S.

7. Show that a relation R on a set A is a preorder (order, equivalence relation) if and only if every restriction of R to subsets of A with at most three elements is a preorder (order, equivalence relation, respectively).

8. Verify that for every ordered set $(A, \leq)$ we have $\mathrm{Aut}(A, \leq) = \mathrm{Aut}(A, \geq)$.

9. Show that the equivalence relations on a set A constitute a closure system on A^2.

10. A nonempty set $\mathcal{C}$ of subsets of a set S is said to have *finite character* if $A \in \mathcal{C}$ is equivalent to the condition that all finite subsets of A belong to $\mathcal{C}$. Show that such a $\mathcal{C}$ must have a maximal member.

11. Let S be a set of sets and let $\lesssim$ be the preorder closure of the relation
$$\{(x,y) \in S^2 : x \in y\}$$
Can you affirm that the descending or the ascending chain condition holds in $(S, \lesssim)$?

12. Write a computer program that for any given order relation R on a finite set of natural numbers
    (a) computes a linear extension of R,
    (b) determines a set of $\mathcal{L}$ of linear extensions such that $\cap \mathcal{L} = R$.

13. Write a program to determine whether two binary relations, each on a finite set of natural numbers, are isomorphic.

14. Write a program that for any binary relation on a finite set of natural numbers finds a path from any given a to any given b or determines that no such path exists.

2. LATTICES AND CLOSURES

An order $(A, \leq)$ is called a *lower* (or *meet*) *semilattice* if every two elements $x, y \in A$ have a glb. We write $x \wedge y$ for $\text{glb}\{x, y\}$. Dually, in an *upper* (or *join*) *semilattice* every two elements x, y have a lub, denoted by $x \vee y$. A *lattice* is both a lower and an upper semilattice. In a lattice, every nonempty finite set has both a glb and a lub. The order $(A, \leq)$ is a *complete lattice* if every subset B of A has both a glb and a lub, denoted by $\wedge B$ and $\vee B$, respectively.

Examples. Consider the set $O(A)$ of all order relations on a set A, ordered by inclusion: $(O(A), \subseteq)$. This is a lower semilattice, but not an upper semilattice. For an example of a lattice that is not a complete lattice, take the well-ordered set $(\omega, \leq)$. Complete lattices are best exemplified by $(\mathcal{P}(S), \subseteq)$, the inclusion-ordered set of subsets of a set. Here $\vee \mathcal{B} = \cup \mathcal{B}$ for all $\mathcal{B} \subseteq \mathcal{P}(S)$ and $\wedge \mathcal{B} = \cap \mathcal{B}$ if $\mathcal{B} \neq \emptyset$.

Obviously every complete lattice $(A, \leq)$ has both a minimum, $\text{glb}\, A$, and a maximum, $\text{lub}\, A$. It is remarkable that if in an ordered set $(A, \leq)$ all subsets of A have a glb, then $(A, \leq)$ is a complete lattice: for $B \subseteq A$,

$$\text{glb}\{x \in A : b \leq x \text{ for every } b \in B\}$$

is a lub of B. In particular:

Proposition 6 *Every closure system, ordered by inclusion, is a complete lattice, with $\wedge K = \cap K$ for each nonempty set K of closed sets.*

Proposition 7 *The inclusion-ordered set of all closure systems on any given set S is a complete lattice, with intersection as glb. The lub of any nonempty set k of closure systems is given by*

$$\vee k = \{\cap a : \emptyset \subset a \subseteq \cup k\}$$

Proof. The first statement follows by applying Proposition 6 to the closure system on $\mathcal{P}(S)$ formed by all closure systems on S. To prove the second statement, observe that

$$\mathcal{G} = \{\cap a : \emptyset \subset a \subseteq \cup k\}$$

is a closure system on S. It is a superset of each member of k, i.e., of $\cup k$. For any closure system $\mathcal{C}$ on S with $\cup k \subseteq \mathcal{C}$ we have necessarily $\mathcal{G} \subseteq \mathcal{C}$ because $\mathcal{C}$ is closed under intersection. □

Corollary. *Let $\mathcal{C}$ and $\mathcal{D}$ be closure systems on a set S. In the lattice of closure systems on S we have*

$$\mathcal{C} \wedge \mathcal{D} = \mathcal{C} \cap \mathcal{D}$$
$$\mathcal{C} \vee \mathcal{D} = \{C \cap D : C \in \mathcal{C}, D \in \mathcal{D}\}$$

Some closure systems play a distinguished role in the description of ordered sets $(A, \leq)$. A subset U of A is called an *upper section* if $x \in U$, $x \leq y$ imply $y \in U$. A *lower section* L is an upper section of the dual order, i.e., $x \in L$, $y \leq x$ imply $y \in L$. The set of upper sections is a closure system $\mathcal{U}$ on A, while the lower sections constitute another closure system $\mathcal{L}$. Consider now the least upper bound $\mathcal{C} = \mathcal{U} \vee \mathcal{L}$ in the lattice of closure systems on A. The closed sets in $\mathcal{C}$ are called *convex* (or *order convex*). Another synonym is *interval*.

It can be deduced from the Corollary of Proposition 7 that a subset C of A is convex if and only if $x, y \in C$ and $x \leq z \leq y$ imply $z \in C$. If $x \leq y$, then the convex closure of $\{x, y\}$ is the set

$$\{z \in A : x \leq z \leq y\}$$

which is called a *segment* and is denoted by $[x, y]$. The upper and lower sections generated by any $x \in A$ are the sets

$$\{z \in A : x \leq z\} \quad \text{and} \quad \{z \in A : z \leq x\}$$

denoted by $[x, \rightarrow)$ and $(\leftarrow, x]$ respectively. For $x \leq y$ we have

$$[x, y] = [x, \rightarrow) \cap (\leftarrow, y]$$

Examples. In the linearly ordered set $(\omega, \leq)$ of natural numbers,

$$5 = \{0, 1, 2, 3, 4\}$$

is a lower section. The complement

$$\omega \setminus 5 = \{5, 6, 7, \ldots\}$$

is an upper section. The segment $[3, 6]$ is precisely

$$\{3, 4, 5, 6\} = [3, \rightarrow) \cap (\leftarrow, 6]$$

In $(\mathcal{P}(\omega), \subseteq)$, the set $\{\{2, 4\}, \{4, 6\}, \{2, 6\}\}$ is convex, and the set

$$\{S \subseteq \omega : \{2, 4, 6\} \subseteq S \subseteq \{2, 4, 6, 7, 8\}\}$$

is a four-element segment.

A closure system $\mathcal{C}$ on a set S and the corresponding closure operator are called *algebraic* if the closure $\overline{A}$ of each $A \subseteq S$ is

$$\cup \{\overline{F} : F \subseteq A, \ F \text{ finite}\}$$

Equivalently, in an algebraic closure system a set C is closed if and only if $\overline{F} \subseteq C$ for all finite $F \subseteq C$. The term "algebraic" will be fully justified only in Chapter XI. Here let us observe only that for any n-ary operation on a set U, the set of subsets of U closed under the operation is an algebraic closure system. Permutation groups, rings of sets on a given set, seen is Chapter I, are examples of algebraic closure systems. Also the lower section, upper section, and convex closure systems on any ordered set are algebraic. On the other hand, the reader can verify that for any infinite ordinal α of the first kind,

$$\{S \subseteq \alpha : S \text{ has a maximum in } (\alpha, \leq)\}$$

is a nonalgebraic closure system on α.

Nested Union Theorem. *A closure system $\mathcal{C}$ on a set S is algebraic if and only if the union of each nonvoid chain in $(\mathcal{C}, \subseteq)$ belongs to $\mathcal{C}$.*

Proof. Assume that $\mathcal{C}$ is algebraic, and let $\mathcal{K}$ be a nonvoid chain in $(\mathcal{C}, \subseteq)$: a set of closed sets "nested by inclusion." Every finite $F \subseteq \cup \mathcal{K}$ is the subset of some $C \in \mathcal{K}$. (Were this not so, take the smallest n such that some F with n elements violates the claim, take any $x \in F$, take $C_1 \in \mathcal{K}$ with $x \in C_1$, and $C_2 \in \mathcal{K}$ with $F \setminus \{x\} \subseteq C_2$; then see whether C_1 is "nested" in C_2 or the other way round.) If $F \subseteq C$ with $C \in \mathcal{K}$, then for the closure of F we must certainly have $\overline{F} \subseteq C$ and hence $\overline{F} \subseteq \cup \mathcal{K}$. By definition of algebraicity, $\cup \mathcal{K}$ is closed.

Conversely, assume that $\cup \mathcal{K} \in \mathcal{C}$ for every nonvoid chain $\mathcal{K}$ in $(\mathcal{C}, \subseteq)$. Let C be a subset of S such that $\overline{F} \subseteq C$ for all finite $F \subseteq C$. Call a subset A of C admissible if

$$\overline{A \cup F} \subseteq C \quad \text{for all finite} \quad F \subseteq C$$

All finite subsets of C are of course admissible. Denote by $\mathcal{A}$ the set of admissible subsets of C. If $\mathcal{K}$ is a chain in $(\mathcal{A}, \subseteq)$, then $\cup \mathcal{K}$ is admissible because for any finite $F \subseteq C$,

$$\bigcup_{K \in \mathcal{K}} \overline{K \cup F} \in \mathcal{C} \quad \text{and} \quad \overline{(\cup \mathcal{K}) \cup F} \subseteq \bigcup_{K \in \mathcal{K}} \overline{K \cup F} \subseteq C$$

Thus Zorn's Lemma applies to $(\mathcal{A}, \subseteq)$. Let A be a maximal admissible subset of C. We must have $A = C$, for if $A \subset C$, then for any $x \in C \setminus A$ the set $A \cup \{x\}$ would be admissible, contradicting the maximality of A. □

An obvious case of algebraicity is at hand when every nonvoid subset of $\mathcal{C}$ possesses a maximal member, i.e., when $(\mathcal{C}, \subseteq)$ satisfies the ascending chain condition.

Proposition 8 *Let $\mathcal{C}$ be a closure system on a set S. The following conditions are equivalent:*

(i) $(\mathcal{C},\subseteq)$ satisfies the ascending chain condition,
(ii) $\mathcal{C}$ is algebraic and every closed set is generated by a finite set.

Proof. Assume (i). Then $\mathcal{C}$ is obviously algebraic. Define the *dimension* of a closed set C as the smallest cardinality of a generating subset of C. If not all closed sets had finite dimension, then let γ be the smallest infinite dimension encountered among all closed sets of infinite dimension. We shall derive a contradiction. Let C be a closed set of dimension γ, let $G \subseteq C$ be a generating set of cardinality γ, and let $f : \gamma \to G$ be a bijection. The cardinal γ has no ordinal predecessor: this may be a matter of course for insiders. Here it follows from the observation that for a predecessor β the closure of

$$\{f(\alpha) : \alpha \in \beta\}$$

would have a finite generating set F, and then the finite set

$$F \cup \{f(\beta)\}$$

would generate C. Thus, $\gamma = \cup \gamma$. For every $\beta \in \gamma$, the closure C_β of

$$\{f(\alpha) : \alpha \in \beta\}$$

has finite dimension, by the very definition of γ. The set

$$\{C_\beta : \beta \in \gamma\}$$

is a chain in $(\mathcal{C},\subseteq)$. By the assumption (i) it has a maximal member $C_\mu, \mu \in \gamma$, and then $C_\beta \subseteq C_\mu$ for all $\beta \in \gamma$. But from $\gamma = \cup \gamma$ it follows that

$$C = \bigcup_{\beta \in \gamma} C_\beta$$

and thus $C = C_\mu$ would have finite dimension, in contradiction with the definition of γ and C. Thus no closed set can have infinite dimension, and (ii) holds.

To establish the converse implication, assume (ii) and let $\mathcal{A}$ be a nonempty set of closed sets. Let $\mathcal{K}$ be a maximal chain in $(\mathcal{A},\subseteq)$. By algebraicity $\cup \mathcal{K}$ is closed. Let F be a finite set of generators for $\cup \mathcal{K}$ and let $c : F \to \mathcal{K}$ be a function with $x \in c(x)$ for all $x \in F$. (Is there any doubt that such a function c exists?) Then

$$\{c(x) : x \in F\}$$

is a subset of $\mathcal{K}$, it is a finite chain in $(\mathcal{A}, \subseteq)$, and as such it has a maximal member $c(y)$, $y \in F$. Since F generates $\cup \mathcal{K}$, we have $c(y) = \cup \mathcal{K}$. Thus $c(y)$ is the maximum of the chain $\mathcal{K}$, and since $\mathcal{K}$ is a maximal chain in $(\mathcal{A}, \subseteq)$, $c(y)$ is a maximal element in the order $(\mathcal{A}, \subseteq)$. □

A closure system satisfying the conditions of the above proposition is called *Noetherian*. An example is provided by the upper section closure system in $(\alpha, \leq)$ for any ordinal α. (What about lower sections?) More complex examples are afforded by rings, vector spaces, and matroids.

EXERCISES

1. Let $(S, \leq)$ be an ordered set. Verify that
   (a) $\emptyset$ has a lub if and only if S has a min,
   (b) $\emptyset$ has a glb if and only if S has a max.

2. Verify that every chain is a lattice.

3. Let k be a nonvoid set of closure systems on a set S. In the complete lattice of all closure systems on S, verify that
$$\vee k = \left\{ \bigcap_{\mathcal{K} \in k} c(\mathcal{K}) : c \in F \right\}$$
where F is the set of all choice functions on k.

4. Let $\mathcal{P}r(A)$ and $\mathcal{E}q(A)$ denote the set of preorders and the set of equivalence relations on a set A. Verify that, ordered by inclusion, these are lattices. Are these lattices complete? Observe that $\mathcal{E}q(A) \subseteq \mathcal{P}r(A)$ and show that for $R, Q \in \mathcal{E}q(A)$, the bounds $R \wedge Q$ and $R \vee Q$ taken in $\mathcal{E}q(A)$ coincide with the bounds taken in $\mathcal{P}r(A)$. Write a computer program to determine $R \vee Q$ for finite $A \subset \omega$.

5. Verify that the closure systems $\mathcal{P}r(A)$, $\mathcal{T}r(A)$ (transitive closure), $\mathcal{E}q(A)$ are algebraic on A^2.

6. Is the closure system of all closure systems on a set S an algebraic system on $\mathcal{P}(S)$?

7. Verify that every finite closure system is algebraic.

8. Verify that, in any ordered set, every antichain is convex.

9. Find all order-convex sets of $(\mathcal{P}(2), \subseteq)$.

3. COVERING RELATIONS

Let us turn our attention to the set $\mathcal{P}r(A)$ of preorders on a set A. Preorders constitute an algebraic closure system on the set A^2 of couples. Order relations on A form a subset $O(A)$ of $\mathcal{P}r(A)$. The preorder generated by any acyclic and antisymmetric binary relation is an order. We shall see that an important class of orders, including all finite ones, are canonically generated by certain acyclic relations having quite remarkable combinatorial properties.

In a preordered set $(A, \lesssim)$ an element y is said to *cover* x if $x < y$ and there is no z with $x < z < y$. The relation

$$\{(x,y) \in A^2 : y \text{ covers } x\}$$

is called the *covering relation* of the preorder, and the corresponding relational structure is denoted by $(A, \prec)$. For $x \prec y$ we also write $y \succ x$. We shall be mainly interested in covering relations of ordered sets. Here $x \prec y$ if and only if $x < y$ and $\{x, y\}$ is convex.

Examples. If α is any ordinal and $\beta, \gamma \in \alpha$, then in $(\alpha, \leq)$ we have $\beta \prec \gamma$ precisely when γ is the successor of β. In the ordered set $(\mathcal{P}(S), \subseteq)$ of subsets of a set S, a set A is covered by a superset B if and only if $B \setminus A$ is a singleton.

Let R be an arbitrary binary relation on a set A. If R is acyclic and so are the relations

$$R_{yx} = [R \setminus \{(x,y)\}] \cup \{(y,x)\}$$

for all $(x, y) \in R$, then R is called *strongly acyclic*.

Proposition 9 *A binary relation R is the covering relation of some order $\leq$ if and only if R is irreflexive, antisymmetric, and strongly acyclic.*

Proof. Acyclicity, irreflexivity, and antisymmetry of covering is obvious as $x \prec y$ implies $x < y$ and orders are acyclic. Further, if xRy in a covering relation R, then for any cycle C in R_{yx},

$$P = C \setminus \{(y,x)\}$$

is a path from x to y in R, and this path excludes the couple (x,y). Therefore, this path P has length greater than 1, $\cup P \neq \{x,y\}$. For any

$$z \in (\cup P) \setminus \{x,y\}$$

P would contain a path from x to z, as well as a path from z to y. But then $x < z < y$, a contradiction with $x \prec y$ proving the nonexistence of cycles in R_{yx}.

Conversely, if R is irreflexive, antisymmetric, and strongly acyclic, then it is acyclic and generates an order. It is straightforward to verify that the covering relation of this order is actually R. □

The covering relation of any order $\leq$ is included in any relation R that generates the order $\leq$: this is because between two elements related by covering there can only be one path in R, a rather short one.

Proposition 10 *For any ordered set $(A, \leq)$ the following are equivalent:*

(i) *among the relations on A that generate the order $\leq$ there is a minimal one, with respect to inclusion between relations,*
(ii) *every segment $[x,y]$ in $(A, \leq)$ has a finite maximal chain,*
(iii) *the order $\leq$ is generated by its covering relation.*

If these conditions hold, then the covering relation is actually the smallest one among all relations that generate the order.

Proof. Assuming (i), let R be a minimal relation generating $\leq$ in the preorder closure system on A^2. If $x < y$, then there is a path P in R from x to y: the restriction of $\leq$ to $\cup P$ is a maximal chain in $[x,y]$. The case $x = y$ being trivial, (ii) is established.

The implication of (iii) by (ii) is obvious.

Finally, by the remark preceding the statement of the proposition, (iii) implies both (i) and the conclusive clause. □

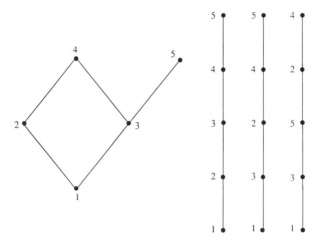

FIGURE 2.3 An order and some of its linear extensions.

In view of this proposition, an order is called *discrete* if it is generated by its covering relation. All finite orders are discrete. A simple example of an infinite discrete order is $(\omega, \leq)$. What about the inclusion-ordered power set $(\mathcal{P}(\omega), \subseteq)$?

Finite orders are usually represented by an arrow diagram of their covering relation. The direction of an arrow is often indicated implicitly only, by the relative position of the endpoints. If the endpoints a and b are joined by a nonhorizontal straight line and a is lower than b, then the arrow is understood to be directed from a to b. Figure 2.3 illustrates in this manner (the covering relation of) an order on five elements and three of the linear extensions.

The concept of covering in finite orders has been used, with a somewhat different terminology, to model the consecutivity of activities in scheduling large-scale industrial projects. The idea is to break down a project, such as building a rocket, into smaller activities, among which some may be design activities, others may involve fabrication or procurement of components from vendors, and there would be numerous activities of assembly, testing, documentation, etc. Obviously, some activities would have to precede others, at least for technical reasons. Here are the basics of the Project Evaluation and Review Technique/Critical Path Method (PERT/CPM) in the language of ordered sets.

Let A be a finite set, the elements of which are called *activities*. Let P be a binary relation on A. Every couple $(x,y) \in P$ is called a *(technical) precedence constraint*. If P fails to be irreflexive, antisymmetric, and acyclic, then the constraints are called *contradictory*, and their meaning should be mulled over once again. Otherwise consider the preorder closure of P: it is an order on the set A of activities, called *precedence order*. By adding harmless kick-off and ribbon-cutting ceremonies to the set of activities, it may be assumed that $(A, \leq)$ has both a minimum ($\min A$) and a maximum ($\max A$). Finally, a *duration function* $d : A \to \omega$ is supposed to be given, the natural number $d(a)$ being called the *duration* of the activity $a \in A$.

We look for a *project schedule*, i.e., a map $s : A \to \omega$ giving for each activity $a \in A$ its *start date* $s(a)$ and such that if aPb, then $s(a) + d(a) \leq s(b)$ in $(\omega, \leq)$: these are the schedule implications of the technical constraints. Since rockets are urgent, we look for a project schedule with the *completion date*

$$s(\max A) + d(\max A)$$

as early as possible: an *optimal schedule* is one whose completion date is less than or equal to that of any other schedule. The following is mathematically straightforward and allows project planners (with or without computers, depending on the number of activities involved) to find an optimal schedule.

Forward Pass Scheduling Theorem. *Given a set of activities A, a set of noncontradictory precedence constraints P generating a precedence order on A, and a duration function d on A, there is an optimal schedule s, and only one, such that*

(i) $s(\min A) = 0$
(ii) *for every $b \in A$*

$$s(b) = \max\{s(a) + d(a) : b \text{ covers } a \text{ in the precedence order}\}$$

The optimal schedule satisfying (i) and (ii) is called the *early start schedule*. An activity a is called *critical* if for any two optimal schedules s, s' we have $s(a) = s'(a)$: starting this activity any time later would delay the completion date.

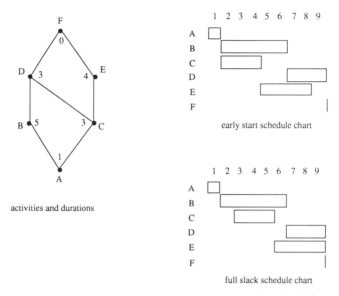

FIGURE 2.4 Activity precedence order and project schedules.

Let C be the completion date of an optimal schedule. Once C is known, we can easily find an optimal schedule σ unique with the following properties (σ is called the *full slack schedule*):

(i) $\sigma(\max A) + d(\max A) = C$
(ii) $\sigma(a) + d(a) = \min\{\sigma(b) : b \text{ covers } a\}$ for every $a \in A$

It is not difficult to show that the knowledge of the early start and full slack schedules (s and σ) are sufficient to determine which activities a are critical: exactly those for which $s(a) = \sigma(a)$. Project managers often focus their attention on *critical paths*: these are paths with critical activities only from $\min A$ to $\max A$ in the covering relation of the activity precedence order. In the project represented in Figure 2.4, there is one critical path, consisting of A, B, D, F.

Historical examples of the use of PERT/CPM include the Polaris missile and the cruiseship *France*.

EXERCISES

1. Verify that a relation is acyclic and antisymmetric if and only if its preorder closure is an order.

2. Write a computer program to determine (i) acyclicity and (ii) strong acyclicity of any relation on a finite set of natural numbers.

3. Let A be any set. Show that there is a bijection between the set of all discrete orders on A and the set of irreflexive, antisymmetric, strongly acyclic relations on A. Write a program to find the corresponding relations for finite $A \subset \omega$.

4. Write a program that for any acyclic and antisymmetric relation on a finite $A \subset \omega$ finds the covering relation of the order it generates.

5. Write a program that, given the covering relation of any order on a finite $A \subset \omega$, finds
   (a) the maximal elements of the order,
   (b) the mimimal elements,
   (c) a maximal chain,
   (d) a maximal antichain,
   (e) the order dimension.
   The program should also determine if the given order is a lattice.

6. Design a PERT/CPM program package, including
   (a) verification of whether a given set of technical constraints is noncontradictory,
   (b) computation of the early start schedule,
   (c) determination of all critical paths.

4. INTERSECTING CONVEX SETS

In a closure system $\mathcal{C}$, what does it take to guarantee that a given finite $\mathcal{F} \subseteq \mathcal{C}$ has nonempty intersection, $\cap \mathcal{F} \neq \emptyset$? Finding a common element of the closed sets given may be viewed as a geometric "common point" version of algebraically specifying "simultaneous solutions." For example, a natural number x satisfying $x \geq 1$, $x \leq 5$, $x \geq 2$, $x \leq 7$ is a common point of the segments [1, 5] and [2, 7].

Given a finite nonempty set $\mathcal{F}$ of closed sets, obviously $\cap \mathcal{F} \neq \emptyset$ if and only if $\cap \mathcal{S} \neq \emptyset$ for all $\emptyset \subset \mathcal{S} \subseteq \mathcal{F}$. But could the knowledge that $\cap \mathcal{S} \neq \emptyset$ for certain $\mathcal{S} \subset \mathcal{F}$ be sufficient to infer that $\cap \mathcal{F} \neq \emptyset$? Before restricting our attention to convex closure in linear orders, we make a general definition, to be called upon again in later chapters. We define the *Helly number of a closure system* $\mathcal{C}$ as the smallest nonzero cardinal h such that every nonempty finite $\mathcal{F} \subseteq \mathcal{C}$ with $\cap \mathcal{F} = \emptyset$ has a subset $\mathcal{S}$ with

$$1 \leq \text{Card}\, \mathcal{S} \leq h \quad \text{and} \quad \cap \mathcal{S} = \emptyset$$

In the worst case $h = \omega$. Helly number $h = 1$, on the other hand, means that every nonempty finite set $\mathcal{F}$ of nonvoid closed sets has nonempty intersection. Helly number $h = 2$ means that $h \neq 1$, and if every pair of members of a nonempty finite set $\mathcal{F} \subseteq \mathcal{C}$ have an element in common, then there is an element common to all members of $\mathcal{F}$.

Helly Theorem for Intervals. *The convex closure system on any linearly ordered set has Helly number at most 2.*

Proof. The convex closure system on a linearly ordered set $(A, \leq)$ has Helly number 1 if and only if A is void or it is a singleton. Suppose this is not the case. Let $\mathcal{F}$ be a finite nonempty set of convex sets and assume that

$$B \cap C \neq \emptyset \quad \text{for } B, C \in \mathcal{F}$$

For every $(B, C) \in \mathcal{F}^2$ let $p(B, C)$ be an element of $B \cap C$. We do not care whether $p(B, C)$ coincides with $p(C, B)$. Let

$$P = \{p(B, C) : (B, C) \in \mathcal{F}^2\}$$

Since $\mathcal{F}$ is finite, so are $\mathcal{F}^2$ and P, and for each $B \in \mathcal{F}$ the set $B \cap P$ is finite and has both a minimum and a maximum. For all $B, C \in \mathcal{F}$, we must have

$$\min(B \cap P) \leq \max(C \cap P)$$

Then

$$\max\{\min(B \cap P) : B \in \mathcal{F}\} \leq \min\{\max(C \cap P) : C \in \mathcal{F}\}$$

and both the max–min and min–max belong to every member of $\mathcal{F}$.
□

We conclude with establishing a lemma that clarifies the meaning of "Helly number 2" and that is often instrumental in proving Helly-type theorems.

Lemma for Helly Number 2. *The following conditions are equivalent for any closure system $\mathcal{C}$:*

(i) *the Helly number of $\mathcal{C}$ is at most 2,*
(ii) *if A, B, C are closed sets, every pair of which intersect, then $A \cap B \cap C \neq \emptyset$.*

Proof. A priori (ii) is weaker than (i). But assume that (ii) holds. If the Helly number of $\mathcal{C}$ is greater than 2, then some finite nonempty set $\mathcal{F}$ of closed sets has empty intersection, even though $A \cap B \neq \emptyset$ for all $A, B \in \mathcal{F}$. Choose such an $\mathcal{F}$ with as few members as possible, say n members $A_1, A_2, A_3, \ldots, A_n$. By (ii), we must have $n \geq 4$. Also by (ii)

$$A_1 \cap A_2 \cap A_i \neq \emptyset$$

for every $3 \leq i \leq n$. Let

$$\mathcal{F}' = \{A_1 \cap A_2\} \cup \{A_i : 3 \leq i \leq n\}$$

Since $\mathcal{F}'$ has fewer members than $\mathcal{F}$, and $X \cap Y \neq \emptyset$ for $X, Y \in \mathcal{F}'$, we must have $\cap \mathcal{F}' \neq \emptyset$. But $\cap \mathcal{F}' = \cap \mathcal{F}$ and thus $\cap \mathcal{F} \neq \emptyset$, a contradiction showing that $\mathcal{C}$ cannot have Helly number greater than 2. □

EXERCISES

1. What is the Helly number of the upper section closure system on a chain?

2. Verify that if $\mathcal{C}$ and $\mathcal{K}$ are closure systems with $\mathcal{C} \subseteq \mathcal{K}$, then the Helly number of $\mathcal{C}$ is at most that of $\mathcal{K}$.

3. What is the Helly number of the closure system $\mathcal{P}(S)$ on an infinite set S?

4. Can you generalize the Lemma for Helly Number 2 to higher Helly numbers?

BIBLIOGRAPHY

Peter C. FISHBURN, *Interval Orders and Interval Graphs: A Study of Partially Ordered Sets*. John Wiley & Sons 1985. A research monograph accessible to the novice. Included are results on linear extensions and order dimension.

Roland FRAÏSSÉ, *Theory of Relations*. North-Holland 1986. Written in the spirit of classical set theory and formal logic, this advanced but accessible monograph offers much material on ordered sets and n-ary relations, $n \geq 2$.

F. HAUSDORFF, Grundzüge einer Theorie der geordneten Mengen. *Math. Annalen* 65, 1908, pp. 435–505. A seminal research paper on linear orders. The concepts of interval and density are explored. An arithmetic of chains is developed, including that of ordinals, based on a lexicographic product construction.

Ferdinand K. LEVY, Gerald D. THOMPSON, and Jerome D. WIEST, The ABCs of the Critical Path Method. *Harvard Business Review*, Sept.–Oct. 1963, pp. 98–108. Short and simple, for managers, with a philosophical footnote by e. e. cummings.

Ferdinand K. LEVY and Jerome D. WIEST, *A Management Guide to PERT/CPM*. Prentice-Hall 1977. Mathematicians interested in PERT/CPM should consult texts designed for engineers and project managers such as this one, in order to gain an understanding of the application context.

Keith LOCKYER and James GORDON, *Critical Path Analysis and Other Project Management Techniques*. Ritman Publishing 1991. A reference to contemporary doctrine. Make sure you cut through some of the mathematically unnecessary notational complexities.

Bertrand RUSSELL, *Introduction to Mathematical Philosophy*. George Allen and Unwin 1970. The notion of order is investigated in Chapter 4.

G. SZÁSZ, *Introduction to Lattice Theory*. Academic Press 1963. Chapter I provides a good introduction to ordered sets in general.

CHAPTER III

GROUPS

1. BINARY OPERATIONS, HOMOMORPHISMS, AND CONGRUENCES

A *groupoid* is a set A together with a binary operation $\odot : A^2 \to A$. Formally it is defined as the couple $(A, \odot)$. When the operation is clearly understood, we speak simply of the "groupoid A." In a way analogous to binary relations, it is convenient to write $x \odot y$ for $\odot(x, y)$. Sometimes the symbol $\odot$ is omitted and the value of the operation on the couple (x, y) is simply denoted by the juxtaposition xy of the two operands.

With the binary operation of cardinal sum $(\omega, +)$ is a groupoid, and so is $(\omega, \cdot)$ with the cardinal product operation. Other examples are $(\mathcal{P}(S), \cup)$ and $(\mathcal{P}(S), \cap)$ for any set S. All these four examples are *associative*, i.e.,

$$(x \odot y) \odot z = x \odot (y \odot z)$$

for all elements x, y, z of the underlying set. Also the *commutative* law

$$x \odot y = y \odot x$$

holds in these groupoids. That this is far from being guaranteed in general is seen as follows. Let m and n be two fixed natural numbers

and define a binary operation on ω by
$$x \odot y = (m \cdot x) + (n \cdot y)$$
Then $\odot$ is commutative only if $m = n$, and it is associative only if neither m nor n is greater than 1: verification of this is an excellent exercise.

Let $(A, \odot)$ be a groupoid. According to an observation made in Chapter I, the set $\mathcal{C}$ of subsets of A closed under the operation $\odot$ is a closure system, and it is indeed an algebraic closure system (Chapter II). Every closed set C, together with the restriction of $\odot$ to C^2, is then a groupoid. It is called a *subgroupoid* of $(A, \odot)$. The operation of the subgroupoid is usually denoted by the same symbol $\odot$, and we often refer to $(C, \odot)$ as the "subgroupoid C."

Let $(A, \odot)$ and $(B, \top)$ be two groupoids. (Later we shall not always bother to use distinct symbols $\odot$, $\top$ for two groupoids if the context clearly indicates the distinctness of the operations.) A *homomorphism* from $(A, \odot)$ to $(B, \top)$ is a function $h : A \to B$ such that
$$h(x \odot y) = h(x) \top h(y) \qquad \text{for all} \quad x, y \in A$$
In shorthand $h : (A, \odot) \to (B, \top)$.

Example. Let S be any set. A homomorphism h from $(\mathcal{P}(S), \cap)$ to the subgroupoid $(\{0,1\}, \cdot)$ of $(\omega, \cdot)$ is given by
$$h(B) = \begin{cases} 0 & \text{if} \quad S \setminus B \text{ is infinite} \\ 1 & \text{if} \quad S \setminus B \text{ is finite} \end{cases}$$
If S is infinite, then this same function h is not a homomorphism from $(\mathcal{P}(S), \cup)$ to $(\{0,1\}, \cdot)$.

Proposition 1 *The composition of two groupoid homomorphisms*
$$h : (A, \circ_1) \to (B, \circ_2) \qquad \text{and} \qquad g : (B, \circ_2) \to (C, \circ_3)$$
is a homomorphism from $(A, \circ_1)$ to $(C, \circ_3)$. The identity mapping on the underlying set of any groupoid is a homomorphism of that groupoid to itself. The inverse of a bijective homomorphism is a homomorphism.

Proof. For the composition, using simplified notation,
$$(gh)(xy) = g[h(x)h(y)] = g[h(x)]g[h(y)]$$

The statement about the identity is obvious. For the inverse, if $h(x)$, $h(y)$ are in Im h, apply the inverse h^* to both sides of the equation $h(x)h(y) = h(xy)$. □

A bijective homomorphism between groupoids is called an *isomorphism*. The groupoids $(A, \odot)$ and $(B, \top)$ are called *isomorphic* if there exists an isomorphism $(A, \odot) \to (B, \top)$.

Example. For any nonzero natural number $n \in \omega$ let

$$\text{Mult}\, n = \{n \cdot x : x \in \omega\}$$

Then $(\text{Mult}\, n, +)$ is a subgroupoid of $(\omega, +)$ and an isomorphism

$$h : (\omega, +) \to (\text{Mult}\, n, +)$$

is given by $h(x) = n \cdot x$.

Proposition 2 *Let $h : A \to B$ be a homomorphism between groupoids. The image by h of any subgroupoid of A is a subgroupoid of B. The inverse image of any subgroupoid of B is a subgroupoid of A.*

A *congruence* relation on a groupoid $(A, \odot)$ is an equivalence relation $\equiv$ on A such that $x \equiv x'$, $y \equiv y'$ together imply $x \odot y \equiv x' \odot y'$.

Congruence Example. For any infinite set S, let $\equiv$ be the equivalence relation on $\mathcal{P}(S)$ where $x \equiv x'$ if and only if either both x and x' are finite or both are infinite. This is a congruence relation on $(\mathcal{P}(S), \cup)$ but not on $(\mathcal{P}(S), \cap)$.

An equivalence relation E on A is a congruence of the groupoid $(A, \odot)$ if and only if for any equivalence classes B, C the *product set*

$$\{b \odot c : b \in B,\ c \in C\}$$

is included in some class D. As such a D is unique, this defines a binary operation $B \odot C = D$ on the quotient set A/E of all equivalence classes: $(A/E, \odot)$ is called the *quotient groupoid of* $(A, \odot)$ *by the congruence E*. The canonical surjection $A \to A/E$ is a homomorphism. Conversely, if h is any homomorphism, then the induced equivalence defined on its domain by

$$x \equiv y \quad \text{if and only if} \quad h(x) = h(y)$$

is a congruence of the domain groupoid.

Example. In the Congruence Example above, the quotient is isomorphic to $(\{0,1\},\cdot)$ where "$\cdot$" is the cardinal product.

A *neutral element* in a groupoid $(A,\odot)$ is an element $u \in A$ such that
$$u \odot x = x \odot u = x \quad \text{for all} \quad x \in A$$
If we had another neutral w, then $u \odot w$ would have to be equal to both u and w, implying $u = w$. Thus a groupoid has at most one neutral element. It may have none: in $(\omega,+)$ consider the subgroupoid
$$[n,\to) = \{x \in \omega : n \leq x\}$$
for any fixed nonzero $n \in \omega$. Zero is of course neutral in $(\omega,+)$.

A *semigroup* is an associative groupoid, i.e., one where
$$(x \odot y) \odot z = x \odot (y \odot z) \quad \text{for all elements} \quad x,y,z$$
The reader will then find the use of brackets superfluous, e.g., we can write $x \odot y \odot z$ instead of $(x \odot y) \odot z$ or $x \odot (y \odot z)$. A semigroup with a neutral element u is called a *monoid*.

A subgroupoid B of a semigroup A is called a *sub-semigroup*. If A is a monoid and B also contains the neutral element of A, then B is a *submonoid*. Not every sub-semigroup of a monoid is a submonoid. On the other hand, every quotient of a semigroup is a semigroup, and every quotient of a monoid is a monoid.

Examples. (1) For any set S, the function set S^S is a monoid with the usual composition of functions as binary operation. The set of constant functions from S to S is a semigroup under composition, but it is not a monoid if S has more than one element. (2) Both groupoids $(\omega,+)$ and $(\omega,\cdot)$ are monoids. (3) Let S be an infinite set. Then $(\mathcal{P}(S),\cup)$ and $(\mathcal{P}(S),\cap)$ are monoids. The set of finite subsets of S constitutes a submonoid of $(\mathcal{P}(S),\cup)$ but only a sub-semigroup in $(\mathcal{P}(S),\cap)$.

By a *semigroup homomorphism* we simply mean a groupoid homomorphism between semigroups. With monoids we must be more careful. If A and B are monoids and $h : A \to B$ is a groupoid homomorphism that maps the neutral element of A to that of B, then h is called a *monoid homomorphism*. [For example, the constant zero function from $(\omega,\cdot)$ to itself is not a monoid homomorphism to $(\omega,\cdot)$.]

A homomorphism from a groupoid A to itself is called an *endomorphism*. The set $\operatorname{End} A$ of all such endomorphisms, together with composition as a binary operation, is a monoid. It is indeed a submonoid of A^A.

For every element a of a monoid $(A,\odot)$ with neutral u, there is a unique function $e_a : \omega \to A$ such that

$$e_a(0) = u \quad \text{and} \quad e_a(n+1) = e_a(n) \odot a \quad \text{for all} \quad n \in \omega$$

For example, if $(A,\odot) = (\omega, +)$, then $e_a(n) = n \cdot a$. If $(A,\odot) = (\omega,\cdot)$, then

$$e_a(n) = \operatorname{Card}(a^n) \quad \text{for all} \quad a, n \in \omega$$

Remember that a^n is the set of all maps $n \to a$. However, it is customary to denote, in any monoid $(A,\odot)$, the element $e_a(n)$ simply by a^n, and we shall follow this tradition. The reader can then verify that the equations

$$(a^n)^m = a^{n \cdot m}$$

and

$$(a^n) \odot (a^m) = a^{n+m}$$

hold for all $a \in A$, $n, m \in \omega$. Furthermore, if $(A,\odot)$ is a commutative monoid, then for all $a, b \in A$ and $n \in \omega$ we have

$$(a \odot b)^n = (a^n) \odot (b^n)$$

Two elements a, b of a not necessarily commutative monoid A are said to *commute* if $a \odot b = b \odot a$. A finite family $(a_i : i \in I)$ of elements of A is called *commutative* if a_i and a_j commute for every $i, j \in I$. Let C be the set of all such commutative families. There is a unique function $p : C \to A$ that maps the empty family to the neutral element of A and such that

$$p(a_i : i \in I) = p(a_i : i \in I\setminus\{j\}) \odot a_j$$

for every family $(a_i : i \in I)$ in C and index $j \in I$. The value of p on $(a_i : i \in I)$ is called the *product* of that family and it is denoted by $\prod_{i \in I} a_i$. Observe that if $I = \{1,2\}$, then this product is the element $a_1 \odot a_2 = a_2 \odot a_1$. On the other hand, let $(a_i : i \in I)$ be a constant

family, with a finite indexing set I. Now all the a_i are equal to the same element a and
$$\prod_{i \in I} a_i = a^{\operatorname{Card} I}$$

If S is a finite set of pairwise commuting elements of the monoid, then the family $(a : a \in S)$ is commutative, and $\prod_{a \in S} a$ is also referred to as the *product of the set S*.

Let B be a set of pairwise commuting elements of a monoid A. Then the submonoid $\overline{B}$ generated by B consists of all products of finite families of elements of S.

Note on Additive Notation. The binary operation of a monoid can be denoted by any symbol you wish. If this symbol is a straight cross "+" or some variant of it, such as in $(\omega, +)$ or $(\mathcal{P}(S), +)$, then the product of a family of monoid elements is customarily denoted by

$$\sum_{i \in I} a_i \quad \text{instead of} \quad \prod_{i \in I} a_i$$

and it is called "sum" rather than "product." Usually this is done only in commutative monoids. We also write na instead of a^n in this case. But never mind the notation—if you never forget what it means.

In a monoid with neutral u, an *inverse* of an element a is an element a' such that
$$a \odot a' = a' \odot a = u$$
If a had another inverse a'', then we would have
$$(a' \odot a) \odot a'' = u \odot a'' = a'' \quad \text{and} \quad a' \odot (a \odot a'') = a' \odot u = a'$$
and by associativity $a' = a''$. Thus every element a has at most one inverse; it shall be denoted by a^*. It may have none: for any set S, $(\mathcal{P}(S), \cap)$ is a monoid with neutral element S, and S is the only element that has an inverse, namely itself, $S^* = S$. Other examples of monoids with a scarcity of inverse elements are $(\omega, +)$ and $(\omega, \cdot)$, the natural numbers with the cardinal sum and product, respectively. What about a map composition monoid S^S?

A *group* is a monoid in which every element has an inverse. The fact that neither $(\omega, +)$ nor $(\omega, \cdot)$ is a group proves to be the grandmother of fundamental inventions in algebra. Groups do exist, however, in nature: with "+" the symmetric difference of sets, $(\mathcal{P}(S), +)$ is always a group. So is every permutation group on any set S, with the composition $f \circ g$ as binary operation, justifying the terminology introduced in Chapter I. Note that on every singleton set there is a *trivial* group structure.

Obviously the groups $(\mathcal{P}(S), +)$ are commutative. On the other hand, the permutation group of all bijections $S \to S$ is not commutative if $\operatorname{Card} S \geq 3$: for three distinct a, b, c in S define bijections f and g by

$$f(a) = b, \qquad f(b) = a, \qquad f(x) = x \qquad \text{for} \quad x \neq a, b$$
$$g(b) = c, \qquad g(c) = b, \qquad f(x) = x \qquad \text{for} \quad x \neq b, c$$

and see if $f \circ g = g \circ f$.

By a *group homomorphism* we simply mean a groupoid homomorphism $h : A \to B$ where A and B are groups. The group structure forces h to be a monoid homomorphim as well between A and B, and to map the inverse x^* of any $x \in A$ to the inverse of $h(x)$ in B,

$$h(x^*) = [h(x)]^*$$

An isomorphism from a groupoid A to itself is called an *automorphism*. The set of all automorphisms is a permutation group on A.

Inverse elements in a group A obey the rules

$$(x^*)^* = x \qquad \text{and} \qquad (xy)^* = y^* x^*$$

For every $a \in A$, the mapping $h_a : A \to A$ given by

$$h_a(x) = axa^*$$

is an automorphism, called *conjugation* or *inner automorphism*. The element axa^* is called a *conjugate* of x in the group A. Inner automorphisms form a permutation group on A. Every inner automorphism may also be viewed as a unary operation on A. Note that if A is commutative, then all inner automorphisms coincide with the identity mapping id_A.

If A is a group and a submonoid B of A contains the inverse b^* for every $b \in B$, then B is called a *subgroup*. If $B \subset A$, then B is called a *proper subgroup* of A. Subgroups form an algebraic closure system on A. A subgroup of A is called *normal* if it is closed under every inner automorphism of the group A. Normal subgroups form an algebraic closure system on A, included in the closure system of all subgroups. Both these closure systems, when ordered by inclusion, are of course complete lattices. The two systems coincide in commutative groups, because here all subgroups are normal.

Note that if B is a proper subgroup of A, then the definition of conjugacy permits two elements of B to be conjugates in A without being conjugates in B.

For any normal subgroup N of a group A, consider the binary relation
$$\mathcal{R}(N) = \{(x,y) \in A^2 : x^*y \in N\}$$
This is quite easily seen to be an equivalence relation. Let us show that $\mathcal{R}(N)$ is indeed a congruence of A. Assume $x\mathcal{R}(N)y$, $v\mathcal{R}(N)w$: we need to show that $xv\mathcal{R}(N)yw$. We have $x^*y \in N$ and $v^*w \in N$, and by conjugation $v^*(x^*y)v \in N$. Observe that
$$v^*w = v^*(y^*y)w$$
By closure under the group operation,
$$[v^*(x^*y)v][v^*(y^*y)w] \in N$$
By associativity this element is equal to $(v^*x^*)(yw) = (xv)^*(yw)$, establishing $xv\mathcal{R}(N)yw$.

Let R be any congruence relation on a group A. Let $\mathcal{N}(R)$ be the congruence class containing the neutral element u of A. It is not difficult to see that $\mathcal{N}(R)$ is a normal subgroup. Let us just verify closure under conjugation: for $x \in \mathcal{N}(R)$, i.e., xRu and any $a \in A$, using aRa and a^*Ra^*, we get
$$(ax)a^* R (au)a^*$$
i.e.,
$$axa^*Ru, \qquad axa^* \in \mathcal{N}(R)$$
It is quite obvious that for every normal subgroup N,
$$\mathcal{N}(\mathcal{R}(N)) = N$$
and for every congruence R,
$$\mathcal{R}(\mathcal{N}(R)) = R$$
The observation that $N \subseteq M$, for any two normal subgroups, is equivalent to $\mathcal{R}(N) \subseteq \mathcal{R}(M)$ allows us to state the following result:

Proposition 3 *Let A be any group. An order isomorphism is established from the set of normal subgroups of A, ordered by inclusion, to the set of congruence relations of A, ordered by inclusion (implication), by associating to each normal subgroup N the congruence*

$$\mathcal{R}(N) = \{(x,y) \in A^2 : x^*y \in N\}$$

The normal subgroup N and the congruence $\mathcal{R}(N)$ are said to *correspond* to each other and $\mathcal{R}(N)$ is also called *congruence modulo* N. For $x\mathcal{R}(N)y$ we usually write

$$x \equiv y \bmod N$$

The quotient groupoid $A/\mathcal{R}(N)$ is actually a group. It is also denoted by A/N and called the *quotient group of A by the normal subgroup N*. For a group homomorphism $h : A \to B$, the class of the neutral element in the congruence $\mathcal{R}_h$ of A induced by h is a normal subgroup, called the *kernel* of h and denoted by $\mathrm{Ker}\,h$. (We have $A/\mathcal{R}_h = A/\mathrm{Ker}\,h$.) On the other hand, $\mathrm{Im}\,h$ is a subgroup of B, but it is not necessarily normal.

Proposition 4 *Let $h : A \to B$ be a group homomorphism. Then $A/\mathrm{Ker}\,h$ is isomorphic to $\mathrm{Im}\,h$.*

Proof. Let γ be a choice function associating to every class C of the congruence induced by h on A an element of C. Consider

$$f : A/\mathrm{Ker}\,h \to \mathrm{Im}\,h$$

given by

$$f(C) = h(\gamma(C)) \qquad \square$$

EXERCISES

1. Let R be any binary relation. Verify that the set of all relation-preserving maps $f : R \to R$ is a monoid under composition. Verify that $\mathrm{Aut}\,R$ is a group.

2. Find all possible group structures on a three-element set.

3. Let S be a set of elements of a group G. Consider the relation R on G where xRy means $x^*y \in S$. Show that

(a) R is an equivalence relation if and only if S is a subgroup,
(b) R is a preorder if and only if S is a submonoid,
(c) R is an order if and only if S is a submonoid in which only the neutral has an inverse.

4. Let $(G, \cdot)$ be a group. Define a groupoid operation $\odot$ on $\mathcal{P}(G)$ by
$$A \odot B = \{a \cdot b : a \in A, b \in B\}$$
Verify that this is a monoid. Verify that every quotient group of G is a subgroupoid of $\mathcal{P}(G)$. Is every quotient of G a submonoid of $\mathcal{P}(G)$? Verify that the set of normal subgroups of G is a submonoid of $\mathcal{P}(G)$. Show that for any two normal subgroups N, M of G, $N \odot M$ coincides with the least upper bound $N \vee M$ in the lattice of normal subgroups of G.

5. Write a computer program to verify if a given equivalence relation, on a group whose underlying set is a finite $A \subset \omega$, is a congruence.

6. For a groupoid structure on a set A, show that the congruence relations form a closure system on A^2, and hence they form an inclusion-ordered lattice.

7. Show that, given a group G, the various cosets of the various subgroups plus the empty set $\emptyset$ constitute a closure system on G.

8. Verify that the kernel of a group homomorphism is the set of elements mapped to the neutral element of the codomain group.

2. PERMUTATION GROUPS

We have noted that for any set A, the set $\Sigma(A)$ of all bijections $A \to A$ constitutes a group under the functional composition $\circ$. The members of $\Sigma(A)$ are called *permutations* of A, and $\Sigma(A)$ is often called the *symmetric group on* A. The permutation groups on A, as defined in Chapter I, are then simply the subgroups of $\Sigma(A)$. If B is any other set equipotent to A, say via a bijection $f : A \to B$, then the function $F : \Sigma(A) \to \Sigma(B)$ given by
$$F(\sigma) = f \circ \sigma \circ f^*$$

is an isomorphism between the two symmetric groups. Informally speaking this means that only the cardinality of A matters if we are interested in the structure of $\Sigma(A)$; the nature of the elements of A does not matter. In particular the symmetric group on any finite set of cardinality $n \in \omega$ would be well understood by studying the symmetric group on the particular set

$$[1, n] = \{x \in n + 1 : 1 \leq x \leq n\}$$

Why $[1, n]$ rather than the cardinal n itself? Tradition and notational convenience. We write Σ_n for $\Sigma([1, n])$. The composition of two premutations is often denoted by juxtaposition.

A permutation σ in Σ_n is traditionally denoted by a two-row table between brackets, the first row being simply the list of the numbers 1 to n and the second row indicating, under each $i \in [1, n]$ of the first row, the image element $\sigma(i)$. For example, for $n = 4$, if $\sigma(1) = 4$, $\sigma(2) = 2$, $\sigma(3) = 1$, and $\sigma(4) = 3$, then σ is written as

$$\begin{pmatrix} 1 & 2 & 3 & 4 \\ 4 & 2 & 1 & 3 \end{pmatrix}$$

In this chapter, permutations may appear responsible for the widely held preconception that mathematics is about counting. Let us first count the permutations that make up a symmetric group.

The *factorial* of a cardinal number n (finite or infinite) is defined as the cardinality of $\Sigma(n)$. The factorial of n is denoted by $n!$.

Observe that if A and B are equipotent sets of common cardinality n, then the set of bijections $A \to B$ has cardinality $n!$ as well. Indeed, let $\text{bi}(A, B)$ be the set of these bijections. Choose any $f \in \text{bi}(A, B)$. The function

$$F : \Sigma(A) \to \text{bi}(A, B)$$

given by

$$F(\sigma) = f \circ \sigma$$

is bijective onto $\text{bi}(A, B)$.

Proposition 5 $0! = 1$ *and for every cardinal number n,*

$$(n + 1)! = n! \cdot (n + 1)$$

Proof. For $0! = 1$ observe that the empty function ("empty set of ordered pairs") is the sole bijection $\emptyset \to \emptyset$.

Let A be any set of cardinality n and $\{s\}$ a singleton disjoint from A, $s \notin A$. By definition, $n+1$ is the cardinality of $B = A \cup \{s\}$. By the Counting Lemma of Chapter I, for every $b \in B$ there is a bijection
$$f_b : A \to B \setminus \{b\}$$
Define a function $F : \Sigma(A) \times B \to \Sigma(B)$ by
$$F(\sigma, b)(s) = b$$
$$F(\sigma, b)(x) = f_b(\sigma(x)) \quad \text{for } x \in A$$
This function F is bijective, establishing the equipotence of the Cartesian product $\Sigma(A) \times B$ and $\Sigma(B)$. But
$$\mathrm{Card}(\Sigma(A) \times B) = n! \cdot (n+1) \quad \text{and} \quad \mathrm{Card}\,\Sigma(B) = (n+1)!$$
$\square$

The equality
$$(n+1)! = n! \cdot (n+1)$$
is often referred to as the "inductive definition" of factorials. It allows the computation of finite factorials, e.g.,
$$1! = 1 \cdot 1 = 1$$
$$2! = 1 \cdot 2$$
$$3! = 1 \cdot 2 \cdot 3$$
$$4! = 1 \cdot 2 \cdot 3 \cdot 4$$
$$5! = 1 \cdot 2 \cdot 3 \cdot 4 \cdot 5$$
and so on, providing in particular a rigorous proof that for finite A, the set $\Sigma(A)$ of permutations is also finite.

Let G be any "abstract" group: a group that is not necessarily a permutation group. Let u be the neutral element of G and let the group operation be denoted by the dot symbol $\cdot$. Let A be any set. An *action* of G on A is a mapping $\alpha : G \times A \to A$ such that
$$\alpha(u, x) = x$$
$$\alpha(h, \alpha(g, x)) = \alpha(h \cdot g, x)$$

for all $h,g \in G$, $x \in A$. Often we write simply gx for $\alpha(g,x)$, and then

$$h(gx) = (h \cdot g)x \quad \text{for all} \quad h,g \in G, \quad x \in A$$

Example. The two-element group $\mathcal{P}(1) = \{0,1\}$ with the symmetric difference operation "+" acts on any power set $\mathcal{P}(S)$ as follows: for $X \in \mathcal{P}(S)$, let

$$0X = X \quad \text{and} \quad 1X = S \setminus X$$

Every group G can be made to act on G itself in a number of ways. *Action by translation* is defined by

$$gx = g \cdot x$$

and *action by conjugation* is given by

$$gx = g \cdot x \cdot g^*$$

for all $g, x \in G$. *Every permutation group G on a set A acts on A by*

$$gx = g(x)$$

for $g \in G$, $x \in A$.

Given an abstract group G acting on a set A, for every $g \in G$ the mapping $\sigma_g : A \to A$ given by

$$\sigma_g(x) = gx$$

is a permutation of A. The function $f : G \to \Sigma(A)$ given by

$$f(g) = \sigma_g$$

is a homomorphism from the group G to $\Sigma(A)$. It is easy to see that different group actions give rise to different homomorphisms this way. Conversely, every homomorphism f from an abstract group G to a symmetric group $\Sigma(A)$ corresponds to a group action on A given by

$$gx = [f(g)](x)$$

for $g \in G$, $x \in A$. Thus the theory of group actions is essentially the theory of homomorphisms from abstract groups to permutation groups.

For an abstract group G acting on itself by translation, the homomorphism f corresponding to the action is injective into $\Sigma(G)$: this is because for the neutral element u of G, obviously

$$gu \neq g'u \quad \text{if} \quad g \neq g'$$

Thus f is an isomorphism between G and the subgroup $\operatorname{Im} f$ of $\Sigma G)$, yielding this century-old result:

Cayley Representation Theorem. *Every group is isomorphic to some permutation group.*

Given a group G acting on a set A, consider the binary relation R on A in which

$$xRy \quad \text{if and only if} \quad gx = y \quad \text{for some} \quad g \in G$$

This relation R is easily seen to be an equivalence. The classes of R are called *orbits*. For $x \in A$, we say that a group element $g \in G$ *stabilizes* x, or *fixes* x, if $gx = x$. Alternatively, x may be called a *fixed point* of g. The set of group elements g fixing a given x is a subgroup of G, called the *stabilizer* of x, denoted by S_x. Assume that x and y are in the same orbit, say $hx = y$ for some $h \in G$. For this group element h consider the conjugation by h, an inner automorphism of G. For $g \in S_x$,

$$h \cdot g \cdot h^* \in S_y$$

and it is not difficult to verify that the restriction of this conjugation to S_x is an isomorphism $S_x \to S_y$. Thus all stabilizers of elements in the same orbit are isomorphic and a fortiori equipotent.

Orbit Counting Theorem. *If O is any orbit of a group G acting on a set A and if S_a is the stabilizer of any $a \in O$, then*

$$\operatorname{Card} O \cdot \operatorname{Card} S_a = \operatorname{Card} G$$

Proof. For each $x \in O$ let g_x be an element of G such that $g_x a = x$. Define a function $f : O \times S_a \to G$ by

$$f(x,h) = g_x \cdot h$$

Is f injective? If $f(x,h) = f(y,k)$, i.e., $g_x \cdot h = g_y \cdot k$, then

$$x = g_x a = g_x(ha) = (g_x \cdot h)a = (g_y \cdot k)a = g_y(ka) = g_y a = y$$

and applying the inverse of $g_x = g_y$,
$$h = g_x^* \cdot (g_x \cdot h) = g_y^* \cdot (g_y \cdot k) = k$$
Thus f is indeed injective. To see that it is also surjective, for $g \in G$ let
$$x = ga \quad \text{and} \quad h = g_x^* \cdot g$$
We have $f(x,h) = g$. Therefore f is bijective, proving the theorem. □

Any group G can be made to act not just on G itself but also on the power set $\mathcal{P}(G)$, by letting
$$gH = \{g \cdot x : x \in H\} \quad \text{for} \quad g \in G, \quad H \in \mathcal{P}(G)$$
Let H be a subgroup of G: what is the orbit O of H? By definition
$$O = \{gH : g \in G\}$$
The sets gH in O are called *(left) cosets* of H. For every coset gH the map associating to each $x \in H$ the element $g \cdot x$ of gH is a bijection from H to gH. Thus every coset is equipotent to H.

If two cosets intersect, $gH \cap g'H \neq \emptyset$, then for some $h, h' \in H$,
$$g \cdot h = g' \cdot h'$$
$$g = g' \cdot h' \cdot h^*$$
and hence $gH \subseteq g'H$. By the same argument we also have $g'H \subseteq gH$, and therefore any two intersecting cosets must coincide. Since every $g \in G$ belongs to some coset (namely to gH), the cosets constitute a partition of G. The cardinality of the set O of cosets is called the *index* of H in G, denoted by $[G : H]$. Observe that if H is a normal subgroup of G, then the cosets are precisely the congruence classes modulo H. In this case $[G : H]$ is equal to the cardinal number of elements of the quotient group G/H.

Lagrange's Subgroup Counting Theorem. *For any subgroup H of a group G we have*
$$\operatorname{Card} H \cdot [G : H] = \operatorname{Card} G$$
If H is a normal subgroup, then
$$\operatorname{Card} H \cdot \operatorname{Card}(G/H) = \operatorname{Card} G$$

Proof. Proposition 2 of Chapter II. (Alternatively, we may recognize here an instance of the Orbit Counting Theorem.) □

Example of Application. Let G be a group of order 5. What are the subgroups of G? Obviously G itself and the trivial subgroup reduced to the neutral element are subgroups. Indeed there is none else. For if H were a subgroup with n elements, $1 < n < 5$, then we would have
$$n \cdot [G : H] = 5$$
This can be ruled out by elementary arguments, without recourse to any number-theoretical concepts.

EXERCISES

1. Design a program that for any subset S of Σ_n
   (a) finds the number of elements of the subgroup $\overline{S}$ of Σ_n generated by S,
   (b) finds the stabilizer of any $i \in [1, n]$ in $\overline{S}$,
   (c) determines the orbits of $\overline{S}$ acting on $[1, n]$.

2. For a finite set S, let $\Sigma(S)$ act on $\mathcal{P}(S)$ by $\sigma A = \sigma[A]$. How many orbits do we have?

3. Show that if S is any set and n, m two cardinals with $n + m = \operatorname{Card} S$, then
$$n! \cdot (m!) \cdot \operatorname{Card}\{X \subseteq S : \operatorname{Card} X = n\} = (\operatorname{Card} S)!$$

4. Let a finite group G acting on a finite set have n different orbits. For every $g \in G$ let $F(g)$ be the number of fixed points of g. Prove the "Burnside equation"
$$n \cdot \operatorname{Card} G = \sum_{g \in G} F(g)$$

5. Let $\mathcal{S}$ be any set of pairwise nonisomorphic binary relations on a finite, n-element set A. Show that
$$\operatorname{Card} \mathcal{S} \cdot n! = \sum_{g \in \Sigma(A)} I(g)$$

where $I(g)$ is the number of relations R on A isomorphic to some member of S and such that $g \in \operatorname{Aut} R$.

3. INTEGERS AND CYCLIC GROUPS

The set-theoretical definition of natural numbers and of ω given in Chapter I should be clear in our minds at this point. Let ω^+ denote the set of nonzero naturals. Consider the set

$$\omega^- = \{\omega \setminus n : n \in \omega^+\}$$

Each $\omega \setminus n$ coincides with the upper section $[n, \rightarrow)$ in $(\omega, \leq)$. Thus ω^- is disjoint from ω. An *integer* is defined as an element of

$$\omega^- \cup \omega = \omega^- \cup \{0\} \cup \omega^+$$

A nonzero integer is *positive* or *negative* according to whether it belongs to ω^+ or ω^-. The set of integers shall be denoted by $\mathbb{Z}$. (*Zahl* means "number" in German.)

Recall that $(\omega, \leq)$ was defined as the inclusion order on ω. The set ω^- can also be ordered by inclusion: $(\omega^-, \subseteq)$ is then isomorphic to the dual of $(\omega^+, \leq)$ via the map associating $\omega \setminus n$ to n. The entire set $\mathbb{Z}$ is ordered by the relation

$$\{(z,v) \in \omega^2 : z \subseteq v\} \cup \{(z,v) \in (\omega^-)^2 : z \subseteq v\}$$
$$\cup \{(z,v) : z \in \omega^-, v \in \omega\}$$

This *integer order* on $\mathbb{Z}$ is usually denoted by $\leq$. Restricted to ω, it obviously yields the natural order on ω, and restricted to ω^- it yields the inclusion order described above. All negative integers are less than any natural number. For any nonzero integer z, the *negative of z* is defined as $\omega \setminus z$, while the *negative of zero* is defined as zero itself. We write $-z$ for the negative of z. Note the double negation rule

$$-(-z) = z$$

for all integers. The negative of a positive integer is negative and vice versa. The map associating to each integer its negative is an order isomorphism from $(\mathbb{Z}, \leq)$ to its dual $(\mathbb{Z}, \geq)$: we have

$$x \leq y \quad \text{if and only if} \quad -y \leq -x$$

FIGURE 3.1 Integers.

The order $(\mathbb{Z},\leq)$ is obviously total. A nonzero integer is positive or negative according to whether it is greater or less than 0. No negative integer is covered by a positive (or vice versa, obviously) because $z < 0 < v$ for all negative z and positive v. For every $n \in \omega$, n is covered by its successor $n+1$ and by nothing else, and the negative $-n$ covers $-(n+1)$ only. Thus the covering relation of $(\mathbb{Z},\leq)$ is

$$\{(n,n+1) : n \in \omega\} \cup \{(-(n+1),-n) : n \in \omega\}$$

Using the fact, already noted in Chapter II, that the natural order $(\omega,\leq)$ is discrete, we quickly realize that the preorder closure of this covering relation on $\mathbb{Z}$ is the full order $(\mathbb{Z},\leq)$, i.e., the integer order on $\mathbb{Z}$ is discrete too. This allows the familiar representation of $\mathbb{Z}$ given in Figure 3.1.

If $x,y \in \mathbb{Z}$, then $\{x,y\}$ generates a convex segment S in $(\mathbb{Z},\leq)$: the *distance* $d(x,y)$ of the two integers is then defined as the number of couples in $(S,\prec)$, the covering relation of $\leq$ restricted to S. For example, if $x = 2$ and $y = -1$, then

$$S = [-1,2] = \{-1,0,1,2\}$$

and the couples in $(S,\prec)$ are $(-1,0),(0,1),(1,2)$. Thus $d(2,-1)$ is 3. Obviously

$$d(x,x) = 0 \quad \text{and} \quad d(x,y) = d(y,x)$$

and less obviously

$$d(x,y) = d(-x,-y)$$

for all $x,y \in \mathbb{Z}$. Note that if $x < y$, then in the covering relation $(\mathbb{Z},\prec)$ there is a unique path from x to y. The length of this path is $d(x,y)$.

Integer addition (*sum*) is defined as the binary operation $\oplus$ on $\mathbb{Z}$ given for $x,y \in \mathbb{Z}$ by

$$x \oplus y = \begin{cases} d(x,-y) & \text{if} \quad -\min(x,y) \leq \max(x,y) \\ -d(x,-y) & \text{if} \quad -\min(x,y) > \max(x,y) \end{cases}$$

The operation $\oplus$ is commutative. Note that if $x, y \in \omega$, then

$$x \oplus y = d(x, -y) = x + y$$

i.e., the restriction of $\oplus$ to ω coincides with the cardinal sum operation. The integer 0 is easily verified to be neutral not only in $(\omega, +)$ but also in $(\mathbb{Z}, \oplus)$. Integer addition is associative as well: verification of this is straightforward, if somewhat tedious. Further,

$$d(x, x) = 0 \quad \text{implies} \quad x \oplus -x = 0$$

for all integers x, and therefore $(\mathbb{Z}, \oplus)$ is a group: the negative of each integer serves as its inverse. Instead of $\oplus$, we shall henceforward use the notation $+$ for integer addition. For $x, y \in \mathbb{Z}$ we shall also write $x - y$ instead of $x + (-y)$, and call $x - y$ *difference*. Note that $y \le x$ if and only if $x - y$ is nonnegative. If A and B are any finite sets such that $B \subseteq A$, then

$$\text{Card}(A \setminus B) = \text{Card}\, A - \text{Card}\, B$$

In the group $(\mathbb{Z}, +)$ all translations are automorphisms of the order $(\mathbb{Z}, \le)$. It follows that

$$\text{if} \quad x \le y \quad \text{and} \quad z \le v, \quad \text{then} \quad x + z \le y + v$$

What are the subgroups of $\mathbb{Z}$? This question will benefit from the introduction of yet another binary operation on $\mathbb{Z}$, called *integer multiplication* (or *product*), denoted by $\odot$, and defined with reference to the cardinal product "$\cdot$" of natural numbers as follows:

$$x \odot y = \begin{cases} x \cdot y & \text{if } x, y \in \omega \\ (-x) \cdot (-y) & \text{if } x, y \in \omega^- \\ -[(-\min(x, y)) \cdot \max(x, y)] & \text{otherwise} \end{cases}$$

It is a matter of simple verification that $(\mathbb{Z}, \odot)$ is a commutative monoid with the integer 1 as neutral element. (It is not a group, however.) The integer 0 also has a special role: $0 \odot x = 0$ for every integer x. Since integer multiplication coincides with the cardinal product when the former is restricted to natural numbers, we shall use the same symbol "$\cdot$" for integer multiplication, instead of $\odot$.

The multiplicative monoid $(\mathbb{Z}, \cdot)$ is linked to the order $(\mathbb{Z}, \le)$ as follows. If $x \le y$ and $z \ge 0$, then $x \cdot z \le y \cdot z$, but $x \cdot z \ge y \cdot z$ if $z \le 0$.

Integer multiplication and addition are linked by *distributivity*:
$$x \cdot (y + z) = (x \cdot y) + (x \cdot z)$$
This is essentially a consequence of the fact that for finite (as well as infinite) sets A, B, C,
$$A \times (B \cup C) = (A \times B) \cup (A \times C)$$
and if $B \cap C = \emptyset$, then
$$(A \times B) \cap (A \times C) = \emptyset$$

What are then the subgroups of $(\mathbb{Z}, +)$? For any fixed $m \in \mathbb{Z}$ the set of multiples
$$m\mathbb{Z} = \{m \cdot x : x \in \mathbb{Z}\}$$
is a subgroup. (Use distributivity to verify closure under addition.) Indeed $m\mathbb{Z}$ is the subgroup of $\mathbb{Z}$ generated by m, and for $m \neq 0$ it is isomorphic to the entire group $\mathbb{Z}$ via the map associating $m \cdot x$ to $x \in \mathbb{Z}$. Let us show that every subgroup G of $\mathbb{Z}$ is of this kind. If G is trivial, then $G = 0\mathbb{Z}$. Otherwise G has some positive elements because for every nonzero $z \in G$ either z or $-z$ is positive. Let m be the smallest positive element of G. Obviously $m\mathbb{Z} \subseteq G$. If we had $m\mathbb{Z} \neq G$, then for every
$$z \in G \setminus m\mathbb{Z}$$
we would also have
$$-z \in G \setminus m\mathbb{Z}$$
and thus $G \setminus m\mathbb{Z}$ would have some positive elements. Let μ be the smallest of these. We cannot have $\mu < m$ because m is the smallest positive element of G. Thus $m < \mu$ and $\mu - m$ is positive. Clearly $\mu - m \in G$ but
$$\mu - m \notin m\mathbb{Z}$$
for otherwise $(\mu - m) + m = \mu$ would belong to $m\mathbb{Z}$. As $0 < \mu - m < \mu$, $\mu - m$ would be a positive element of $G \setminus m\mathbb{Z}$ smaller than μ, which is impossible by the definition of μ. Thus $m\mathbb{Z} = G$. Observe that m and $-m$ are the only generators of $m\mathbb{Z} = -m\mathbb{Z}$. Note that $1\mathbb{Z} = \mathbb{Z}$.

A group G is called *cyclic* if in the subgroup closure system on G the entire group G is generated by some element of G. We have shown that $\mathbb{Z}$ as well as all subgroups of $\mathbb{Z}$ are cyclic. On the other hand it is clear that every quotient of a cyclic group is cyclic (the congruence class of any generator generates the quotient). Thus we get more examples of cyclic groups by taking the various quotients $\mathbb{Z}/m\mathbb{Z}$. If $m > 0$, then we sometimes use the notation $\mathbb{Z}_m$ for $\mathbb{Z}/m\mathbb{Z}$. The group $\mathbb{Z}_m$ has exactly m elements:

(i) the m nonnegative integers less than m are all in distinct cosets of $m\mathbb{Z}$ because if two such i, j (with say $i < j$) were in the same coset, then $j - i$ would belong to $m\mathbb{Z}$, but $0 < j - i < m$ would contradict the minimality of m among the positive elements of $m\mathbb{Z}$,

(ii) every coset C of $m\mathbb{Z}$ contains a nonnegative integer less than m because for any $z \in C$ one of

$$z + z \cdot m \quad \text{or} \quad z - z \cdot m$$

is nonnegative, and the smallest nonnegative element μ of C must be less than m or else look at $\mu - m \in C$.

Thus the function associating to each $i \in m$ its coset

$$\bar{i} = \{i + m \cdot x : x \in \mathbb{Z}\}$$

is a bijection onto $\mathbb{Z}/m\mathbb{Z}$ whose inverse function

$$c : \mathbb{Z}/m\mathbb{Z} \to m$$

is a choice function on the set of cosets. Using this bijective correspondence, a group structure $\oplus$ isomorphic to $\mathbb{Z}/m\mathbb{Z}$ is defined on $m = \{0, \ldots, m - 1\}$ itself by

$$i \oplus j = c(\bar{i} + \bar{j})$$

This group structure is called *addition modulo m* $(\bmod\, m)$. Note that if $m \geq 3$, then the binary relation on m given by

$$R = \{(i, i \oplus 1) : i \in m\}$$

is a cycle (Figure 3.2). For $i, j \in m$, $j \geq 1$, there is in R a path of length j from i to $i \oplus j$: indeed there is no other $z \in m$ such that a path of length j would exist from i to z.

For $m = 2$, addition $\bmod 2$ on $\{0, 1\}$ coincides with the symmetric difference on the power set $\mathcal{P}(\{\emptyset\}) = \mathcal{P}(1) = 2$. Every integer is

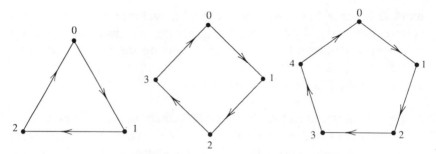

FIGURE 3.2 Integers modulo 3, 4, and 5.

congruent modulo $2\mathbb{Z}$ either to 0 or to 1, and it is called either *even* or *odd* accordingly.

Let G be any cyclic group generated by an element g. Define a function $e : \mathbb{Z} \to G$ by

$$e(n) = \begin{cases} g^n & \text{for } n \in \omega, \\ (g^*)^{-n} & \text{for } n \in \omega^- \end{cases}$$

This function e is a surjective homomorphism from the additive group $\mathbb{Z}$ to G. Let $\operatorname{Ker} e = m\mathbb{Z}$, $m \geq 0$. If $m = 0$, then e is an isomorphism from $\mathbb{Z}$ to G. If $m > 0$, then G is isomorphic to $\mathbb{Z}/m\mathbb{Z}$. In this case the restriction of e to $m = \{0, \ldots, m-1\}$ is an isomorphism from $(m, \oplus)$ to G:

Cyclic Group Structure Theorem. *A cyclic group is isomorphic either to $\mathbb{Z}$ or to the group $(m, \oplus)$ of integers modulo m for some $m \geq 1$.*

Of course $(m, \oplus)$ is isomorphic to $\mathbb{Z}/m\mathbb{Z} = \mathbb{Z}_m$. Let G be a subgroup of $\mathbb{Z}/m\mathbb{Z}$. Remember that G is a set of cosets. Clearly $\cup G$ is then a subgroup $n\mathbb{Z}$ of $\mathbb{Z}$, with $m\mathbb{Z} \subseteq n\mathbb{Z}$. In fact the quotient group

$$n\mathbb{Z}/m\mathbb{Z}$$

coincides with G. Since every quotient of a cyclic group is cyclic, G must be cyclic as well. Thus we have shown that every subgroup of a cyclic group is cyclic.

Any element g of any group H generates a cyclic subgroup G of H. Let $e : \mathbb{Z} \to G$ be the surjective group homomorphism defined above. It is actually the only homomorphism $e : \mathbb{Z} \to G$ with $e(1) = g$. It is customary to denote $e(z)$ by g^z for any integer z. For $z \in \omega$ this is compatible with the exponential notation introduced earlier in this chapter. Note the rules

$$(g^z)(g^y) = g^{z+y}$$
$$(g^z)^y = g^{z \cdot y}$$

as well as

$$(gh)^z = g^z h^z \quad \text{if} \quad gh = hg$$

Note also that the inverse g^* of g coincides with g^{-1}. From now on we shall use this latter notation g^{-1} in preference to g^*. By analogy, we shall even take the notational liberty of designating the inverse map of any bijection $f : A \to B$ between two sets by the symbol f^{-1}. [If $A = B$, then this makes good sense in the group $\Sigma(A)$ to which f belongs.] Further, for any map $f : A \to B$, not necessarily bijective, and any $T \subseteq B$, the inverse image $f^{\text{inv}}[T]$ is customarily denoted by $f^{-1}[T]$.

EXERCISES

1. Design a program that for any two integers m, n finds a generator of the subgroup $m\mathbb{Z} \cap n\mathbb{Z}$ in $(\mathbb{Z}, +)$ as well as a generator of the lub subgroup $m\mathbb{Z} \vee n\mathbb{Z}$.

2. Describe a submonoid of $(\mathbb{Z}, \cdot)$ that is not generated by any single integer.

3. Show that $\text{Aut}(\mathbb{Z}, \leq)$ is a group isomorphic to $\mathbb{Z}$. Is there an analogous result for finite cyclic groups?

4. Which are the generators of the group $\mathbb{Z}$?

5. Describe $\text{Aut}(\mathbb{Z}, +)$ and $\text{Aut}(5, \oplus)$.

6. Let $x, y \in \mathbb{Z}$, $x \leq y$. Verify that

$$\text{Card}[x, y] = y - x + 1$$

7. Show that if g is any member of a finite group G with neutral element u, then $g^{\text{Card } G} = u$.

4. ALTERNATING GROUPS

For $k \in \omega$, $k \geq 2$, a permutation $\sigma \in \Sigma_n$ is called a k-*cycle* if there are k distinct elements $n_1, \ldots, n_k \in [1, n]$ such that

$$\sigma(n_i) = n_{i+1} \quad \text{for all} \quad 1 \leq i < k, \qquad \sigma(n_k) = n_1$$

and $\sigma(m) = m$ for all other elements m of $[1, n]$. For example, the permutation

$$\begin{pmatrix} 1 & 2 & 3 & 4 \\ 4 & 2 & 1 & 3 \end{pmatrix}$$

is a 3-cycle; it is more commonly written as $(1, 4, 3)$. Generally, since any k-cycle σ is determined by the k-tuple

$$(n_1, \ldots, n_k)$$

this sequence-in-brackets notation of the k-tuple is used, somewhat abusively, to denote the permutation σ.

Note that a k-cycle is not a k'-cycle for any $k' \neq k$. A *cycle* is a permutation σ that is a k-cycle for some k, and k is then called the *length* of σ. If σ is a k-cycle, $k \geq 3$, then

$$\{(i, \sigma(i)) : 1 \leq i \leq n, \ \sigma(i) \neq i\}$$

is a relational cycle of length k. A 2-cycle is called a *transposition*. Two cycles σ and τ in Σ_n are called *disjoint* if

$$\{i \in [1, n] : \sigma(i) \neq i\} \cap \{i \in [1, n] : \tau(i) \neq i\} = \emptyset$$

In this case σ and τ commute: $\sigma \circ \tau = \tau \circ \sigma$.

Proposition 6 *The set of transpositions of $[1, n]$ generates, in the subgroup closure system, the entire symmetric group Σ_n.*

Proof. The proposition is obviously true for $n = 1, 2$. Assuming it failed to hold for all n, let m be the first natural number such that Σ_m is not generated by transpositions. Then $m \geq 3$ and Σ_{m-1} is generated by the transpositions of $[1, m-1]$. This means that the stabilizer S_m of m in Σ_m is generated by those transpositions of $[1, m]$ that leave m fixed. (Why?) Let now $\sigma \in \Sigma_m$ such that σ is not in the allegedly proper subgroup S of Σ_m generated by the transpositions of $[1, m]$. Then $\sigma \notin S_m$. Let τ be the transposition of $[1, m]$ such that

$$\tau(m) = \sigma(m), \qquad \tau(\sigma(m)) = m$$

Obviously $\tau \circ \sigma \in S_m$ and therefore $\tau \circ \sigma \in S$. But also $\tau \in S$ and thus

$$\tau \circ (\tau \circ \sigma) \in S$$

As $\tau \circ (\tau \circ \sigma) = \sigma$, this contradicts $\sigma \notin S$. □

Cycle Representation Theorem. *Every permutation $\sigma \in \Sigma_n$ is the product of a unique set of pairwise disjoint cycles.*

Proof. Let $\sigma \in \Sigma_n$. Let

$$(\sigma) = \{\sigma^z : z \in \mathbb{Z}\}$$

This is the subgroup of Σ_n generated by σ. Consider the action of the permutation group (σ) on $[1, n]$. For every nonsingleton orbit C define $\sigma_C \in \Sigma_n$ by

$$\sigma_C(x) = \begin{cases} \sigma(x) & \text{if } x \in C, \\ x & \text{if } x \notin C \end{cases}$$

Then the various σ_C form a set of pairwise disjoint, commuting cycles and their product is σ. □

A cycle appearing in such a product representation of σ is called a *cycle factor* of σ. The *cycle structure* of a permutation $\sigma \in \Sigma_n$ is the equivalence relation on $[1, n]$ whose nonsingleton classes are the sets

$$\{i \in [1, n] : \tau(i) \neq i\}$$

corresponding to the various cycle factors τ of σ. For example, if

$$\sigma = \begin{pmatrix} 1 & 2 & 3 & 4 & 5 & 6 & 7 & 8 & 9 \\ 6 & 7 & 3 & 1 & 2 & 4 & 5 & 9 & 8 \end{pmatrix}$$

then the cycle factors are $(1, 6, 4)$, $(2, 7, 5)$, and $(8, 9)$. We can write

$$\sigma = (1, 6, 4)(2, 7, 5)(8, 9)$$

and the cycle structure has four classes:

$$\{1, 6, 4\}, \{2, 7, 5\}, \{8, 9\}, \{3\}$$

Two permutations $\sigma, \pi \in \Sigma_n$ are said to have *similar cycle structures* if some bijection $\varphi : [1, n] \to [1, n]$ is an isomorphism from the cycle structure (equivalence relation) of σ to that of π. Almost obviously, this means precisely that for every $2 \leq k \leq n$ the number of k-cycle factors of σ equals the number of k-cycle factors of π.

Example. In Σ_5, $\sigma = (1,3,5)(2,4)$ and $\pi = (1,2,4)(3,5)$ have distinct but similar cycle structures. This σ and $(1,5,3)(2,4)$ have the same cycle structure. What about σ and $(2,4)(3,5,1)$? The latter is just an equivalent notation for σ.

In Σ_n any two k-cycles are conjugates. Indeed, let

$$\sigma = (n_1,\ldots,n_k) \quad \text{and} \quad \pi = (n'_1,\ldots,n'_k)$$

be two k-cycles. There is some $\varphi \in \Sigma_n$ such that

$$\varphi(n_i) = n'_i \quad \text{for} \quad i = 1,\ldots,k$$

We have $\sigma = \varphi^{-1}\pi\varphi$.

The above argument extends to arbitrary permutations $\sigma, \pi \in \Sigma_n$ having similar cycle structures. There exists in this case some $\varphi \in \Sigma_n$ such that for every cycle factor $(n_1,\ldots,n_k)$ of σ and a corresponding cycle factor $(n'_1,\ldots,n'_k)$ of π having the same length k,

$$\tau(n_i) = n'_i \quad \text{for} \quad i = 1,\ldots,k$$

Clearly for such φ we have

$$\sigma = \varphi^{-1}\pi\varphi$$

A simple converse verification completes the proof of the following:

Proposition 7 *Two permutations σ and π of $[1,n]$ have similar cycle structures if and only if they are conjugates in the symmetric group Σ_n.*

Let $S(n,2)$ be the set of two-element subsets of $[1,n]$. A pair

$$\{i,j\} \in S(n,2)$$

is *inverted* by a $\sigma \in \Sigma_n$ if

$$i < j \quad \text{and} \quad \sigma(j) < \sigma(i)$$

or if

$$j < i \quad \text{and} \quad \sigma(i) < \sigma(j)$$

The number of pairs inverted by σ is a nonnegative integer $\nu(\sigma)$, and σ is called *odd* or *even* according to whether $\nu(\sigma)$ is odd or even. For example, all transpositions are odd, and the identity permutation is even.

The function $p : \Sigma_n \to \mathbb{Z}/2\mathbb{Z}$ mapping each $\sigma \in \Sigma_n$ to the congruence class modulo $2\mathbb{Z}$ of the number $\nu(\sigma)$ of pairs inverted by σ is called the *parity function*.

Proposition 8 *The parity function $p : \Sigma_n \to \mathbb{Z}_2$ is a group homomorphism.*

Proof. First, verify the following auxiliary fact: for any finite set S, the function

$$P : \mathcal{P}(S) \to \mathbb{Z}_2$$

mapping each $A \subseteq S$ to the congruence class modulo $2\mathbb{Z}$ of the integer Card A is a group homomorphism from the symmetric difference group $(\mathcal{P}(S), +)$ to $\mathbb{Z}_2$. Call this P the *subset parity function* of S.

Consider any $\sigma_1, \sigma_2 \in \Sigma_n$ and their product

$$\sigma_3 = \sigma_2 \circ \sigma_1$$

Let r_1 denote the subset of $S(n,2)$ consisting of the pairs inverted by σ_1. Let

$$r_2 = \{\{i,j\} \in S(n,2) : \{\sigma_1(i), \sigma_1(j)\} \text{ is inverted by } \sigma_2\}$$

and let r_3 be the set of pairs inverted by σ_3. Then r_3 is the symmetric difference of r_1 and r_2. Applying the subset parity function P of $S(n,2)$ we have in $\mathbb{Z}_2$ the equality

$$P(r_3) = P(r_2) + P(r_1)$$

Finally, observe that the equality

$$p(\sigma_k) = P(r_k)$$

holds, obviously for $k = 1, 3$, and less obviously for $k = 2$. □

The kernel of the parity function $\Sigma_n \to \mathbb{Z}_2$, consisting of all even permutations, is called the *alternating group* A_n. By Lagrange's Subgroup Counting Theorem we have, for $n \geq 2$,

$$2 \cdot \operatorname{Card} A_n = \operatorname{Card} \Sigma_n = n!$$

Example. We have $2 \cdot \operatorname{Card} A_4 = 4! = 2 \cdot 3 \cdot 4$. Therefore
$$\operatorname{Card} A_4 = 3 \cdot 4$$
Consider the four-element subset N of A_4 consisting of the identity permutation, $(1,2)(3,4)$, $(1,3)(2,4)$, and $(1,4)(2,3)$. The set N constitutes a normal subgroup of Σ_4 and a fortiori of A_4. (The Simplicity Theorem for Alternating Groups will show that this situation is quite exceptional.)

All 3-cycles are even. Indeed any 3-cycle (a,b,c) can be written as the product
$$(a,b)(b,c)$$
of two transpositions and we can apply the parity function. [Alternative proof: $(1,2,3)$ is even and all other 3-cycles are its conjugates.]

The set of 3-cycles generates the alternating group A_n. Proof: Let A be the subgroup of Σ_n generated by the set of 3-cycles. Obviously $A \subseteq A_n$. Assuming strict inclusion will lead to a contradiction. Let μ be an element of $A_n \setminus A$ with the greatest number of fixed points. No cycle factor σ of μ has length greater than 2, for otherwise take $i \in [1,n]$ with $\sigma(i) \neq i$; let τ be the 3-cycle
$$(i, \sigma(i), \sigma^2(i))$$
and observe that $\tau^{-1}\mu \in A_n \setminus A$ has more fixed points than μ. Thus all cycle factors of μ are transpositions and obviously μ has at least two such factors (a,b) and (c,d). Let
$$\tau = (a,b)(c,d) = (a,b,c)(b,c,d)$$
Again $\tau^{-1}\mu \in A_n \setminus A$ has more fixed points than μ, which is absurd.

For $n \geq 5$, all 3-cycles are conjugates in A_n. (Do not say, "of course.") Proof: Let
$$\sigma = (n_1, n_2, n_3) \quad \text{and} \quad \pi = (n_1', n_2', n_3')$$
be 3-cycles. Let $\alpha \in \Sigma_n$ be any permutation with
$$\alpha(n_i) = n_i' \quad \text{for} \quad i = 1,2,3$$
Let a and b be two distinct elements of $[1,n] \setminus \{n_1, n_2, n_3\}$. If α is even, $\alpha \in A_n$, then
$$\sigma = \alpha^{-1}\pi\alpha$$

If α is odd, then let β be the product (composition) of α and (a,b),
$$\beta = \alpha(a,b)$$
Clearly β is even, $\beta \in A_n$, and $\sigma = \beta^{-1}\pi\beta$.

If $\sigma = (a_1a_2)(a_3a_4)$ and $\pi = (b_1b_2)(b_3b_4)$ are permutations in Σ_n, each being the product of two disjoint transpositions, then σ and π are conjugates in A_n. Indeed, take any permutation α of $[1,n]$ with
$$\alpha(a_i) = b_i \quad \text{for} \quad i = 1,2,3,4$$
If α is even, then $\sigma = \alpha^{-1}\pi\alpha$ proves our claim. Otherwise $\beta = \alpha(a_1a_2)$ is even and $\sigma = \beta^{-1}\pi\beta$.

A group G is called *simple* if it has no normal subgroups other than G itself and the trivial subgroup reduced to the neutral element.

Example. The two-element group $\mathbb{Z}_2$ is simple. The alternating group A_4 is not simple.

Simplicity Theorem for Alternating Groups. *The alternating group A_n is simple if $n \geq 5$.*

Proof. We shall show that every nontrivial normal subgroup N of A_n contains a 3-cycle. It would then follow that N contains all 3-cycles (since these are all conjugates in A_n) and $N = A_n$ because the 3-cycles generate A_n.

For each permutation $\sigma \in \Sigma_n$ the cycle structure of σ is an equivalence relation on $[1,n]$. Call the number of equivalence classes the *cycle count* of σ. It is equal to the number of cycle factors plus the number of fixed points of σ. The cycle count is n for the identity permutation, and it is less than n for all other permutations in Σ_n.

Among the elements of N other than the identity permutation, let μ have the highest possible cycle count. We may assume that μ is not a 3-cycle, for otherwise the proof is finished.

First of all, μ has no cycle factor of length 5 or more. For if it had one, say σ, then we would take any $i \in [1,n]$ with $\sigma(i) \neq i$, form the product of transpositions
$$\tau = (i,\sigma(i))(\sigma^2(i),\sigma^3(i))$$
and compare the cycle count of $\tau\mu\tau^{-1}\mu$ to that of μ. This construction also shows that if N contains a 5-cycle, then it contains a 3-cycle. We shall use this fact at the end of the proof.

Second, μ has no cycle factor of length 4 either. For otherwise the cycle count of μ^2 would be higher than that of μ.

Third, μ cannot have cycle factors of length 2 and 3 simultaneously, for μ^2 would contradict the choice of μ in this case as well.

Fourth, the cycle factors of μ cannot all be 3-cycles. For in that case, by assumption, μ would have at least two distinct cycle factors σ and π. Then choose $i,j \in [1,n]$ with $\sigma(i) \neq i$, $\pi(j) \neq j$. Let

$$\tau = (i,\sigma(i))(j,\pi(j))$$

Compare the cycle counts of $\tau\mu\tau^{-1}\mu$ and μ.

Therefore all the cycle factors of μ are transpositions, and obviously there are at least two of these, say (a,b) and (c,d). Let

$$\rho = (a,b)(c,d)\mu$$

We have $\rho \in A_n$ and

$$\mu = (a,b)(c,d)\rho$$

Let

$$\tau_1 = (a,c)(b,d), \qquad \tau_2 = (a,d)(b,c)$$

These two permutations are conjugates of $(a,b)(c,d)$ in A_n, and therefore both $\tau_1\rho$ and $\tau_2\rho$ are conjugates of μ in A_n. Since N is normal, both $\tau_1\rho$ and $\tau_2\rho$ belong to N. Then

$$\tau_1\rho\mu\tau_2\rho = \tau_1\mu\tau_2 = \rho$$

also belongs to N and it has a higher cycle count than μ. Therefore ρ must be the identity permutation, i.e., $\mu = (a,b)(c,d)$.

Let now

$$e \in [1,n] \setminus \{a,b,c,d\}$$

Composing μ with $(b,c)(d,e)$, its conjugate in A_n, we obtain the 5-cycle

$$(b,c)(d,e)\mu = (a,c,e,d,b)$$

as a member of N. By a previous observation, it follows that N also contains a 3-cycle. $\square$

From the simplicity of A_n it follows that for $n \geq 5$, Σ_n has no normal subgroups other than Σ_n itself, A_n, and the trivial subgroup. For if N were some other normal subgroup, then $N \cap A_n$ would be

a normal subgroup of A_n, implying that either $N \cap A_n$ is trivial or that
$$A_n \subset N \subset \Sigma_n$$
In the former case, if
$$\sigma, \pi \in N \setminus A_n$$
then $\sigma\pi$ is even, $\sigma\pi \in N \cap A_n$, and $\sigma\pi$ is the identity, from which we conclude that N has only two elements. This is impossible because every nonidentity permutation in Σ_n has distinct conjugates belonging to the same normal subgroups. In the latter case, $A_n \subset N \subset \Sigma_n$, we have
$$\operatorname{Card} A_n < \operatorname{Card} N < n!$$
and by Lagrange's Subgroup Counting Theorem,
$$2 \cdot \operatorname{Card} A_n = \operatorname{Card} N \cdot \operatorname{Card}(\Sigma_n/N) = n!$$
Since $\operatorname{Card} N < n!$, we have
$$1 < \operatorname{Card}(\Sigma_n/N)$$
Since $\operatorname{Card} A_n < \operatorname{Card} N$, we have
$$\operatorname{Card}(\Sigma_n/N) < 2$$
But there is no integer between 1 and 2.

Historical Comment. In historical order this third chapter should have been the first. If modern algebra is the arithmetic of non-numbers, then modern algebra is 200 years old and it was born as group theory. While there was considerable algebraic manipulation of imaginary and unknown quantities in earlier times, this was more an extension of, rather than abstraction from, numerical mathematics. Permutation groups, on the other hand, are nonnumerical structures par excellence. It is puzzling that they were invented not to study the symmetry of geometric shapes, but to investigate the solvability of equations in numerical fields.

EXERCISES

1. Write a computer program that from the two-row representation of any $\sigma \in \Sigma_n$ determines the cycle factors of σ.

2. Write a program that, given cycles $\sigma_1,\ldots,\sigma_k \in \Sigma_n$, produces a two-row representation of the product $\sigma_1\cdots\sigma_k$.

3. Write a program to compute the parity function $\Sigma_n \to Z_2$.

4. Show that there is a bijection between Σ_n and the set of linear orders on $\{1,\ldots,n\}$.

5. For a positive integer n, let S be the set of $n-1$ transpositions $(i,i+1)$, $1 \leq i \leq n-1$. Show that S generates Σ_n.

6. Verify that if $\sigma \in \Sigma_n$ is a k-cycle, then σ^k is the identity permutation.

7. Show that k-cycles are even or odd permutations according to whether k is an odd or even integer.

8. Verify that the equivalence classes of the cycle structure of any $\sigma \in \Sigma_n$ are precisely the orbits of the permutation group generated by σ acting on $\lfloor 1,n \rfloor$.

9. Show that A_4 has a quotient isomorphic to Z_3. What are the quotient groups of A_n for $n \geq 5$?

10. Let n be a nonzero natural member. Show that Z_n is not simple if and only if $n = m \cdot k$ for some $m,k \in n$.

BIBLIOGRAPHY

Richard A. DEAN, *Classical Abstract Algebra*. Harper & Row 1990. Chapters 3 to 6 and Chapter 11 contain a patient yet lively development of group theory, from the beginning, with many examples and exercises, to the point where structural theorems such as Sylow's can be proved.

William J. GILBERT, *Modern Algebra with Applications*. John Wiley & Sons 1976. Groups are introduced here before semigroups. The latter are motivated by automata theory, to which a very concise introduction is given in Chapter 7. The example of a pushbutton elevator should convince the reader of the everyday reality of finite state automata. Pólya-Burnside enumeration, an application of group theory to counting, is discussed in Chapter 6.

Nathan JACOBSON, *Basic Algebra I*. Freeman and Company 1985. Chapter 1, some 60 pages long, is devoted to monoids and groups. Including most of the basic theory, it is suitable for a first course on the subject. Chapter 6 contains more specialized material on the structure of geometric groups.

Serge LANG, *Algebra*. Addison-Wesley 1971. Chapter I is a classical exposition of basic group theory. The development is self-contained but rapid: 50 pages include Sylow's theorem, free groups and relations, and the classification of finitely generated commutative groups. What the integers are is assumed to be known. A minimum of categorical language is developed in order to discuss free groups.

George E. MARTIN, *Transformation Geometry: An Introduction to Symmetry*. Springer 1982. An undergraduate text on groups that occur in nature. No abstract group theory is needed. Translations, rotations, and reflections of the Euclidean plane and space are discussed in an elementary language. Mosaics, Platonic solids, and more.

Mario PETRICH, *Inverse Semigroups*. John Wiley & Sons 1984. These are semigroups in which for every element x there is a unique y (called the inverse of x) such that $x = xyx$ and $y = yxy$. The presentation is self-contained, thorough, and comprehensive. The first chapter, over 70 pages, is an introduction to semigroups in general.

Hala O. PFLUGFELDER, *Quasigroups and Loops: Introduction*. Heldermann Verlag 1990. A short and accessible text on structures intermediate between groups and groupoids, on their combinatorial aspects, and how they appear in discrete geometry.

H. WIELANDT, *Finite Permutation Groups*. Academic Press 1968. The researcher's ideal primer on permutation groups. Concise and self-contained.

CHAPTER IV

RINGS

1. IDEALS

A group is a set with a binary operation satisfying certain conditions. Similarly a ring can be defined as a set A with two binary operations, $+$ and $\cdot$, satisfying certain conditions. Formally, a *ring* is a triple $(A, +, \cdot)$ such that

  (i) $(A, +)$ is a commutative group, with $+$ called *sum* or *addition*, and with a neutral element denoted by 0_A or simply 0,
  (ii) $(A, \cdot)$ is a commutative semigroup, with "$\cdot$" called *product* or *multiplication*,
  (iii) the product is *distributive* over the sum,

$$x \cdot (y + z) = x \cdot y + x \cdot z \quad \text{for all} \quad x, y, z \in A$$

The product $x \cdot y$ is often denoted simply by juxtaposition of the factors as xy. When the context is clear, we can refer to the ring $(A, +, \cdot)$ as "the ring A." The additive neutral 0_A is called the *zero element* of the ring. In the group $(A, +)$ the inverse of any $x \in A$ is denoted by $-x$. We write $y - x$ for $y + (-x)$. In the case where $(A, \cdot)$ is a monoid, the multiplicative neutral is generally denoted by 1_A or simply 1, and it is called the ring's *identity* element.

Examples. The set $\mathbb{Z}$ of integers, together with the integer sum and product operations defined in the previous chapter, is a ring $(\mathbb{Z},+,\cdot)$. Another important example is $A = \mathcal{P}(S)$, the power set of any set, with $+$ the symmetric difference and the product $X \cdot Y$ being defined as the intersection $X \cap Y$. Here 0 is the empty set $\emptyset$ and 1 is the whole set S.

Remark. Many texts consider rings with noncommutative multiplication.

We should point out at the outset that requiring the multiplicative semigroup to be also a group would not take us anywhere. Indeed, since in any ring $0x = 0$ for all elements x (verify), the product of 0 with its multiplicative inverse would be simultaneously the identity 1_A and 0, implying $1_A = 0$. But then every element x would be zero,

$$x = 1_A x = 0x = 0$$

and such a ring A is *trivial*, $A = \{0\}$. However, requiring $A \setminus \{0\}$ to be closed under multiplication and to form a multiplicative group is meaningful, and such rings, to be studied in the next chapter, are called *fields*. Here let us just take note that the ring of integers is not a field, and $(\mathcal{P}(S), +, \cap)$ is not a field if Card $S \neq 1$. The two-element field

$$(\mathcal{P}(1), +, \cap) = (2, +, \cap)$$

on the other hand, will be seen to be of recurring importance.

A *subring* of $(A, +, \cdot)$ is any subgroup of $(A, +)$ that is also a subsemigroup of $(A, \cdot)$. The set of subrings constitutes an algebraic closure system on A. For example, the subrings of $(\mathcal{P}(S), +, \cap)$ are precisely the rings of sets on S defined in Chapter I. Note that the identity element of the ring, if it has one, does not have to belong to every subring.

Every commutative group $(A, +)$ can be made into a ring by defining the product $x \cdot y$ as always 0. We then refer to $(A, +, \cdot)$ as the *null product ring* structure on $(A, +)$. Here every additive subgroup is a subring.

A *homomorphism from a ring A to a ring B* is a group homomorphism $h : (A, +) \to (B, +)$ that is also a multiplicative semigroup homomorphism from $(A, \cdot)$ to $(B, \cdot)$. For example, the constant function

$A \to B$ mapping everything to the 0 of B is a ring homomorphism. As for groupoids, we have mutatis mutandis:

Proposition 1 *The composition of any two ring homomorphisms $h : A \to B$ and $g : B \to C$ is a ring homomorphism from A to C. The identity mapping on each ring is a ring homomorphism. The inverse of a bijective ring homomorphism is a ring homomorphism.*

A bijective ring homomorphism is called a *ring isomorphism*. Rings A and B are *isomorphic* if there is an isomorphism $A \to B$. Any isomorphism $A \to A$ is called an *automorphism* of A. The set of all such automorphisms is a permutation group on A, denoted by Aut A. If A is a field, then Aut A is the cornerstone of Galois theory.

An *ideal* is a subgroup I of $(A, +)$ such that $a \in I$, $c \in A$ imply $ac \in I$. For example, for every ring element a, the set

$$\{ac : c \in A\}$$

is an ideal. An ideal is a fortiori a subring. Not all subrings are ideals: in the power set ring $(\mathcal{P}(S), +, \cap)$ of a set S with at least two elements, consider the subring $\{\emptyset, S\}$. The full ring A is always an ideal; all other ideals are called *proper*. The ideal reduced to the ring's zero element is called *trivial*. Ideals play a role similar to normal subgroups in group theory. Of course, all ideals are normal subgroups of $(A, +)$ because $(A, +)$ is commutative. *Ideal cosets* are the cosets of ideals viewed as subgroups of $(A, +)$. If I is any ideal and $a \in A$, then the coset of I containing a is

$$\{a + i : i \in I\}$$

This coset is denoted by $a + I$.

The key observation to make is that every congruence relation of the additive group modulo any subgroup that is an ideal is actually also a congruence of the multiplicative semigroup. Indeed, if $a \equiv a'$ and $b \equiv b'$ modulo an ideal I, then

$$(a - a')b \in I \quad \text{and} \quad a'(b - b') \in I$$

Hence

$$(a - a')b + a'(b - b') \in I$$

that is, $ab \equiv a'b' \bmod I$. Thus we have a multiplicative quotient semigroup structure on the additive quotient group A/I, where

$$(a + I) \cdot (b + I) = ab + I$$

The multiplicative quotient is distributive over the additive quotient because

$$\begin{aligned}
(a + I) \cdot [(b + I) + (c + I)] &= (a + I) \cdot [(b + c) + I] \\
&= [a(b + c)] + I \\
&= (ab + ac) + I \\
&= (ab + I) + (ac + I) \\
&= [(a + I) \cdot (b + I)] + [(a + I) \cdot (c + I)]
\end{aligned}$$

Therefore the set of cosets of every ideal I has a *quotient ring* structure

$$(A/I, +, \cdot)$$

The foregoing also shows that the canonical surjection g from A to A/I given by

$$q(a) = a + I \quad \text{for} \quad a \in A$$

is a ring homomorphism. Let now $h : A \to B$ be an arbitrary ring homomorphism. The *kernel* of h,

$$\operatorname{Ker} h = \{x \in A : h(x) = 0\}$$

is not only an additive subgroup of A but also an ideal. Exactly as in the case of groups, we can see that $\operatorname{Im} h$ is a subring of B, and it is isomorphic to the quotient ring $A/\operatorname{Ker} h$.

Example. We have seen that every subgroup of $(\mathbb{Z}, +)$ is of the form

$$m\mathbb{Z} = \{mc : c \in \mathbb{Z}\}$$

for some nonnegative integer m. Thus every additive subgroup of $\mathbb{Z}$ is actually an ideal of the ring $\mathbb{Z}$, and $\mathbb{Z}_m = \mathbb{Z}/m\mathbb{Z}$ is not only a quotient group but also a quotient ring.

On any ring A, the set $\mathcal{I}(A)$ of ideals of A is an algebraic closure system. The ideal generated by a single element a of A is called a

principal ideal, denoted by (a). If the ring has an identity element, then
$$(a) = \{ca : c \in A\}$$
The ideal generated by a nonempty finite set $\{a_1,\ldots,a_n\}$ of elements of A is denoted by $(a_1,\ldots,a_n)$. In a ring with identity
$$(a_1,\ldots,a_n) = \{c_1 a_1 + \cdots + c_n a_n : c_1,\ldots,c_n \in A\}$$
If every ideal of A is principal, A is called a *principal ring*. The ring A is called *Noetherian* if the ideal closure system $\mathcal{I}(A)$ is Noetherian, i.e., if every ideal is generated by a finite set of elements.

Examples. The ring $\mathbb{Z}$ is a principal ring. For any finite set S, the ring $(\mathcal{P}(S), +, \cap)$ is principal. Every ideal I of $\mathcal{P}(S)$ has a unique generator $S' \subseteq S$, and
$$I = \mathcal{P}(S')$$
This is no longer true for infinite S, for in that case the set of finite subsets of S constitutes an ideal which is not principal. This ideal is not even finitely generated, so $\mathcal{P}(S)$ is not Noetherian if S is infinite.

Since $\mathcal{I}(A)$ is a closure system, $\mathcal{I}(A)$ ordered by inclusion is a complete lattice. Given two ideals $I, J \in \mathcal{I}(A)$, their greatest lower bound is their intersection $I \cap J$, while their least upper bound is the set
$$I + J = \{x \in A : x = a + b \text{ for some } a \in I, b \in J\}$$
We also refer to this as *ideal addition* or *sum*. Let us point out that, for all $a, b \in A$, we have
$$(a) + (b) = (a, b)$$
Two proper ideals I and J are called *coprime*, or *comaximal*, if $I + J$ is A, the maximum of the lattice $\mathcal{I}(A)$. An ideal I is called *maximal* if it is covered by A in $\mathcal{I}(A)$, i.e., if it is a maximal member of the inclusion-ordered set of all proper ideals. Equivalently, this means that I is comaximal with every other proper ideal J not contained in I. In particular, every two distinct maximal ideals are comaximal.

Examples. Let S be a finite set. In the ring $\mathcal{P}(S)$ consider two proper ideals I, J generated by $A, B \in \mathcal{P}(S)$. Then $I \cap J$ is generated by $A \cap B$ and the sum $I + J$ is generated by $A \cup B$. Thus I and

J are comaximal if and only if $A \cup B = S$. Clearly I is maximal if and only if $S \setminus A$ is a singleton.

Maximal Ideal Theorem. *In a ring with identity, each proper ideal is contained in some maximal ideal.*

Proof. Observe that an ideal I is proper if and only if the identity element does not belong to I. Then apply Zorn's Lemma to the inclusion-ordered set of all proper ideals. □

Proposition 2 *Two proper ideals I and J of a ring A are coprime if and only if every coset of I intersects every coset of J.*

Proof. If $I + J = A$, then consider two cosets $a + I$ and $b + J$. As the ring element $a - b$ belongs to $I + J$,

$$a - b = i + j \quad \text{for some } i \in I, \ j \in J$$

Then $a - i = b + j$ is in both $a + I$ and $b + J$.

Conversely, assume that $I + J \neq A$. Let $a \in A \setminus (I + J)$. Then the coset $a + I$ is disjoint from $I + J$, and a fortiori disjoint from the coset $0 + J$. □

Corollary. *Two proper ideals I and J of a ring A are coprime if and only if for every $a, b \in A$ there is a solution $x \in A$ to the two simultaneous congruences*

$$x \equiv a \bmod I$$
$$x \equiv b \bmod J$$

Proof. Clearly x is a solution if and only if $x \in (a + I) \cap (b + J)$. □

For the proof of the next theorem it will be convenient to define the *product IJ of two ideals* I and J as the ideal generated by the set of all products ab, $a \in I$, $b \in J$. It is easy to see that IJ consists of all sums of the form

$$a_1 b_1 + \cdots + a_n b_n$$

where $a_i \in I$ and $b_i \in J$ for $1 \leq i \leq n$. Clearly

$$IJ \subseteq I \cap J$$

Example. In a power set ring $(\mathcal{P}(S), +, \cap)$ the product of any two ideals happens to coincide with their intersection, $IJ = I \cap J$. This is not so in $\mathbb{Z}$, where, for example, $(2)(4) = (8)$ and $(2) \cap (4) = (4)$.

The set $\mathcal{I}(A)$ of ideals of any ring A is a commutative semigroup under ideal multiplication. Indeed $\mathcal{I}(A)$ is a monoid, with A as the neutral element, as long as the ring A itself is assumed to have an identity. Ideal multiplication is distributive over ideal addition, that is,
$$I(J + K) = IJ + IK$$
for all ideals I, J, K.

Lemma. *Let J and $K_1, \ldots, K_m$ be $m + 1$ ideals of a ring A. Then*
$$(J + K_1)(J + K_2) \cdots (J + K_m) \subseteq J + (K_1 K_2 \cdots K_m)$$

Proof. By induction on m. For $m = 2$, we have, by distributivity,
$$(J + K_1)(J + K_2) = JJ + JK_2 + K_1 J + K_1 K_2$$
which is contained in $J + K_1 K_2$. Let $m > 2$. Assuming the statement true, by induction hypothesis, for J and $K_2, \ldots, K_m$, we have
$$(J + K_1)(J + K_2) \cdots (J + K_m) \subseteq (J + K_1)[J + (K_2 K_3 \cdots K_m)]$$
But the latter product is equal, by distributivity again, to
$$JJ + J(K_2 K_3 \cdots K_m) + K_1 J + (K_1 K_2 K_3 \cdots K_m)$$
which is contained in $J + (K_1 K_2 \cdots K_m)$. $\square$

The following generalizes a 2000-year-old result of Sun Tse.

Chinese Remainder Theorem. *Let $I_1, \ldots, I_n$ be pairwise coprime ideals of a ring A with identity, $I_j + I_k = A$ for $j \neq k$. Then for every $a_1, \ldots, a_n \in A$ there is a solution $x \in A$ to the n simultaneous congruences*
$$x \equiv a_1 \bmod I_1$$
$$\vdots$$
$$x \equiv a_n \bmod I_n$$

Proof. Trivial if $n = 1$. If $n = 2$, apply the Corollary of Proposition 2. If $n > 2$, then for each index j fixed, $1 \leq j \leq n$, consider the ideal product of $n - 1$ sums,

$$\prod_{k \neq j}(I_j + I_k)$$

Since each sum $I_j + I_k$ is equal to A, so is their product. But also, by the lemma, this product is contained in

$$I_j + \prod_{k \neq j} I_k$$

Thus

$$I_j + \prod_{k \neq j} I_k = A$$

which means that I_j and $\prod_{k \neq j} I_k$ are coprime. There is then a solution x_j to the two simultaneous congruences

$$x_j \equiv 1 \bmod I_j$$

$$x_j \equiv 0 \bmod \prod_{k \neq j} I_k$$

Such an x_j exists for each $1 \leq j \leq n$. Let

$$x = a_1 x_1 + \cdots + a_n x_n$$

Clearly the n desired congruences hold. □

The Chinese Remainder Theorem can be restated by saying that if $\mathcal{R}$ is a set of pairwise coprime ideals of a ring with identity, and if $\mathcal{C}$ is the set of all cosets of the various ideals in $\mathcal{R}$, then for each finite set F of cosets in $\mathcal{C}$ we have $\cap F \neq \emptyset$ as soon as every pair of members of F intersects. This is reminiscent of the Helly Theorem for Intervals.

On any ring A, all ideal cosets plus the empty set $\emptyset$ constitute a closure system, called the *ideal coset closure system* on A. For a broad class of rings we shall determine the Helly number of this closure system.

In the remainder of this section, we shall assume that A is a ring with identity. The reader can verify that if I and J are coprime ideals, then $IJ = I \cap J$. Further, if we form the *power ideals* I^n and J^m

in the multiplicative monoid $\mathcal{I}(A)$, then the ideals I^n and J^m remain coprime for all positive integer exponents n, m. [To see this, write $1 = i + j$ for some $i \in I, j \in J$, and consider $(i + j)^{n+m}$. Alternatively, just consider $(I + J)^{n+m}$.] Of course I^n and I^m are never coprime because $I^n \supseteq I^m$ if $n \leq m$.

Suppose there is a set of ideals $\mathcal{M} \subseteq \mathcal{I}(A)$ such that any two distinct members of $\mathcal{M}$ are coprime and the ideals in $\mathcal{M}$ plus the trivial ideal generate the full multiplicative monoid $\mathcal{I}(A)$. This means that all proper nontrivial ideals of the ring A are of the product form

$$I_1^{k_1} \cdots I_n^{k_n}$$

where the $I_1, \ldots, I_n$ are $n \geq 1$ distinct members of $\mathcal{M}$ and the k_i are positive integers. It is easily deduced from the Maximal Ideal Theorem that $\mathcal{M}$ cannot be anything else than the set of all maximal ideals of A. The ring A is now said to have the *ideal decomposition property*.

Example. For any finite set S, the power set ring $\mathcal{P}(S)$ has the ideal decomposition property. Other natural examples will appear in Section 3 of this chapter.

Helly Theorem for Ideal Cosets. *If a ring R has the ideal decomposition property, then the ideal coset closure system of R has Helly member at most 2.*

Proof. First some local definitions. For any maximal ideal M of R, call the ideals of the form M^n, $n \geq 1$, M-*primary* ideals. Call an ideal I *primary* if it is M-primary for some M; such an M is then called a *radical* of I. We know that primary ideals with distinct radicals are coprime, and primary ideals with the same radical are comparable by inclusion.

Let $\mathcal{F}$ be a nonempty finite set of ideal cosets. Assume that we have $A \cap B \neq \emptyset$ for all $A, B \in \mathcal{F}$. We shall show that $\cap \mathcal{F} \neq \emptyset$.

Let c be a choice function on $\mathcal{F}$, $c(A) \in A$ for $A \in \mathcal{F}$.

Each member A of $\mathcal{F}$ is the coset of precisely one ideal $I(A)$. For distinct $A, B \in \mathcal{F}$, the nondisjointness of A and B implies that $I(A)$ and $I(B)$ are distinct. Each $I(A)$ is of the product form

$$P_1 \cdots P_n$$

where the P_i are primary ideals such that P_i and P_j are coprime for $i \neq j$. We also have
$$I(A) = P_1 \cap \cdots \cap P_n$$
and
$$A = (c(A) + P_1) \cap \cdots \cap (c(A) + P_n)$$
Call each P_i a *primary factor* of $I(A)$. Obviously if P is a primary factor of $I(A)$ and Q a primary factor of $I(B)$, then
$$(c(A) + P) \cap (c(B) + Q) \neq \emptyset$$
Convince yourself that if P and Q have the same radical, then the cosets $c(A) + P$ and $c(B) + Q$ are comparable by inclusion. Thus for any given maximal ideal M for which M-primary factors appear, the inclusion-ordered set
$$\{c(A) + P : A \in \mathcal{F} \text{ and } P \text{ is an } M\text{-primary factor of } I(A)\}$$
is a chain, indeed a finite chain whose smallest member we denote by $J(M)$. Clearly $\cap \mathcal{F}$ is nothing else but the intersection of the various $J(M)$. Since $J(M)$ and $J(N)$ are cosets of coprime ideals if $M \neq N$, the Chinese Remainder Theorem applies. □

Historical Note. Ideals were invented by Richard Dedekind, a contemporary and friend of Georg Cantor. The reader should note the analogy between cardinals and ideals: these creatures are sets that can be added and multiplied in a manner that generalizes and explains the arithmetic of integers. Are mathematicians getting carried away to an artificial paradise of abstract concepts? According to the down-to-earth theology of Kronecker (another contemporary and countryman of Cantor), "God created the integers" only. We could engage in a Byzantine argumentation about whether God created positive integers only and left it to the engineers to invent zero and the negatives or if He created zero only and left to us the repetitious task of manufacturing successor ordinals.

EXERCISES

1. Verify that if x, y, z are elements of a ring, then $x(-y) = (-x)y = -(xy)$ and $x(y - z) = xy - xz$.

2. For $n \in \omega$, let $\binom{n}{k}$ denote the number of k-element subsets of n. Let a ring A have an identity element. Show that for any $x, y \in A$
$$(x+y)^n = \sum_{k=0,\ldots,n} \binom{n}{k} x^k y^{n-k}$$

3. Verify that if the closure system generated by the ideal cosets of a ring with identity has Helly number 1, then the ring is a field or it consists of a single element.

4. Verify that $\mathrm{Aut}(\mathbb{Z}, +, \cdot)$ is trivial.

5. Show that in any ring with identity the product of any two principal ideals is principal.

6. In a power set ring $(\mathcal{P}(S), +, \cap)$, if $\mathcal{A} \subseteq \mathcal{P}(S)$, then what can you say about the ideal generated by $\mathcal{A}$?

7. Verify that if $h : A \to B$ is a ring homomorphism and I is an ideal of B, then the inverse image $h^{-1}[I]$ is an ideal of A. What if I is only a subring?

8. Verify that every quotient of a principal ring is principal.

9. Verify that a ring is Noetherian if and only if the additive monoid of all ideals is generated by the set of principal ideals.

10. Consider a simultaneous congruence system (modulo ideals in a ring that has the ideal decomposition property):
$$x \equiv a_1 \bmod I_1$$
$$\vdots$$
$$x \equiv a_n \bmod I_n$$
Under what condition is there a unique solution?

11. Write a computer program that for any finite set $S \subset \omega$ determines all solutions of any given simultaneous congruence system in the power set ring $(\mathcal{P}(S), +, \cap)$.

12. Write a computer program that determines all solutions of any given simultaneous congruence system in the ring $\mathbb{Z}_6$.

13. Show that infinite power set rings $\mathcal{P}(S)$ do not have the ideal decomposition property.

2. POLYNOMIALS

Our definition of polynomials may not at first sight correspond to what many users of mathematics have in mind. Kindly bear with us if this is your case—the abstraction is for simplicity's sake. All rings discussed in this section shall possess an identity element 1. A *(formal) polynomial (in one indeterminate) over a ring* A is a family $p = (p_i)_{i \in \omega}$ of elements of A indexed by the ordinal ω and such that, for some $d \in \omega$, $p_i = 0$ whenever $i > d$. The first such index d is called the *degree* of the polynomial p, denoted by $\deg p$. The set P, or $P(A)$, of all polynomials over A can itself be made into a ring by defining the sum of two polynomials p and q by

$$(p+q)_i = p_i + q_i \quad \text{for all} \quad i \in \omega$$

and their product pq by

$$(pq)_i = \sum_{0 \leq k \leq i} p_k q_{i-k} \quad \text{for all} \quad i \in \omega$$

The reader should verify that the ring axioms are indeed satisfied. Observe that the *zero polynomial* $(p_i)_{i \in \omega}$ specified by

$$p_i = 0_A \quad \text{for all} \quad i \in \omega$$

is the additive zero of the ring P, while the degree zero polynomial p for which $p_0 = 1_A$ is the identity in P. Further, all degree zero polynomials form a subring of P isomorphic to A, the isomorphism being given by mapping each $a \in A$ to the degree zero polynomial p^a for which $p_0^a = a$. Usually we refer to p^a as the "polynomial corresponding to a" or simply the *"polynomial a."* This abuse of language is quite harmless as long as we know from the context whether the symbol a refers to an element of A or to the corresponding degree zero polynomial. (Many texts call these constant polynomials.) Polynomials of degree 1 are more commonly called *linear*. The linear polynomial p given by

$$p_0 = 0_A \quad \text{and} \quad p_1 = 1_A$$

is referred to as an *indeterminate* over A, and it is denoted by the capital letter X. Observe that for every natural number m, X^m is equal to the degree m polynomial p such that

$$p_m = 1 \quad \text{and} \quad p_i = 0 \quad \text{for} \quad i \neq m$$

Proposition 3 *Let $c = (c_i)_{i \in \omega}$ be a polynomial of degree n. Then we have, in the ring P of polynomials,*

$$c = c_n X^n + c_{n-1} X^{n-1} + \cdots + c_1 X + c_0 \tag{1}$$

Proof. The product of the degree zero polynomial (corresponding to) c_i with X^i is the polynomial p given by

$$p_i = c_i, \qquad p_j = 0 \quad \text{for} \quad j \neq i \qquad \square$$

Proposition 3 should explain why c is called a polynomial. The various c_i are called the *coefficients* of c, each c_i being the ith *coefficient* and c_n the *leading coefficient*. Assume now that for some natural number m and zero-degree polynomials $a_0, a_1, \ldots, a_m$ such that $a_m \neq 0$, we have in P an expression

$$c = a_m X^m + a_{m-1} X^{m-1} + \cdots + a_1 X + a_0$$

as well as the expression (1). Then $m = n$ and $a_i = c_i$ for all $i \leq n$. The reason is that, by definition of polynomial addition, each coefficient c_i of c is obtained as the sum of the respective ith coefficients of the $m + 1$ polynomials

$$a_m X^m, \ a_{m-1} X^{m-1}, \ \ldots, \ a_1 X, \ a_0$$

But the ith coefficient of $a_j X^j$ is a_j if $i = j$; otherwise it is zero.

A polynomial with at most one nonzero coefficient is called a *monomial*. According to Proposition 3, every polynomial is the sum of monomials, and according to the above remarks, there is only one way to write a polynomial as a sum of monomials $a X^i$ of different degrees, except for the order of terms of the addition and the omission of zero terms.

Example. Over the ring $\mathbb{Z}$ of integers,

$$X^3 - 2X^2 = 1 \cdot X^3 + (-2) \cdot X^2 + 0 \cdot X + 0$$

is a polynomial with leading coefficient 1.

On two occasions in this volume we shall need polynomials in more than one indeterminate, but not more than five. The *ring of polynomials in two indeterminates* over a ring A is defined as the ring $P(P(A))$ of polynomials in one indeterminate over $P(A)$. The indeterminate of $P(P(A))$ over $P(A)$ is denoted by Y. [The letter X is already used to denote the indeterminate of $P(A)$ over A.] Alternatively, we may use X_1 and X_2 instead of X and Y, respectively. As every $a \in A$ is identified with a polynomial in $P(A)$, it is further identified with a polynomial in $P(P(A))$. With this in mind, we can write every $p \in P(P(A))$ in the form

$$p = \sum_{(i,j) \in I} a_{ij} X_1^i X_2^j$$

where $(a_{ij} : (i,j) \in I)$ is a uniquely determined family of nonzero elements of A indexed by a finite subset I of ω^2. Let us go a few steps further. Abbreviate $P(P(A))$ as P_2. Let

$$P_3 = P(P_2), \qquad P_4 = P(P_3), \qquad P_5 = P(P_4)$$

Let X_3 be the indeterminate of P_3 over P_2, X_4 the indeterminate of P_4 over P_3, and X_5 the indeterminate of P_5 over P_4. The members of P_5 are called *polynomials in five indeterminates* over the original ring A. Every such polynomial $p \in P_5$ can be written in the form

$$p = \sum_{(i,j,k,l,m) \in I} a_{ijklm} X_1^i X_2^j X_3^k X_4^l X_5^m$$

where $(a_{ijklm} : (i,j,k,l,m) \in I)$ is a unique family of nonzero elements of A indexed by a finite subset I of ω^5. For example, over $A = \mathbb{Z}$, we have among the polynomials in five indeterminates

$$3X_1 X_2 X_3^3 + 2X_3 X_4 X_5^2 \quad \text{and} \quad 3X_1 X_2 X_3^3 - 2X_3 X_4 X_5^2$$

Their sum is $6X_1 X_2 X_3^3$ and their product is

$$9X_1^2 X_2^2 X_3^6 - 4X_3^2 X_4^2 X_5^4$$

In the remainder of this chapter we shall only consider polynomials in a single indeterminate X.

Let A be a subring of a ring B. Let b be any element of B. Denote by $A[b]$ the subring of B generated by $A \cup \{b\}$. Clearly $A[b] = A$ if and only if $b \in A$. For every polynomial

$$c = c_n X^n + \cdots + c_1 X + c_0$$

the element

$$c_n b^n + \cdots + c_1 b + c_0$$

of B is called the *value of the polynomial c at b*. Clearly this value belongs to $A[b]$. Conversely, the set of all polynomial values at b contains $A \cup \{b\}$ (why?), it is closed under addition and multiplication (verify!), and hence it coincides with $A[b]$:

Proposition 4 *If A is a subring of a ring B and $b \in B$, then the set of values at b of all the polynomials over A constitute the subring $A[b]$ of B generated by A and the element b.*

As a particular instance, if we abusively denote by A the set of corresponding zero-degree polynomials and if we let $B = P(A)$, the set of all polynomials over A, then $X \in B$ and $P(A)$ coincides with $A[X]$. This explains the traditional notation $A[X]$ for the ring of polynomials over A.

Proposition 5 *If A is a subring of B and $b \in B$, then $A[b]$ is isomorphic to a quotient ring of $A[X]$.*

Proof. The function from $A[X]$ to $A[b]$ that maps each polynomial to its value at b is a ring homomorphism surjective onto $A[b]$. □

Let $p = p_n X^n + \cdots + p_1 X + p_0$ be a polynomial over a ring A. The function $f : A \to A$ mapping each $a \in A$ to the value of p at a,

$$f(a) = p_n a^n + \cdots + p_1 a + p_0$$

is called the *polynomial function defined by*, or *corresponding to*, p. Accordingly, we shall write f_p for f. (Many texts refer to f_p simply as "the polynomial function p.") For example, all constant functions $A \to A$ are polynomial functions; they are defined by the degree zero polynomials. Also, the identity function id_A is a polynomial function, defined by the linear polynomial X. Polynomial functions constitute a subset of the set A^A of all functions from A to A, and indeed a most noteworthy subset. To be more precise about this, let us define a ring structure $(A^A, +, \cdot)$ on A^A by letting, for $f, g \in A^A$ and $a \in A$,

$$(f + g)(a) = f(a) + g(a)$$
$$(f \cdot g)(a) = f(a)g(a)$$

The constant functions mapping all elements of A to 0 and to 1 are the additive and multiplicative neutrals, respectively, of the ring A^A. (Verify that the ring axioms hold.)

Proposition 6 *Polynomial functions form a subring of A^A. This subring is generated by the set of all constant functions $A \to A$ plus the identity function id_A. The subring of polynomial functions is also closed under composition of functions.*

Proof. The subring property follows from the facts that for polynomials p and q,

$$f_{p+q} = f_p + f_q \quad \text{and} \quad f_{pq} = f_p \cdot f_q$$

in A^A. Also $f_{-p} = -f_p$. The second statement then becomes clear, as every polynomial function defined by a monomial is either constant or the product of a constant function and a function of the form

$$id_A^k = id_A \cdots id_A \qquad (k \text{ factors}, \ k \geq 1)$$

If a polynomial function f is not defined by a monomial, then it is the sum of two or more functions, each defined by some monomial.

Finally, to establish closure under composition, let f_p and f_q be defined by $p, q \in A[X]$. The polynomial p can be viewed as a polynomial over the ring $A[X]$. As such, it defines a polynomial function F_p from $A[X]$ to $A[X]$. The value of F_p at $q \in A[X]$ is a polynomial $F_p(q)$ in $A[X]$. The polynomial function $A \to A$ defined by $F_p(q)$ is then nothing else but the composition $f_p \circ f_q$. The verification of details is left as an exercise. $\square$

Examples. Every function $\mathbb{Z}_2 \to \mathbb{Z}_2$ is a polynomial function. There is actually only one function $f : \mathbb{Z}_2 \to \mathbb{Z}_2$ that is neither constant nor the identity, given by

$$f(0) = 1, \qquad f(1) = 0$$

This function f is defined by the polynomial $X + 1$ as well as by many other polynomials, such as

$$X^3 + X^2 + X + 1$$

Thus different polynomials can define the same function. (We shall show in the next chapter that, in general, not all functions $A \to A$ are polynomial functions.)

Given some property of a ring A, the question arises whether the polynomial ring $A[X]$ also has that property. For example, if the

equation
$$a = -a$$
holds for every element of A, then this property is transferred to $A[X]$ because we shall also have
$$p = p_n X^n + \cdots + p_1 X + p_0 = -p_n X^n - \cdots - p_1 X - p_0 = -p$$
for every polynomial p over A. On the other hand, the property of being finite is clearly not transferred from A to $A[X]$. One of the best known transferable properties is the Noetherian condition:

Hilbert's Transfer Theorem. *If A is a Noetherian ring, then the polynomial ring $A[X]$ is also Noetherian.*

Proof. We shall prove that every ideal J of $A[X]$ is finitely generated. For every natural number n, let the set I_n consist of 0 plus those elements of A that appear as leading coefficients of some degree n polynomial in J. For $n < m$, we have $I_n \subseteq I_m$ because if $p \in J$, $p \neq 0$, and p has degree n, then $X^{m-n} p$ also belongs to J, it has degree m, and its leading coefficient is the same as that of p. Since
$$\deg(p + q) \leq \max(\deg p, \deg q)$$
and for $a \in A$,
$$\deg(ap) \leq \deg p$$
it is not difficult to see that every I_n is an ideal of A. As A is Noetherian, the chain of ideals $\{I_n : n \in \omega\}$ has a maximal member I_m. For each $n = 0, 1, \ldots, m$ let G_n be a finite set generating the ideal I_n. For each $c \in G_n$ let $p(n,c)$ be a degree n polynomial in J with leading coefficient c. Let us show that the finite set
$$P = \bigcup_{0 \leq n \leq m} \{p(n,c) : c \in G_n\}$$
generates the ideal J. If this is not so, let $q \in J$ be a polynomial of least possible degree d that is not in the ideal J_P generated by P. For some n between 0 and m, we have $I_d = I_n$, and let the choice of such n be minimal. Then $n \leq d$ and G_n generates I_d. Since I_d is nontrivial, G_n is nonempty. Let $c_1, \ldots, c_k$ be the elements of G_n, and let a be the leading coefficient of q. As $a \in I_d$,
$$a = a_1 c_1 + \cdots + a_k c_k$$

for some elements $a_1,\ldots,a_k$ of A. For $i = 1,\ldots,k$ let the polynomial $t(i)$ be defined by
$$t(i) = X^{d-n} p(n, c_i)$$
All the $t(i)$ belong to J_P and the leading coefficients of $t(1),\ldots,t(k)$ are $c_1,\ldots,c_k$, respectively. The polynomial
$$q - a_1 t(1) - \cdots - a_k t(k)$$
belongs to J, and since its degree is less than d, it must belong to J_P. But then, adding back to it the polynomials $a_i t(i)$ belonging to J_P, we would get $q \in J_P$, a contradiction completing the proof. $\square$

Corollary. *If A is a subring of B and $b \in B$, then $A[b]$ is Noetherian provided that A is Noetherian.*

Proof. The quotient of a Noetherian ring is always Noetherian. (Why?) And according to Proposition 5, $A[b]$ is isomorphic to a quotient of $A[X]$. $\square$

We can now easily produce an example of a Noetherian ring that is not principal. By Hilbert's Transfer Theorem, $Z[X]$ is Noetherian. However, the ideal generated by 2 and X is not generated by any single polynomial, and thus $Z[X]$ is not principal. This also shows that the property of being a principal ring is not transferred from A to $A[X]$.

Finding ring elements x at which a given polynomial p has zero value, $p(x) = 0$, has traditionally been a central concern of algebra. Such elements x are called *roots* of the polynomial p. The word *algebra* itself, used by Mohammed ibn-Musa al-Khwarizmi in the title of his book published around the year 830, refers to an essentially combinatorial approach to finding roots of polynomials.

EXERCISES

1. Write a computer program for adding and multiplying polynomials, in one as well as several indeterminates, over
   (a) Z,
   (b) Z_n for any positive integer n,
   (c) $(\mathcal{P}(S), +, \cap)$ for any finite subset S of ω.

Write a program that, for any polynomial in one indeterminate over one of these rings, searches for roots of that polynomial in the ring being considered.

2. Let A be a subring of a ring B. Show that those polynomials over A whose value is 0 at some fixed element b of B form an ideal I_b of $A[X]$. Can we have $I_b = A[X]$? Show that if I_b is trivial, then $A[b]$ is isomorphic to $A[X]$.

3. Give examples of polynomial functions that are injective, bijective, or surjective from a ring A to itself and examples of polynomial functions that are not.

4. Is it possible that the value of some polynomial over a ring A is not zero at any element of A?

5. Can a linear polynomial have more than one root in a ring?

6. Is a ring A necessarily Noetherian if $A[X]$ is Noetherian?

3. FACTORIZATION AND THE EUCLIDEAN ALGORITHM

The object of our interest in this section is the multiplicative structure of a ring A. We assume, throughout the section, that this multiplicative structure is a monoid, with the neutral element denoted by 1. We define the binary relation $a|b$, in words "*a divides b*" or "*b is a multiple of a*," to mean that $b = qa$ for some $q \in A$. Due to associativity, this *divisibility relation* is transitive, and it is reflexive since $a = 1 \cdot a$. Thus it is a preorder on A, with the usual notation $\lesssim$, $<$ ($\lesssim$ but not $\gtrsim$), and $\sim$ for the associated equivalence ($\lesssim$ and $\gtrsim$), as introduced in Chapter II. Obviously $1 \lesssim a \lesssim 0$ for every ring element a, and $a < 0$ if $a \neq 0$.

Examples. Divisibility is an order relation on any power set ring $(\mathcal{P}(S), +, \cap)$; here $a|b$ is equivalent to $a \supseteq b$. It is not an order in $\mathbb{Z}$, because $1|-1$ and $-1|1$, i.e., $-1 \sim 1$. However, divisibility restricted to positive integers is an order, of which the usual integer order $\leq$ is a linear extension. A lower section of the divisibility order of positive integers is displayed in Figure 4.1. The divisibility order on the power set ring $\mathcal{P}(3)$ is displayed in Figure 4.2.

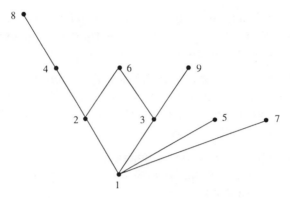

FIGURE 4.1 Divisibility order of positive integers.

Those elements u of a ring A for which an inverse exists in the multiplicative monoid of A form a group U under multiplication, with 1_A as the neutral element. These elements are called the *units* of A. The inverse of a unit u is denoted by u^{-1}. If u is a unit, then so is $-u$ and

$$(-u)^{-1} = -(u^{-1})$$

The binary relation R defined on A by the condition that

aRb if and only if $a = ub$ for some unit u

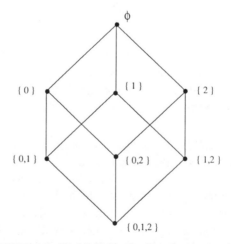

FIGURE 4.2 Divisibility order in a power set ring.

is an equivalence relation, called *unit equivalence*. Each element a of A is unit equivalent to $-a$. Any two unit-equivalent elements are also equivalent in the divisibility preorder. It follows that a nonzero element a covers 1 in the divisibility preorder of A if and only if

a is not a unit and $a = bc$ implies $a \sim b$ or $a \sim c$

In this case a is called a *prime* element of the ring A.

Examples. In $\mathbb{Z}$, 2, 3, and -3 are prime elements; 0, 1, and 6 are not. In $\mathbb{Z}[X]$, $2X + 1$ is prime and $2X + 2$ is not. In a power set ring $(\mathcal{P}(S), +, \cap)$ a nonzero element (subset of S) is prime if and only if it is the complement of a singleton in S. In the subring of $\mathcal{P}(S)$ formed by S itself and its finite subsets, no element is prime if S is infinite.

In any ring, the principal ideal (a) generated by a ring element a consists precisely of the multiples of a. Thus $a|b$ is equivalent to $(a) \supseteq (b)$. In a principal ring a nonzero element a is prime if and only if (a) is a maximal ideal.

If two elements a and b of a ring have a greatest lower bound d in the divisibility preorder, i.e.,

$d \lesssim a$, $d \lesssim b$ and $c \lesssim d$ for every $c \lesssim a$, $c \lesssim b$

then d is called a *greatest common divisor* (gcd) of a and b. Every other element unit equivalent to d is of course also a gcd. Dually, a *least common multiple* (lcm) is a least upper bound of a and b. Observe that the ideal $(a) \cap (b)$ consists precisely of the common multiples of a and b. An element d of this ideal generates $(a) \cap (b)$ if and only if d is an lcm of a and b. Thus no lcm exists if $(a) \cap (b)$ is not principal. A gcd of a and b, on the other hand, does not have to belong, in general, to the ideal $(a) + (b)$ which is the lub of (a) and (b) in the lattice of ideals. For example, in $\mathbb{Z}[X]$, the gcd's of the zero-degree polynomial 2 and the monomial X are 1 and -1, which do not belong to the proper ideal $(2, X)$. However:

Proposition 7 *If A is a principal ring and $a, b \in A$, then the greatest common divisors of a and b are precisely the generators of the ideal $(a) + (b)$.*

Proof. Let d be a generator of $(a) + (b)$,

$$(d) = (a) + (b) = (a, b)$$

Obviously d divides both a and b. But d is also of the form $xa + yb$ for some $x, y \in A$, and thus every common divisor of a and b divides d as well.

Conversely, let d be a gcd of a and b. Then (d) contains the ideals (a) and (b); hence
$$(d) \supseteq (a) + (b)$$
Let g be a generator of $(a) + (b)$. Since g is a common divisor of a and b, it must also divide d, that is, $(g) \supseteq (d)$. But
$$(g) = (a) + (b)$$
by definition, so $(d) = (g)$ and d also belongs to and generates the ideal $(a) + (b)$. $\square$

Corollary. *The following conditions are equivalent for any two non-unit elements a, b of a principal ring A:*

(i) *1 is a greatest common divisor of a and b,*
(ii) *the ideals (a) and (b) are coprime, $(a) + (b) = A$.*

Such two elements a and b of a principal ring A are called *coprime* elements. We then have, for some elements $x, y \in A$, the equality
$$xa + yb = 1$$
This result was first reported for the ring $\mathbb{Z}$ by Bachet de Méziriac in 1624. No other rings were studied at that time.

Returning to the general case of a not necessarily principal ring A, observe that $a < b$ by divisibility if and only if the ideal (b) is strictly contained in (a), $(b) \subset (a)$. Thus the divisibility preorder of every Noetherian ring satisfies the descending chain condition. Particular examples are $\mathbb{Z}$, $\mathbb{Z}[X]$, and of course every finite ring.

Proposition 8 *If the divisibility preorder of a ring A satisfies the descending chain condition, then every nonzero nonunit element a of A is a product of prime elements, $a = a_1 \cdots a_n$, $n \geq 1$.*

Proof. Call a nonzero nonunit element a of the ring *nonfactorizable* if a does not obey the proposition. If the set of nonfactorizable elements of A is not empty, let a be one of its elements minimal in the divisibility preorder. As a cannot be prime, $a = bc$ where none of b,c is a unit or zero and both $b < a$ and $c < a$ by divisibility. By the minimal choice of a,

$$b = b_1 \cdots b_n \quad \text{and} \quad c = c_1 \cdots c_m$$

where all the factors b_i and c_i are prime. Thus

$$a = b_1 \cdots b_n c_1 \cdots c_m$$

contradicting the assumption that a is nonfactorizable. □

A ring A is called *Euclidean* if there is a function $\delta : A \to W$ to a well-ordered set W with the following properties:

(i) $a < b$ in the divisibility preorder of A implies $\delta(a) < \delta(b)$ in the well-ordered set W,
(ii) for every $a, b \in A$, there is a $q \in A$ such that $a - qb = 0$ or $\delta(a - qb) < \delta(b)$.

Such a function δ is called a *Euclidean norm*. Observe that in every Euclidean ring the divisibility preorder satisfies the descending chain condition.

Example. A power set ring $(\mathcal{P}(S), +, \cap)$ is Euclidean if and only if S is finite. In that case a Euclidean norm $\delta : \mathcal{P}(S) \to \omega$ is defined by

$$\delta(T) = \operatorname{Card}(S \setminus T)$$

Proposition 9 *Every Euclidean ring is principal.*

Proof. Let I be an ideal of a Euclidean ring A. Let $b \in I$ be minimal in the divisibility preorder, i.e., there is no $c \in I$ with $c < b$. If $b = 0$, then I is trivially principal, $I = \{0\}$. So let $b \neq 0$. We claim that every $a \in I$ is a multiple of b. Let $q \in A$ be such that

$$a - qb = 0 \quad \text{or} \quad \delta(a - qb) < \delta(b)$$

By the definition of b we cannot have

$$\delta(a - qb) < \delta(b)$$

for the element $a - qb$ of I. Thus $a - qb = 0$ and a is a multiple of b, as claimed. □

A ring A is said to be *entire* provided that $ab = 0$ only if $a = 0$ or $b = 0$. A nontrivial entire ring, i.e., where $1 \neq 0$, is called an *integral domain*. This means that $A \setminus \{0\}$ is a multiplicative submonoid.

Examples. The ring $\mathbb{Z}$ is entire. Every field is entire. The power set ring $(\mathcal{P}(S), \dotplus, \cap)$ is not entire as soon as S has more than one element: the product (intersection) of two nonzero (nonempty) ring elements (subsets of S) can be zero (the null set).

Let us observe that in an entire ring any two elements a, b equivalent by divisibility are also unit equivalent. Indeed, if we have

$$a = qb \quad \text{and} \quad b = ra$$

for some ring elements q and r, then $a = (qr)a$ and $a(1 - qr) = 0$. If $a = 0$, then $b = 0$. If $a \neq 0$, then we must have

$$1 - qr = 0$$

and both q and r are units.

Factorization Theorem. *Let A be a principal entire ring. Let $a \in A$ be a nonzero element that is not a unit. Then there exist prime elements $a_1, \ldots, a_n$, $n \geq 1$, whose product is a,*

$$a = a_1 \cdots a_n$$

Further, this prime factorization is unique in the sense that if

$$a = b_1 \cdots b_m$$

is any other factorization with m prime factors, then $m = n$ and there is a permutation σ of $\{1, \ldots, n\}$ such that for each $i = 1, \ldots, n$, b_i and $a_{\sigma(i)}$ are unit equivalent.

Proof. Every principal ring is Noetherian, and therefore its divisibility preorder satisfies the descending chain condition. The existence of a prime factorization

$$a = a_1 \cdots a_n$$

is therefore assured by Proposition 8.

To prove uniqueness, let us first show that if a prime p divides a product cd, then it divides c or d. Suppose it does not divide c. Thus the ideal I generated by $\{p, c\}$ contains the ideal (p) strictly.

As A is principal, for some $g \in A$ we have $I = (g)$. Then $g|p$ but p does not divide g. Since p is prime, g must be a unit and therefore $I = A$. But then every element of A can be expressed as $xp + yc$ with suitable $x, y \in A$, in particular

$$1 = xp + yc$$

Multiplying with d, we get

$$d = xpd + ycd$$

As $p|cd$, we must have $p|d$.

It follows by induction that if a prime p divides a product $c_1 \cdots c_k$, then it divides one of the factors c_i.

Suppose now that we do not have unique prime factorization in A. Let $n \geq 1$ be the smallest natural number such that some product of n prime factors,

$$a = a_1 \cdots a_n$$

is not unique in the sense that a also has some other prime factorization

$$a = b_1 \cdots b_m$$

not obeying the theorem. Clearly $m \geq n$. Also, $m > 1$, for otherwise we would have

$$a = a_1 = b_1$$

Consider the prime element a_1. It divides the product $b_1 \cdots b_m$ and therefore it divides one of the b_i. Without loss of generality we can suppose that $a_1|b_1$. Since b_1 is also prime, b_1 divides a_1 as well. As A is entire, a_1 and b_1 are unit equivalent, $a_1 = ub_1$ for some unit u. We cannot have $n = 1$, for then we would have

$$a = ub_1 = b_1(b_2 \cdots b_m)$$
$$b_1(u - b_2 \cdots b_m) = 0$$

and hence, since $b_1 \neq 0$,

$$u = b_2 \cdots b_m$$

which is impossible since no prime can divide a unit. Thus $n > 1$. Let

$$q = a_2 \cdots a_n$$

We have

$$ub_1 q = a = b_1(b_2 \cdots b_m)$$

and thus
$$b_1(uq - b_2 \cdots b_m) = 0$$

Hence $uq - b_2 \cdots b_m$ is zero, that is,
$$uq = b_2 \cdots b_m$$

Now
$$uq = (ua_2) \cdots a_n$$

is a product of $n - 1$ prime factors. By the minimal choice of n, this factorization of uq is unique, so that $m = n$ and there is some permutation τ of $\{2, \ldots, n\}$ such that for each $2 \leq i \leq n$, b_i is unit equivalent to $a_{\tau(i)}$. But then, since b_1 is unit equivalent to a_1, the factorizations
$$a = a_1 \cdots a_n \quad \text{and} \quad a = b_1 \cdots b_m$$

do not violate the uniqueness condition of the theorem, contrary to the hypothesis that they do. □

Corollary. *Let A be a principal entire ring. Then A possesses the ideal decomposition property and the ideal coset closure system of A has Helly number at most 2.*

Proof. For a prime factorization $a = a_1 \cdots a_n$ we have
$$(a) = (a_1) \cdots (a_n) \qquad \square$$

Proposition 9 implies the unique factorization property in every Euclidean entire ring. Even without this we already knew that the ring $\mathbb{Z}$ of integers is principal and entire, and therefore it has unique factorization. However, let us restate this very important result as a consequence of the Euclidean property:

Integer Factorization Theorem. *The ring $\mathbb{Z}$ of integers is Euclidean and entire. Therefore $\mathbb{Z}$ is principal with unique prime factorization: every integer n greater than 1 has a unique expression*
$$n = a_1^{k_1} \cdots a_t^{k_t}$$

where $a_1 < \cdots < a_t$ are positive prime numbers and $k_1, \ldots, k_t \geq 1$.

Proof. The function δ mapping each nonzero integer a to its *absolute value*
$$|a| = \max(a, -a)$$
and 0 to ω is a Euclidean norm on $\mathbb{Z}$. Let $a, b \in \mathbb{Z}$. If $a = 0$, then we have
$$a - qb = 0$$
for $q = 0$. If $a \neq 0$ but $b = 0$, then for $q = 0$ we have
$$\delta(a - qb) < \delta(b)$$
Otherwise $ab \neq 0$. Let q_1 be the smallest natural number such that $|a| < q_1|b|$. If $ab > 0$, then for $q = q_1 - 1$ we have
$$a - qb = 0 \quad \text{or} \quad \delta(a - qb) < \delta(b)$$
If $ab < 0$, then for $q = 1 - q_1$ we have
$$a - qb = 0 \quad \text{or} \quad \delta(a - qb) < \delta(b) \qquad \square$$

Let $n = a_1^{k_1} \cdots a_t^{k_t}$ and $m = b_1^{h_1} \cdots b_s^{h_s}$ be two integers greater than 1 factorized according to this theorem. Then n divides m if and only if $t \leq s$ and there is an injection σ from $[1, t]$ to $[1, s]$ such that
$$a_i = b_{\sigma(i)} \quad \text{and} \quad k_i \leq h_{\sigma(i)}$$
for all $1 \leq i \leq t$. The integers n and m are coprime if and only if
$$\{a_1, \ldots, a_t\} \cap \{b_1, \ldots, b_s\} = \emptyset$$
In that case $n \cdot m$ is an lcm of n and m.

Group Theory Revisited. (1) It is customary to call the cardinality of a group G the *order* of G. By the *order of a group element* $a \in G$ we mean the order of the subgroup generated by a. If it is finite, then it is the smallest positive integer n such that a^n is the neutral element of the group. Indeed the multiples in $\mathbb{Z}$ of the order of a are precisely the exponents z for which a^z is the neutral of G. For finite groups, Lagrange's Subgroup Counting Theorem says that the order of every subgroup divides the order of the entire group. Thus the order of every group element divides the order of the group. In particular, a finite group of prime order has no nontrivial proper subgroups, and it is generated by each of its nonneutral elements. (2) Let G be any group and let $a, b \in G$ be commuting elements with finite coprime orders n and m. Then the order of ab is $n \cdot m$.

Polynomial Factorization Theorem. *If A is a field, then the polynomial ring $A[X]$ is Euclidean and entire. Therefore $A[X]$ is principal with unique prime factorization: every polynomial $p \in A[X]$ of nonzero degree has an expression*

$$p = u p_1 \cdots p_n$$

where u is of degree zero and the p_i are prime polynomials with leading coefficient 1. No other prime polynomial with leading coefficient 1 divides p.

Proof. We claim that the function $\delta : A[X] \to \omega \cup \{\omega\}$ given by

$$\delta(p) = \deg p \quad \text{if} \quad p \neq 0 \quad \text{and} \quad \delta(0) = \omega$$

is a Euclidean norm. Assume this is not so. We shall derive a contradiction. Let $p \neq 0$ be a polynomial of least possible degree such that for some polynomial $t \neq 0$ no difference $p - qt$, $q \in A[X]$, is zero or of degree less than $\deg t$. Clearly

$$\deg p \geq \deg t$$

for else we take $q = 0$. Let a and b denote the leading coefficients of p and t, respectively. Let

$$q_0 = a b^{-1} X^{\deg p - \deg t}$$

Then $p - q_0 t \neq 0$ and $\deg(p - q_0 t) < \deg p$. But then there must exist, by the minimal degree choice of p, a polynomial q_1 such that

$$(p - q_0 t) - q_1 t$$

is zero or of degree less than t. Letting

$$q = q_0 + q_1$$

this means that $p - qt$ is zero or of degree less than t, contradicting the definitions of p and t. This proves that δ is a Euclidean norm. $\square$

FACTORIZATION AND THE EUCLIDEAN ALGORITHM

In a Euclidean ring A with norm δ, for nonzero elements a, b we call an element of the form $a - qb$ a *remainder of a by b* if

$$a - qb = 0 \quad \text{or} \quad \delta(a - qb) < \delta(b)$$

A *Euclidean algorithm* is a sequence $(a_i)_{i<\gamma}$ of elements of A indexed by an ordinal γ, $3 \leq \gamma \leq \omega$, such that

(i) for all $2 \leq i < \gamma$, a_{i-2} and a_{i-1} are not zero and a_i is a remainder of a_{i-2} by a_{i-1},
(ii) if the sequence $(a_i)_{i<\gamma}$ is finite, then its last term $a_{\gamma-1}$ is zero.

The elements a_0 and a_1 are called *inputs*, and if the sequence is finite, $\gamma < \omega$, then $a_{\gamma-2}$ is called the *output*.

Euclidean Algorithm Theorem. *Every Euclidean algorithm is finite. The output is a greatest common divisor of the inputs.*

Proof. An infinite Euclidean algorithm $(a_i)_{i<\omega}$ would yield a set

$$\{\delta(a_i) : 1 \leq i < \omega\}$$

that would not be well ordered in the codomain W of the norm δ, so every Euclidean algorithm is finite. For $i \geq 2$, every common divisor of a_{i-2} and a_{i-1} is also a common divisor of the remainder

$$a_i = a_{i-2} - q a_{i-1}$$

It easily follows, by induction, that all terms a_i, $i \geq 2$, are multiples of the greatest common divisors of a_0 and a_1. In particular the gcd's must divide the output. On the other hand, we claim that the output $a_{\gamma-2}$ divides every term a_i. If this is not so, let i be the largest index such that the output does not divide a_i. Obviously $i \leq \gamma - 3$. But

$$a_i - q a_{i+1} = a_{i+2} \quad \text{for some } q$$

and as the output divides a_{i+1} and a_{i+2}, it must also divide a_i, a contradiction proving our claim. It follows, in particular, that the output divides the inputs, and hence it divides every gcd of the inputs. The output is therefore equivalent by divisibility to the gcd's, which means that the output itself is a greatest common divisor of the inputs. □

Historical Note. The idea of this algorithm is over 2000 years old. It was first presented in Euclid's *Elements*, Book VII.

EXERCISES

1. In a power set ring $(\mathcal{P}(S), +, \cap)$ what is the gcd and the lcm of two elements $A, B \in \mathcal{P}(S)$?

2. For any finite set S, show that for every nonempty $A, B \in \mathcal{P}(S)$ there is a Euclidean algorithm of length $\gamma \leq 4$ in the power set ring $(\mathcal{P}(S), +, \cap)$, with inputs A and B. Show that this is no longer true in the ring $\mathbb{Z}$.

3. Verify that two elements of a ring are equivalent in the divisibility preorder if and only if they generate the same ideal. Show that the inclusion-ordered set of principal ideals is order isomorphic to the dual of the order associated with the divisibility preorder of the ring.

4. Show that integer divisibility restricted to the set of natural numbers is order isomorphic to $(\mathcal{I}(\mathbb{Z}), \subseteq)$ where $\mathcal{I}(\mathbb{Z})$ denotes the set of ideals of $\mathbb{Z}$.

5. Show that a nonzero integer b covers a in the divisibility preorder of $\mathbb{Z}$ if and only if $b = pa$ for some prime integer p.

6. Show that $\mathbb{Z}$ has an infinity of prime elements.

7. Verify that two integers are coprime if and only if no prime integer divides both of them. (Disregard $0, 1, -1$.)

8. Describe the units of the ring $\mathbb{Z}_m$. Show that if $x \in \mathbb{Z}_m$ is not a unit, then for some $a \in \mathbb{Z}_m$, $a \neq 0$, we have $xa = 0$.

9. Show that the ring $\mathbb{Z}_m$ is a field if and only if m is prime.

10. Show that if a ring A is not trivial, then A^A is not entire.

11. Show that if A is an entire ring, then the units of $A[X]$ are the zero-degree polynomials corresponding to the units of A.

12. Is the property of being entire transferred from A to $A[X]$?

13. Verify that two nonunit nonzero elements of a principal ring are coprime if and only if their only common divisors are the units.

14. Show that if A is a principal ring and F a nonempty finite subset of A, then the greatest lower bounds of F in the divisibility preorder are precisely the generators of the ideal closure of F.

15. Show that a ring A is Euclidean if and only if the following holds: the divisibility pre-order has a well-ordered linear order extension where for every $a, b \in A$ there is a $q \in A$ such that either $a - qb = 0$ or $a - qb$ is less than b in the linear order extension.

16. Write a computer program to execute the Euclidean algorithm, for inputs in the ring
    (a) $\mathbb{Z}$,
    (b) $\mathbb{Z}_p[X]$, p prime integer.

17. Write a computer program to determine divisibility between ring elements in $\mathbb{Z}$ and $\mathbb{Z}_p[X]$ where p is prime, as well as to find gcd's and lcm's, and to express any given gcd of ring elements a, b in the form $xa + yb$.

18. Write a computer program that performs prime factorization in the rings $\mathbb{Z}$ and $\mathbb{Z}_p[X]$, p prime.

BIBLIOGRAPHY

M. F. ATIYAH and I. G. MACDONALD, *Introduction to Commutative Algebra*. Addison-Wesley 1969. An advanced text on rings and modules. Modules can be defined in a way similar to vector spaces; here they truly appear as generalized ideals.

Richard A. DEAN, *Classical Abstract Algebra*. Harper & Row 1990. Contains an accessible and rigorous development of the basic theory of rings. The presentation is not limited to commutative rings.

Jean ITARD, *Arithmétique et théorie des nombres*. "Que sais-je?" Series No. 1093. Presses Universitaires de France 1967. Written in the best encyclopedic tradition, this pocket-size booklet is both rigorous and easy to read. The oldest questions of number theory are introduced with a modicum of abstract algebra.

Jean ITARD, *Les nombres premiers*. "Que sais-je?" Series No. 571. Presses Universitaires de France 1969. More algebra is used in this sequel to Itard's "Arithmétique." The focus is on historically more recent results and problems, specifically concerning prime numbers.

Nathan JACOBSON, *Basic Algebra I*. Freeman 1985. Chapter 2 is a classical introduction to ring theory, including noncommutative rings. Matrix rings and quaternions are discussed as examples of the latter. (This assumes some familiarity with matrices, determinants, and real and complex numbers.) There is much further advanced material on rings and fields in Jacobson's *Basic Algebra II* (Freeman 1989).

Irving KAPLANSKY, *Commutative Rings*. Allyn and Bacon 1970. A selection of advanced topics accessible with basic background knowledge of algebra.

Serge LANG, *Algebra*. Addison-Wesley 1984. Polynomials are defined at once so as to allow for several variables. Factorization of polynomials is treated in this general context. Also included is the construction of formal power series. A generalization of Hilbert's transfer theorem is shown to hold for these.

Oystein ORE, *Number Theory and Its History*. Dover Publications 1988. The good old real world behind abstract ring theory. Ore's presentation is both mathematically and historically precise.

Rodney Y. SHARP, *Steps in Commutative Algebra*. Cambridge University Press 1990. A medium level expertise-builder on rings, ideals, and modules.

David SHARPE, *Rings and Factorization*. Cambridge University Press 1987. A short and very readable undergraduate text on Euclidean domains, primes, and factorization. The factorization of polynomials is discussed in fair detail.

B. L. VAN DER WAERDEN, *Modern Algebra*. Frederick Ungar Publishing Co. 1949. Much of this seminal account of abstract algebra is devoted to rings and fields.

CHAPTER V

FIELDS

1. RATIONAL AND REAL NUMBERS

Fields were defined in the preceding chapter as rings of a particular kind. The nonzero elements of a field F form a group F^* under multiplication. A field is necessarily an integral domain, and every subring of a field is entire, but some integral domains such as the ring $\mathbb{Z}$ of integers are not fields.

Fraction Field Theorem. *Every integral domain D is contained in some field Q as a subring. Moreover, D is contained in some field Q which is minimal in the sense that no proper subring T of Q containing D is a field. Any two such minimal fields Q and Q' are isomorphic.*

Proof. Let $S = \{(a,b) \in D^2 : b \neq 0\}$. Define two binary operations $+$ and $\cdot$ on S by

$$(a,b) + (c,d) = (a \cdot d + b \cdot c, b \cdot d)$$

$$(a,b) \cdot (c,d) = (a \cdot c, b \cdot d)$$

where "+" and "·" on the right-hand-side stand for sum and product in D. The reader should verify that $(S, +)$ and $(S, \cdot)$ are commutative monoids, with $(0,1)$ and $(1,1)$ as respective neutral elements. Define a binary relation R on S by

$$(a,b)R(c,d) \quad \text{if and only if} \quad a \cdot d = b \cdot c \quad \text{in } D$$

This relation R is obviously reflexive and symmetric. Indeed it is also transitive: $(a,b)R(c,d)$ and $(c,d)R(e,f)$ mean

$$(ad)f = (bc)f, \qquad (cf)b = (de)b$$

implying

$$(af)d = (be)d, \qquad (af - be)d = 0$$

and since D is an integral domain and $d \neq 0$,

$$af - be = 0, \qquad af = be$$

that is, $(a,b)R(e,f)$. Thus R is an equivalence relation on S. Furthermore, it is not difficult to verify that R is a congruence of both monoid structures $(S, +)$ and $(S, \cdot)$. We can then form two corresponding quotient monoids $(S/R, +)$ and $(S/R, \cdot)$. The same symbols "+" and "$\cdot$" are used to denote both the original and the quotient operations. We claim that $(S/R, +, \cdot)$ is a ring, and indeed a field.

First, $(S/R, +)$ is a commutative group because

$$(a,b) + (-a,b) = (0, b^2)$$

in S is always congruent to $(0,1) \bmod R$.

Second, we note that $(S/R, \cdot)$ is a commutative monoid whose neutral element is the R-class of $(1,1)$.

Third, product is distributive over sum in S/R, because in S we have

$$[(a,b) + (c,d)] \cdot (e,f) = ((aed + bce), bfd)$$
$$(a,b) \cdot (e,f) + (c,d) \cdot (e,f) = ((aedf + cebf), bfdf)$$

and in D we have

$$(aed + bce) \cdot bfdf = bfd \cdot (aedf + cebf)$$

Fourth, for nonzero $a, b \in D$, in S the product $(a,b) \cdot (b,a)$ is always congruent to $(1,1) \bmod R$, proving that S/R is a field as claimed.

The function $f : D \to S/R$ that maps $x \in D$ to the R-class of $(x,1)$ is easily seen to be an injective ring homomorphism, establishing an isomorphism between D and the subring $\mathrm{Im} f$ of S/R. By the

Injection–Extension Theorem, there is a set $Q \supseteq D$ and a bijection $g : Q \to S/R$ such that $g|D = f$. Define a field structure on Q as follows. For $x, y \in Q$ let the sum of x and y be defined as

$$g^{-1}(g(x) + g(y))$$

and let their product be defined as

$$g^{-1}(g(x) \cdot g(y))$$

If x and y are both in D, then this sum and product coincide with the sum and product in the original ring structure of D. Therefore the symbols "+" and "·" shall be used to denote the ring operations both in D and in Q. The multiplicative inverse of any nonzero $q \in Q$ is denoted by q^{-1}. Obviously $g : Q \to S/R$ is an isomorphism between these fields, and D is a subring of Q.

If $q \in Q$ and (a, b) is any element of the R-class $g(q)$, then

$$(a, b) = (a, 1) \cdot (1, b)$$

in S, which implies that $q = a \cdot b^{-1}$ in Q. If T is a subring of Q containing D, and if T is a field, then $a \cdot b^{-1} \in T$ for every $a, b \in D$, $b \neq 0$, and thus every $q \in Q$ belongs to T. This proves that Q is minimal in the sense that no proper subring T of Q containing D is a field.

Finally, let

$$(Q, +, \cdot) \quad \text{and} \quad (Q', \oplus, \odot)$$

be two fields containing D as a subring and both minimal in the above sense. Define $h : Q \to Q'$ as follows. Every $q \in Q$ is of the form $a \cdot b^{-1}$ for some $a, b \in D$. We can form the element $a \odot b^{-1}$ in Q'. Note that if

$$q = a \cdot b^{-1}$$

has another expression in Q as

$$q = c \cdot d^{-1}$$

with $c, d \in D$, then

$$a \cdot d = b \cdot c$$

in D, which can also be written as

$$a \odot d = b \odot c$$

in Q' since D is also a subring of $(Q', \oplus, \odot)$, and multiplying with $b^{-1} \odot d^{-1}$ in Q' yields

$$a \odot b^{-1} = c \odot d^{-1}$$

We may therefore define $h(q)$ as the element $a \odot b^{-1}$ in Q', independently of the choice of $a, b \in D$ to represent q in the form $a \cdot b^{-1}$ in Q. The verification that h is an isomorphism from $(Q, +, \cdot)$ to $(Q', \oplus, \odot)$ is straightforward gymnastics. □

A field Q satisfying the conditions of the above theorem is called a *field of fractions* of the integral domain D. In the proof we have seen that every $q \in Q$ is of the form $a \cdot b^{-1}$ with $a, b \in D$: we shall also write a/b instead of $a \cdot b^{-1}$. Note that

$$a/b = -a/-b$$

and if a/b is not 0, then

$$(a/b)^{-1} = b/a$$

Let us choose once and for all a field of fractions of the integers $\mathbb{Z}$, and let us denote it by $\mathbb{Q}$ (for "quotients"). The elements of $\mathbb{Q}$ shall henceforward be called *rational numbers*.

Another important example of fraction field is based on the following transfer property: the ring of polynomials of an integral domain A is entire as well. (Proof: The leading coefficient of the product of two polynomials is the product of the respective leading coefficients of the two polynomials.) A field of fractions of such a ring $A[X]$ is called a field of *rational fractions* over A.

The order relation $\leq$ on $\mathbb{Z}$ that we ordinarily use to compare integers is far from being arbitrary. The relation $\leq$ is the only order on $\mathbb{Z}$ for which $0 < 1$ and such that

$$x \leq y \quad \text{implies} \quad x + z \leq y + z \quad \text{for all} \quad x, y, z \in \mathbb{Z}$$

Generally if $(G, \otimes)$ is a group, then an order $\leq$ on G is said to be *compatible* with the group structure if

$$x \leq y \quad \text{implies} \quad x \otimes z \leq y \otimes z \quad \text{and} \quad z \otimes x \leq z \otimes y$$

It is equivalent to require that

$$x \leq y, \ x' \leq y' \quad \text{imply} \quad x \otimes x' \leq y \otimes y'$$

The ordinary comparison order of the integers is thus compatible with the additive group $(\mathbb{Z},+)$. As a particular aspect of the rational field $(\mathbb{Q},+,\cdot)$ consider now the additive group $(\mathbb{Q},+)$.

Proposition 1 *There is a unique linear order on $\mathbb{Q}$ compatible with the additive group structure $(\mathbb{Q},+)$ and in which 0 is less than 1.*

Proof. Let

$$\mathbb{Q}^+ = \{q \in \mathbb{Q} : q = a/b \text{ for some positive integers } a,b\}$$

Observe that P is closed under addition. Verify that for every nonzero $q \in \mathbb{Q}$, exactly one of q or $-q$ belongs to $\mathbb{Q}^+$. It follows that

$$R = \{(x,y) \in \mathbb{Q}^2 : x = y \text{ or } y - x \in \mathbb{Q}^+\}$$

is a linear order on $\mathbb{Q}$. For $x, y \in \mathbb{Z}$,

$$x \leq y \quad \text{if and only if} \quad xRy$$

Obviously $0R1$ and it is not difficult to verify that R is compatible with $(\mathbb{Q},+)$.

Uniqueness is proven by showing that if T is another compatible linear order in which $0T1$, then

$$\{q \in \mathbb{Q} : 0Tq,\ 0 \neq q\}$$

coincides with the set $\mathbb{Q}^+$ defined above and $T = R$. □

Since the linear order on $\mathbb{Q}$ specified by the above proposition coincides on $\mathbb{Z}$ with the ordinary order of integers, it shall be denoted by the same symbol $\leq$. It is called the *standard order* on the rational numbers. Extending the notion of positive and negative integers, we say that $q \in \mathbb{Q}$ is *positive* if $0 < q$ and *negative* if $q < 0$. Positive rationals are exactly the members of the set $\mathbb{Q}^+$ used in the proof of the above proposition. The set $\mathbb{Q}^+$ of positive rationals constitutes a subgroup of the multiplicative group $(\mathbb{Q}^*, \cdot)$ of all nonzero rationals. The order $\leq$ restricted to $\mathbb{Q}^+$ is compatible with the group $(\mathbb{Q}^+, \cdot)$. Moreover, if a,b,c,d are rationals and b,d are positive, then

$$ab^{-1} < cd^{-1} \quad \text{if and only if} \quad ad < bc$$

This provides a particularly convenient criterion for comparing rationals expressed as quotients of integers.

For a chain $(S, \leq)$ a *cut* is a couple (S_1, S_2) of disjoint nonempty subsets of S such that
$$S_1 \cup S_2 = S$$
and
$$s_1 < s_2 \quad \text{for every} \quad s_1 \in S_1, \quad s_2 \in S_2$$
Saying that for every such cut S_1 has a maximum and S_2 has a minimum means precisely that $(S, \leq)$ is discrete. If for each cut either S_1 has a maximum or S_2 has a minimum but not both, then $(S, \leq)$ is called *continuous*. The chain $(\mathbb{Z}, \leq)$ is discrete, not continuous. The chain $(\mathbb{Q}, \leq)$ is not discrete: consider the cut
$$(\mathbb{Q} \setminus \mathbb{Q}^+, \mathbb{Q}^+)$$
Let us show that $(\mathbb{Q}, \leq)$ is not continuous either. Consider the cut
$$(S, \mathbb{Q} \setminus S)$$
with
$$S = (\mathbb{Q} \setminus \mathbb{Q}^+) \cup S_1 \quad \text{where} \quad S_1 = \{q \in \mathbb{Q}^+ : q^2 < 2\}$$
First, let us verify that S has no maximum, i.e., that S_1 has no maximum. Let $q \in S_1$. Write $q = a/b$ with a, b positive integers. As $2 - q^2$ is positive rational, let
$$2 - q^2 = c/d$$
with c, d positive integers. Let n be any integer larger than $3ad$. The reader can verify, by elementary manipulations, that
$$q + (1/n) \in S_1$$
Consequently S has no maximum. Second, let us show that $\mathbb{Q} \setminus S$ has no minimum. Observe that
$$\mathbb{Q} \setminus S = \{q \in \mathbb{Q}^+ : q^2 > 2 \text{ or } q^2 = 2\}$$
If we knew that there is no rational q with $q^2 = 2$, then for every $q \in \mathbb{Q} \setminus S$ we could easily produce an even smaller element of $\mathbb{Q} \setminus S$: just take any positive integer n greater than
$$2q/(q^2 - 2)$$
and consider $q - (1/n)$. We shall therefore establish that there is no rational q with $q^2 = 2$. This fact, already known to Pythagoras and

Plato, is often referred to as the "irrationality of the square root of 2." Seeking a contradiction, suppose we had a rational q with $q^2 = 2$. Let
$$q = a/b$$
where $a, b \in \mathbb{Z}$. Then the integer a^2 would be equal to the integer $2b^2$. Note that 2 is prime in the ring $\mathbb{Z}$. Let
$$a = p_1 \cdots p_n \quad \text{and} \quad b = r_1 \cdots r_m$$
be prime factorizations of a and b in $\mathbb{Z}$. Obviously
$$a^2 = p_1 \cdots p_n \cdot p_1 \cdots p_n \quad \text{and} \quad 2b^2 = 2r_1 \cdots r_m \cdot r_1 \cdots r_m$$
i.e., $a^2 = 2b^2$ would be a product of $2n$ prime factors on the one hand and of $2m + 1$ prime factors on the other hand. The Factorization Theorem implies
$$2n = 2m + 1$$
which is absurd. This completes the proof that in the cut $(S, \mathbb{Q} \setminus S)$ the set S has no maximum and $\mathbb{Q} \setminus S$ has no minimum: $(\mathbb{Q}, \leq)$ is neither discrete nor continuous.

Proposition 2 *There is a continuous linearly ordered set $(R, \leq)$ such that*

(i) *$\mathbb{Q} \subseteq R$ and the restriction of $\leq$ to $\mathbb{Q}$ is the standard order of rational numbers,*
(ii) *for every $r_1 < r_2$ in R there are q_1, q_2, q_3 in $\mathbb{Q}$ such that*
$$q_1 < r_1 < q_2 < r_2 < q_3$$

Any two such continuous ordered sets are order isomorphic.

Proof. Let C_0 be the set of those cuts (S_1, S_2) of the standard rational chain for which neither S_1 has a maximum nor S_2 has a minimum. Perhaps C_0 is disjoint from $\mathbb{Q}$, but our elementary development of set theory does not permit to conclude this. This is why we have the Disjoint Copy Lemma: take any set C equipotent to C_0 and disjoint from $\mathbb{Q}$. Let $f : C \to C_0$ be a bijection. For $x \in C$ write the cut $f(x)$ as
$$(f_1(x), f_2(x))$$

132 FIELDS

Let $R = Q \cup C$. Define the order $\leq$ on R by letting $a \leq b$ mean that

	$a, b \in Q$	and	$a \leq b$ in Q
or	$a \in Q,\ b \in C$	and	$a \in f_1(b)$
or	$a \in C,\ b \in Q$	and	$b \in f_2(a)$
or	$a, b \in C$	and	$f_1(a) \subseteq f_1(b)$

The rest of the proof is elementary. □

Choose, once and for all, an ordered set $(R, \leq)$ such as specified by the above proposition. Call the elements of R *real numbers*. The more distinctive notation for the set of real numbers is R; it will be used henceforth. Condition (i) of the proposition supports the continued use of the symbol $\leq$ to compare, not just rationals, but real numbers as well. Consistently with the terminology for rationals, this order is called *standard* real order. A real number x is called *positive* if $x > 0$, *negative* if $x < 0$. However, we still do not know how to add and multiply real numbers! Behold:

Real Field Structure Theorem. *There is a unique field structure* $(R, +, \cdot)$ *on the set of real numbers such that*

(i) *the field of rationals is a subring of* $(R, +, \cdot)$,
(ii) $(R, \leq)$ *is compatible with the additive group* $(R, +)$,
(iii) *the set* R^+ *of positive real numbers is a subgroup of* $(R^*, \cdot)$.

Remark. Condition (iii) implies that $(R^+, \leq)$ is compatible with $(R^+, \cdot)$.

Proof. For $r, t \in R$ let

$$S(rt) = \{a \in R : a \leq x + y \text{ (sum in } Q)$$
$$\text{for some } x, y \in Q,\ x \leq r,\ y \leq t\}$$

Clearly $S(rt)$ is a nonempty lower section of $(R, \leq)$ and

$$R \setminus S(rt) \neq \emptyset$$

Thus

$$(S(rt), R \setminus S(rt))$$

is a cut in $(\mathbb{R}, \leq)$. By continuity of $(\mathbb{R}, \leq)$ either $S(rt)$ has a maximum, or $\mathbb{R} \setminus S(rt)$ has a minimum. Whichever is the case, let the real sum $r + t$ be defined as this max-or-min. We have

$$r + t = \text{lub}\, S(rt) = \text{glb}(\mathbb{R} \setminus S(rt))$$

For positive real numbers r, t let

$$P(rt) = \{a \in \mathbb{R} : a \leq x \cdot y \text{ (product in } \mathbb{Q})$$
$$\text{for some } x, y \in \mathbb{Q},\ 0 < x \leq r,\ 0 < y \leq t\}$$

Then

$$(P(rt), \mathbb{R} \setminus P(rt))$$

is a cut in $(\mathbb{R}, \leq)$. Let the real product $r \cdot t$ be defined as the max of $P(rt)$ or the min of $\mathbb{R} \setminus P(rt)$, whichever exists—in either case

$$r \cdot t = \text{lub}\, P(rt) = \text{glb}(\mathbb{R} \setminus P(rt))$$

For r, t negative, the real numbers

$$r' = \text{lub}\{a \in \mathbb{R} : a \leq x \text{ for some } x \in \mathbb{Q} \text{ with } r \leq -x\}$$

and

$$t' = \text{lub}\{a \in \mathbb{R} : a \leq x \text{ for some } x \in \mathbb{Q} \text{ with } t \leq -x\}$$

are positive, and we define $r \cdot t$ as equal to $r' \cdot t'$. For r negative, t positive, let $r \cdot t$ be the real number

$$\text{lub}\{a \in \mathbb{R} : a \leq x \text{ for some } x \in \mathbb{Q} \text{ with } (-r) \cdot t \leq -x\}$$

Finally, if r or t is 0, then let $r \cdot t$ be 0.

The rest of the proof is bearable tedium and low-level ingenuity. □

The final result of this section will establish an important algebraic fact about the real field by making use of the continuous linear order structure of $\mathbb{R}$. Observe that in any continuous linear order $(R, \leq)$, if (A, B) is a cut, then both $\text{lub}\, A$ and $\text{glb}\, B$ exist and they are equal. It can be deduced that in any continuous linear order $(R, \leq)$, if a set $S \subseteq R$ has an upper bound, then it has a least upper bound: let $\overline{S}$ be the lower section generated by S and consider the cut

$$(\overline{S}, R \setminus \overline{S})$$

Similarly, in any continuous linear order, if a set S has a lower bound, then it has a greatest lower bound.

The result that we wish to prove concerns real numbers of the form x^2. Clearly x^2 is positive if $x > 0$, but if $x < 0$, then x^2 is still positive because

$$(-x)^2 = (-1)^2 x^2 = x^2$$

We shall establish that all nonnegative real numbers are of the form x^2. In the proof we shall rely on the elementary observation that

$$\text{if } 0 \leq x < y, \text{ then } x^2 < y^2$$

Somewhat less elementary is the fact called "continuity of the square root operation" in real analysis. [Real analysis proceeds from the joint study of the structures $(\mathbb{R}, +, \cdot)$ and $(\mathbb{R}, \leq)$.] If S is any nonempty set of nonnegative real numbers, then so is

$$S^2 = \{s^2 : s \in S\}$$

Therefore both S and S^2 have a glb, and obviously

$$(\text{glb } S)^2 \leq \text{glb}(S^2)$$

Could the inequality be strict? If it were, let

$$a = (\text{glb } S)^2, \quad b = \text{glb}(S^2)$$

If $\text{glb } S = 0$, then the reader can easily show that $\text{glb}(S^2) = 0$ as well. Thus $\text{glb } S > 0$. By condition (ii) of Proposition 2 we would have a (rational) q such that

$$0 < q < \min\{\text{glb } S, (b-a)/3 \cdot \text{glb } S\}$$

Since $\text{glb } S + q$ is not a lower bound of S, there is an $s \in S$ such that

$$s < \text{glb } S + q$$

We have

$$s^2 < (\text{glb } S + q)^2 \leq b = \text{glb}(S^2)$$

which is impossible. Thus for every set S of nonnegative real numbers $(\text{glb } S)^2 = \text{glb}(S^2)$.

Square Root Theorem. *For every positive real number y there exists a positive real x such that $x^2 = y$.*

Proof. Let

$$S = \{t \in \mathbb{R}^+ : y \leq t^2\}$$

Obviously $y \leq \mathrm{glb}(S^2)$, i.e.,
$$y \leq (\mathrm{glb}\, S)^2$$
Can the inequality be strict? In that case, let
$$z = (\mathrm{glb}\, S)^2$$
and let q be any rational number (indeed any real will do) such that
$$0 < q < \min\{1, \mathrm{glb}\, S, (z-y)/2 \cdot \mathrm{glb}\, S\}$$
We would have
$$y < (\mathrm{glb}\, S - q)^2 < z = (\mathrm{glb}\, S)^2$$
which is contradictory. Thus $(\mathrm{glb}\, S)^2 = y$. □

For every positive real a there can only be one positive real x with
$$x^2 = a$$
for if also
$$z^2 = a$$
for some $z \in \mathbb{R}^+$, then
$$0 = x^2 - z^2 = (x-z)(x+z)$$
which implies $x - z = 0$ by integrality, i.e., $x = z$. The unique positive real x with $x^2 = a$ is called the *square root* of a, and it is denoted by $\sqrt{a}$. We also write $\sqrt{0}$ for 0. Observe that for every $r \in \mathbb{R}$,
$$\sqrt{r^2} = \max(r, -r)$$
We call $\sqrt{r^2}$ the *absolute value* of r and denote it by $|r|$.

Every degree 1 polynomial $aX + b$ over a field A has a (unique) root in A, namely the element $-b/a$. This is clearly not true if A is only a ring: the polynomial
$$2X - 1 \in \mathbb{Z}[X]$$
has no integer root. Indeed, a ring A is a field if and only if every degree 1 polynomial over A has a root in A: consider the polynomials $aX - 1$ for $a \in A$. Degree 2 polynomials may fail to have roots even in a field: witness
$$X^2 - 2 \in \mathbb{Q}[X]$$

However, $X^2 - 2$ also belongs to $\mathbb{R}[X]$ and in $\mathbb{R}$ it does possess the roots $\sqrt{2}$ and $-\sqrt{2}$. So just because a polynomial over a field F does not have a root in F, this does not mean that it has no roots in some larger field E. No real number fits as a root of

$$X^2 + 1 \in \mathbb{R}[X]$$

but would not roots in some larger field containing $\mathbb{R}$ be imaginable? In mathematics, wishful thinking can pay off, just like in philosophy, theoretical physics, economics, and everyday life.

EXERCISES

1. Adapt the computer programs specified in exercise 1 in Section 2, Chapter IV, and exercises 16 and 17 in Section 3, Chapter IV, to handle polynomials over $\mathbb{Q}$.

2. Write a program to add and multiply rational fractions over $\mathbb{Q}$.

3. (a) Verify that every segment of $(\mathbb{Z}, \leq)$ or $(\mathbb{R}, \leq)$ is a complete lattice. What about the segments of $(\mathbb{Q}, \leq)$?
   (b) Verify that any two nonsingleton segments of $(\mathbb{R}, \leq)$ are order isomorphic and that the same is true for $(\mathbb{Q}, \leq)$ but not for $(\mathbb{Z}, \leq)$.

4. Verify that if (S_1, S_2) is a cut in a chain, then S_1 is a lower section and S_2 is an upper section.

5. Let p be a positive prime integer. Show that $\sqrt{p} \notin \mathbb{Q}$. Does $\sqrt{6}$ belong to $\mathbb{Q}$?

6. What can you say about roots of monomials in $\mathbb{Q}$ or $\mathbb{R}$?

7. Show that every finite integral domain is a field.

8. Let D be an integral domain. What can you say about the number of roots in D of a polynomial $p \in D[X]$?

9. Describe the quotient rings of a field.

10. Let a be a nonzero element of a ring with identity. Verify that a is prime if it generates a maximal ideal.

2. GALOIS GROUPS AND IMAGINARY ROOTS

Throughout this section, all rings considered are still assumed to have an identity element. In any field E, the notation a/b will refer to the element $a \cdot b^{-1}$ where $a, b \in E$, $b \neq 0$. By a *subfield* of E we mean any subring of E that is a field. Subfields form an algebraic closure system on E, contained in the closure system of subrings. Recall that by a maximal ideal we mean a maximal member of the inclusion-ordered set of all proper ideals.

Proposition 3 *For any ideal I of a ring A the following are equivalent:*

(i) *I is a maximal ideal,*
(ii) *the quotient ring A/I is a field.*

Proof. If I is maximal, then the difference set $A \setminus I$ is closed under multiplication, because for every $a \in A \setminus I$ the proper ideal

$$\{b \in A : ab \in I\}$$

includes I and therefore equals I. For every $a \in A \setminus I$, the ideal generated by $I \cup \{a\}$ is the set

$$\{i + ca : i \in I, c \in A\}$$

As it strictly includes I, it must be equal to A. In particular

$$1 = i + ca \quad \text{for some} \quad i \in I, \quad c \in A$$

The coset of I containing c is then the multiplicative inverse, in the quotient ring A/I, of the coset containing a.

If I is not maximal, let J be an ideal with $I \subset J \subset A$. Then

$$A/I \cap \mathcal{P}(J)$$

is a nontrivial proper ideal of A/I. But a field cannot have any nontrivial proper ideals. (Why not?) □

Example. The ring $\mathbb{Z}/p\mathbb{Z}$, containing p elements, is a field for every positive prime integer p.

In any principal ring A, the ideal (p) generated by a nonzero element p of A is maximal if and only if p is prime. We shall take a particular interest in rings of polynomials over a field F. (Recall the Polynomial Factorization Theorem.) A polynomial $p \in F[X]$ is usually called *irreducible* over F if it is a prime element of the ring $F[X]$. Every degree 1 polynomial is irreducible. The converse is false: witness $X^2 - 2$ over $\mathbb{Q}$. Note that if a polynomial f divides a polynomial p in $F[X]$, then every root that f may have in F is also a root of p.

Proposition 4 *Let F be a field, p be a nonzero polynomial over F, and $\alpha \in F$. Then α is a root of p if and only if the polynomial $X - \alpha$ divides p in $F[X]$.*

Proof. The "if" part is obvious. For the converse, recall that $F[X]$ is Euclidean (Polynomial Factorization Theorem). Let

$$r = p - q \cdot (X - \alpha)$$

be a remainder of p by $X - \alpha$. Obviously the value of r at α is 0. By definition of the remainder, either $r = 0$ or $r \neq 0$ and

$$\deg r < \deg(X - \alpha)$$

i.e., $r \neq 0$ and $\deg r = 0$. As the latter would contradict $r(\alpha) = 0$, we have $r = 0$, and

$$p = q \cdot (X - \alpha) \qquad \square$$

If p is a polynomial over a field F, $\alpha \in F$ a root of p, and $p = f \cdot g$ in $F[X]$, then α must be a root of f or g: this follows from

$$p(\alpha) = f(\alpha) \cdot g(\alpha)$$

How many roots can p have? Since α is a root,

$$p = q \cdot (X - \alpha)$$

for some $q \in F[X]$, and if $p \neq 0$, then $q \neq 0$ and

$$\deg p = 1 + \deg q$$

The roots of p are α plus the roots of q. Using this as an inductive argument, one obtains the following:

Proposition 5 *A nonzero polynomial p over a field F has at most $\deg p$ roots in F.*

As one consequence of this, we can see now that being a polynomial function can be quite a restrictive condition for a mapping from a ring to itself. Consider functions such as $f : \mathbb{Q} \to \mathbb{Q}$ given by

$$f(x) = \begin{cases} 0 & \text{if } x \leq 0 \\ 1 & \text{otherwise} \end{cases}$$

Another important consequence concerns the structure of a field's multiplicative group.

Proposition 6 *Let F^* be the multiplicative group of a field F. Every finite subgroup of F^* is cyclic.*

Proof. Let G be a finite subgroup of F^*. Let

$$\text{Card}\, G = p_1^{k_1} \cdots p_m^{k_m}$$

where for $i = 1, \ldots, m$ the p_i are distinct positive prime numbers and each k_i is a positive integer. (Recall the Integer Factorization Theorem.) Let $g \in G$ have largest possible order n. By Lagrange's Subgroup Counting Theorem, n divides $\text{Card}\, G$, and therefore

$$n = p_1^{h_1} \cdots p_m^{h_m}, \qquad 0 \leq h_i \leq k_i \quad \text{for all } i$$

Let (g) denote the subgroup generated by g. Suppose $(g) \neq G$: we shall obtain a contradiction. Let

$$b \in G \setminus (g)$$

By Lagrange again, the cyclic subgroup (b) has

$$\nu = p_1^{e_1} \cdots p_m^{e_m}$$

elements, with $0 \leq e_i \leq k_i$ for all i.

Consider the degree n polynomial $X^n - 1$ over F. All the n elements of (g) are roots of $X^n - 1$. Since $X^n - 1$ can only have n roots, b is not a root. Thus ν does not divide n and for some $1 \leq j \leq m$ we have $e_j > h_j$. Let

$$r = \prod_{i \neq j} p_i^{e_i}, \qquad s = \prod_{i \neq j} p_i^{h_i}$$

Write e for e_j, h for h_j, p for p_j. Then the order of b^r is p^e, while the order of $g^{(p^h)}$ is s. It follows that the order of $b^r g^{(p^h)}$ is $p^e \cdot s > n$, contradicting the choice of g. $\square$

Suppose now that a polynomial $p \in F[X]$ has no root in the field F and $\deg p \geq 1$. Using the Factorization Theorem in the principal entire ring $F[X]$, write p as a product $p_1 \cdots p_n$ of n irreducible polynomials. Clearly, none of the factors p_i has a root in F. What if we are allowed to search for roots in some larger field E containing F as a subfield? The roots of p in E would still coincide with the roots of the various p_i taken together. To find a field E in which p has a root, we must find a field in which an irreducible polynomial p_i has a root.

For any irreducible polynomial p over a field F, the ideal (p) of $F[X]$ generated by p is maximal. (This ideal consists of all product polynomials $p \cdot q$, $q \in F[X]$.) The quotient ring $F[X]/(p)$ is a field. The map

$$f : F \to F[X]/(p)$$

associating to each $a \in F$ the coset of (p) containing the zero-degree polynomial corresponding to a is injective, and it is a ring homomorphism: call it the *canonical injection* of F into $F[X]/(p)$. This injection establishes an isomorphism between the field F and the subring $\operatorname{Im} f$ of $F[X]/(p)$, which subring must then itself be a field. Let α be the coset of (p) containing X, $\alpha \in F[X]/(p)$. With

$$p = c_n X^n + \cdots + c_1 X + c_0$$

in $F[X]$, let

$$p_f = f(c_n) X^n + \cdots + f(c_1) X + f(c_0)$$

This is a polynomial over $\operatorname{Im} f$. Obviously α is a root of p_f in the larger field $F[X]/(p)$. Apply now the Injection–Extension Theorem to the canonical injection f. There is a set $E \supset F$ and a bijection

$$g : E \to F[X]/(p)$$

with $g|F = f$. Define a field structure $(E, +, \cdot)$ on E by

$$a + b = g^{-1}(g(a) + g(b))$$
$$a \cdot b = g^{-1}(g(a) \cdot g(b))$$

Then F becomes a subfield of E, and $\rho = g^{-1}(\alpha)$ is a root of p in E: we have constructed a field E larger than F in which p has a root. Of course, E is isomorphic to $F[X]/(p)$:

Imaginary Root Theorem. *For every polynomial p over a field F, there is some field E containing F as a subfield such that p has a root ρ in E.*

As a concrete example of the construction leading to this result, let

$$F = \mathbb{R}, \qquad p = X^2 + 1$$

The field E isomorphic to $F[X]/(p)$ is then called the field of *complex numbers*, denoted by $\mathbb{C}$. It is also customary to denote a selected root of $X^2 + 1$ in $\mathbb{C}$ by the letter i. Using the isomorphism

$$g : \mathbb{C} \to F[X]/(p)$$

it is then straightforward to verify that every complex number can be written, in a unique fashion, as

$$a + bi \quad \text{with} \quad a, b \in \mathbb{R}$$

An element α of a field E is called *root of unity* if $\alpha^n = 1$ for some positive integer n. We also say that α is an nth *root of unity* in this case. For given n, the set of nth roots of unity forms a subgroup of the multiplicative group E^*. This subgroup has at most n elements, since these are exactly the roots of the polynomial $X^n - 1$. The various roots of unity are precisely the finite order elements of E^*. For every root of unity α, its order in E^* is the smallest positive integer n with $\alpha^n = 1$. For example, $\mathbb{C}$ has four fourth roots of unity:

$$1, i, -1, -i$$

Of these, only i and $-i$ are of order 4.

Let K be a subfield of a field E. An element α of E is called a *radical* over K if for some positive integer n and $b \in K$, we have

$$\alpha^n = b$$

Such an n is called an *exponent* of α over K. Every nth root of unity is obviously a radical over K with exponent n. On the other hand, $\sqrt{2}$ is a radical in $\mathbb{R}$ over $\mathbb{Q}$, but it is not a root of unity.

If K is a subfield of E, then the couple (E, K) is called a *field extension*, and it is more commonly denoted by $E : K$. The extension $E : K$ is called *simple* if for some $\alpha \in E$, called *primitive element*, the subfield of E generated by $K \cup \{\alpha\}$ is the whole of E. (The simplest case is $K = E$.) If some primitive element is the root of some polynomial $p \in K[X]$, then $E : K$ is called *simple algebraic*. A simple algebraic extension can be obtained from any field K and any polynomial $p \in K[X]$ as follows: apply the Imaginary Root Theorem, and consider the subfield generated by K and the new root. This is essentially the only way to obtain simple algebraic extensions $E : K$. Indeed, for a primitive element α of such an extension, let

$$I = \{f \in K[X] : \alpha \text{ is a root of } f\}$$

Clearly I is an ideal of the principal ring $K[X]$. Any generator of I is irreducible, and it is called an *irreducible polynomial* of α over K. Let p be such an irreducible polynomial. (Obviously p is linear if and only if $E = K$.) The function h associating to each polynomial in $K[X]$ its value at α is a surjective ring homomorphism from $K[X]$ onto E, and its kernel is $I = (p)$. The function g associating to each coset $C \in K[X]/(p)$ the value of any one member of C at α is a ring isomorphism $K[X]/(p) \to E$:

Proposition 7 *For every simple algebraic extension $E : K$ with a primitive element α there is an irreducible polynomial p over K and an isomorphism*

$$g : K[X]/(p) \to E$$

mapping the class of X to α and such that for the canonical injection

$$j : K \to K[X]/(p)$$

we have $g \circ j = id_K$.

For any set A of elements of a field E, those automorphisms δ of E that fix every element of A constitute a subgroup $\mathcal{G}(A)$ of $\text{Aut}\,E$:

$$\mathcal{G}(A) = \{\delta \in \text{Aut}\,E : \delta(a) = a \text{ for all } a \in A\}$$

is called the *fixing group* of A. [In the full symmetric group $\Sigma(E)$, $\mathcal{G}(A)$ is the intersection of $\text{Aut}\,E$ with all the stabilizers S_a, $a \in A$.]

For any subset B of $\text{Aut}\,E$, those elements a of E that are fixed by every automorphism in B constitute a subfield $\mathcal{F}(B)$ of E. It is called the *fixed field* of B. For the functions

$$\mathcal{G} : \mathcal{P}(E) \to \mathcal{P}(\text{Aut}\,E) \quad \text{and} \quad \mathcal{F} : \mathcal{P}(\text{Aut}\,E) \to \mathcal{P}(E)$$

we have

$$\mathcal{F} \circ \mathcal{G} \circ \mathcal{F} = \mathcal{F} \quad \text{and} \quad \mathcal{G} \circ \mathcal{F} \circ \mathcal{G} = \mathcal{G}$$

Quite obviously also
$$\mathcal{G}(A) \supseteq \mathcal{G}(A') \quad \text{if} \quad A \subseteq A'$$
and
$$\mathcal{F}(B) \supseteq \mathcal{F}(B') \quad \text{if} \quad B \subseteq B'$$
It follows that $\mathcal{G} \circ \mathcal{F}$ is a closure operator on Aut E and $\mathcal{F} \circ \mathcal{G}$ is a closure operator on E. The corresponding closure systems are
$$\operatorname{Im}(\mathcal{G} \circ \mathcal{F}) = \operatorname{Im} \mathcal{G} \quad \text{and} \quad \operatorname{Im}(\mathcal{F} \circ \mathcal{G}) = \operatorname{Im} \mathcal{F}$$
These consist precisely of all possible fixing groups, and of all fixed fields, respectively. For any subfield K of E, the fixing group of K is called the *Galois group* of the extension $E : K$, and it is also denoted by $\mathcal{G}(E : K)$.

Let $E : K$ be a simple field extension. We claim that if two automorphisms
$$\sigma, \tau \in \mathcal{G}(E : K)$$
have the same value on a primitive element α, then $\sigma = \tau$. Clearly our claim is equivalent to saying that if $\delta \in \mathcal{G}(E : K)$ fixes α, then δ is the identity. Since the fixed field of $\{\delta\}$ includes $K \cup \{\alpha\}$, this follows from the definition of a primitive element.

Proposition 8 *If a primitive element α of a simple field extension $E : K$ is a root of unity of prime order p, then the Galois group of $E : K$ is cyclic.*

Proof. Let R be the multiplicative group of pth roots of unity in E, and let $n = \operatorname{Card} R$. The case $E = K$ being trivial, assume that $E \neq K$. Then $\alpha \neq 1$ and $1 < n \leq p$. If $n < p$, then n and p are relatively prime:
$$1 = an + bp$$
for some integers a, b, and
$$\alpha = (\alpha^n)^a \cdot (\alpha^p)^b$$
would belong to K, which is not the case since $E \neq K$. Thus $n = p$ and the element α must generate R. (A group of prime order has no nontrivial proper subgroups, by Lagrange.)

For every $\sigma \in \mathcal{G}(E : K)$, $\sigma(\alpha)$ is in R, and therefore the set

$$h(\sigma) = \{z \in \mathbb{Z} : \alpha^z = \sigma(\alpha)\}$$

is not empty. Indeed $h(\sigma)$ must be a coset of the ideal $p\mathbb{Z}$ of $\mathbb{Z}$— other than $p\mathbb{Z}$ itself because $\alpha^0 = 1$ cannot be $\sigma(\alpha)$ for any automorphism σ of E. Thus we have defined an injective function h from $\mathcal{G}(E : K)$ to the set of nonzero elements of the ring $\mathbb{Z}/p\mathbb{Z} = \mathbb{Z}_p$. Since $(\mathbb{Z}_p, +, \cdot)$ is a field, every nonzero element is a unit, and h is an injection to $\mathbb{Z}_p^*$. But $(\mathbb{Z}^*, \cdot)$ is a cyclic group, and we shall show that h is a group homomorphism from $\mathcal{G}(E : K)$ to $(\mathbb{Z}_p^*, \cdot)$. This will imply that $\mathcal{G}(E : K)$, being isomorphic to a subgroup of a cyclic group, is itself cyclic.

Let

$$\tau, \sigma \in \mathcal{G}(E : K), \qquad t \in h(\tau), \qquad s \in h(\sigma)$$

From

$$(\tau \circ \sigma)(\alpha) = \tau(\alpha^s) = (\tau(\alpha))^s = \alpha^{ts}$$

we conclude that $ts \in h(\tau \circ \sigma)$, that is

$$h(\tau) \cdot h(\sigma) = h(\tau \circ \sigma) \quad \text{in} \quad \mathbb{Z}_p^*$$

This proves that h is a group homomorphism as claimed. □

Proposition 9 *If a primitive element α of a simple extension $E : K$ is a radical with exponent n over K, and if $E \setminus K$ contains no nth root of unity, then the Galois group of $E : K$ is cyclic.*

Proof. The result being trivial for $E = K$, assume $E \neq K$. For every σ in $\mathcal{G}(E : K)$, we have

$$\sigma(\alpha)^n = \sigma(\alpha^n) = \alpha^n$$

therefore $\sigma(\alpha)/\alpha$ is an nth root of unity, and as such it must belong to K. Let R_n be the multiplicative group of all nth roots of unity in K: this is a cyclic group. Define an injective map h from $\mathcal{G}(E : K)$ to R_n by

$$h(\sigma) = \sigma(\alpha)/\alpha$$

This is indeed a group homomorphism because

$$h(\tau \circ \sigma) = \tau(\sigma(\alpha))/\alpha = \tau(\alpha \cdot [\sigma(\alpha)/\alpha])/\alpha$$

and since τ fixes $\sigma(\alpha)/\alpha \in K$,
$$h(\tau \circ \sigma) = \tau(\alpha) \cdot \sigma(\alpha)/\alpha^2 = h(\tau)h(\sigma)$$
The Galois group, being isomorphic to the subgroup $\operatorname{Im} h$ of R_n, is then cyclic. □

Let $E : K$ be a field extension and let $p \in K[X]$. Let $R(p, E)$ be the set of all roots of p in E. The *relative splitting field* of p in $E : K$ is the subfield F of E generated by
$$K \cup R(p, E)$$
The term "splitting" is justified by the observation that if $\deg p = n > 0$ and p has n roots in E (the maximum number it can have), then the Polynomial Factorization Theorem in the ring $F[X]$ yields
$$p = c \prod_{\alpha \in R(p, E)} (X - \alpha)$$
with some $c \in K$, i.e., p "splits into linear factors" in $F[X]$.

Example. The relative splitting field of $X^4 - 4$ in $\mathbb{C} : \mathbb{Q}$ is strictly larger than in $\mathbb{R} : \mathbb{Q}$:
$$X^4 - 4 = (X - \sqrt{2})(X + \sqrt{2})(X - i\sqrt{2})(X + i\sqrt{2})$$
and
$$X^4 - 4 = (X - \sqrt{2})(X + \sqrt{2})(X^2 + 2)$$
are prime factorizations over the two relative splitting fields.

Galois Quotient Theorem (For Relative Splitting Fields). *Let $E : K$ be a field extension. If F is the relative splitting field in $E : K$ of some polynomial $p \in K[X]$, then $\mathcal{G}(E : F)$ is a normal subgroup of $\mathcal{G}(E : K)$ and*
$$\mathcal{G}(E : K)/\mathcal{G}(E : F)$$
is isomorphic to a subgroup of $\mathcal{G}(F : K)$.

Proof. Since the coefficients of p are fixed by every $\sigma \in \mathcal{G}(E : K)$, $\alpha \in E$ is a root of p if and only if $\sigma(\alpha)$ is a root. It follows that $\sigma[F] = F$. Thus the restriction of every $\sigma \in \mathcal{G}(E : K)$ to F is a member of $\mathcal{G}(F : K)$. The function h from $\mathcal{G}(E : K)$ to $\mathcal{G}(F : K)$ mapping each $\sigma \in \mathcal{G}(E : K)$ to its restriction to F is quite obviously a group

homomorphism with kernel $\mathcal{G}(E:F)$. The quotient
$$\mathcal{G}(E:K)/\mathcal{G}(E:F)$$
is isomorphic to the subgroup $\operatorname{Im} h$ of $G(F:K)$. □

A field extension $E:K$ is a *radical extension* if there are subfields $E_0, E_1, \ldots, E_n$ of $E, n \geq 1$, with
$$E = E_0 \supseteq E_1 \supseteq \cdots \supseteq E_n = K$$
such that for every $i = 0, \ldots, n-1$, $E_i : E_{i+1}$ is a simple extension with a primitive element α_i that is a radical over E_{i+1}.

A group G is called *solvable* if there are subgroups
$$G_0 \subseteq \cdots \subseteq G_n = G, \qquad n \geq 1$$
where G_0 is trivial and for every $i = 0, \ldots, n-1$, G_i is a normal subgroup of G_{i+1} with G_{i+1}/G_i cyclic. It is an excellent exercise in group theory to verify that every subgroup of a solvable group is solvable.

Examples. (1) $\mathbb{C} : \mathbb{R}$ is a radical extension. If E is the subfield of $\mathbb{C}$ generated by $\mathbb{Q} \cup \{\sqrt{2}, i\}$, then $E : \mathbb{Q}$ is a radical extension. (2) Every cyclic group is solvable. There are noncyclic solvable groups: for any finite set S with at least two elements, consider the symmetric difference group $(\mathcal{P}(S), +)$.

Recall the Simplicity Theorem for Alternating Groups. Can A_n be solvable if $n \geq 5$? Obviously a simple group is solvable only if it is cyclic. We leave it to the reader to rule out this possibility for A_n, $n \geq 5$. Thus the alternating group A_n is not solvable if $n \geq 5$. On the other hand, it is easy to verify that for $n < 5$, A_n is solvable.

Solvability Theorem for Radical Extensions. *The Galois group of every radical field extension $E : K$ is solvable.*

Proof. Let
$$E = E_0 \supseteq E_1 \supseteq \cdots \supseteq E_n = K$$
be subfields, $E_i : E_{i+1}$ simple for every $i = 0, \ldots, n-1$, with a radical $\alpha_i \in E_i$ over E_{i+1} as primitive element. For each i let $m(i)$ be an exponent with
$$\alpha_i^{m(i)} \in E_{i+1}$$

For subfields F, H of E containing K, suppose that we have intermediate subfields

$$F = F_0 \supseteq F_1 \supseteq \cdots \supseteq F_k = H, \quad k \geq 1$$

such that, for $i = 0, \ldots, k-1$,

(i) the Galois group $\mathcal{G}(F_i : F_{i+1})$ is cyclic,
(ii) F_i is a relative splitting field in $E : F_{i+1}$.

Then let us call $(F_0, \ldots, F_k)$ a *cyclic normal extension sequence from* $H = F_k$ *to* $F = F_0$. Obviously (ii) implies that F_i is a relative splitting field (of some appropriate polynomial) in $F : F_{i+1}$. Consider the subgroups

$$\mathcal{G}(F : F_0) \subseteq \cdots \subseteq \mathcal{G}(F : F_k) = \mathcal{G}(F : H)$$

The Galois Quotient Theorem implies that $\mathcal{G}(F : H)$ is a solvable group!

To prove our theorem, we shall seek to establish that there is a cyclic normal extension sequence from K to the whole of E. Let us introduce some notation. For any set $S \subseteq E$, let $K(S)$ be the subfield of E generated by $K \cup S$; for the case $S = \{a_1, \ldots, a_j\}$ write simply

$$K(a_1, \ldots, a_j)$$

As in the proof of Proposition 9, let R_i denote the multiplicative group of ith roots of unity in E. Each R_i is finite, and therefore cyclic. Any generator g of R_i is a primitive element of the extension $K(R_i) : K$ and

$$K(R_i) = K(g)$$

is the relative splitting field of $X^i - 1$ in $E : K$. Let us now verify two claims.

First Claim: *For every positive integer m, there is a cyclic normal extension sequence from K to $K(R_m)$.* Let m be the smallest integer allegedly violating this claim. Obviously $m > 1$. Let p be any positive prime factor of m, and let $q = m/p$. Let γ be a generator of R_m, β a generator of R_q. Clearly $\gamma^p \in R_q$. If the order t of γ were less than m, then the claim would be true for $K(R_t)$. But $R_t = R_m$: thus m is the order of γ. Two cases are possible:

Case 1. There is a pth root of unity

$$\rho \in K(\gamma) \setminus K(\beta)$$

Then $\rho\beta$ is in R_m and since γ is a generator, $\gamma^e = \rho\beta$ for some positive integer e. The prime p cannot divide e because $\rho\beta$ is not in the subset R_q of $K(\beta)$. Thus p and e are relatively prime,

$$ap + be = 1$$

for some integers a, b. We have

$$\gamma = \gamma^{ap}\gamma^{be} = (\gamma^p)^a(\gamma^e)^b = (\gamma^p)^a(\rho\beta)^b$$

Since $\gamma^p \in K(\beta)$, this expression for γ shows that γ is in the subfield generated by $K(\beta)$ and ρ, i.e., ρ is a primitive element of

$$K(\gamma) : K(\beta)$$

By Proposition 8, the Galois group $\mathcal{G}(K(\gamma) : K(\beta))$ is cyclic.

Case 2. The set $K(\gamma) \setminus K(\beta)$ contains no pth root of unity. Since $\gamma^p \in K(\beta)$, the group

$$\mathcal{G}(K(\gamma) : K(\beta))$$

is cyclic in this case too, by Proposition 9.

In both cases, by the minimal choice of m, there is a cyclic normal extension sequence

$$(K(\beta), \ldots, K)$$

from K to $K(\beta) = K(R_q)$. But then

$$(K(\gamma), K(\beta), \ldots, K)$$

is a cyclic normal extension sequence from K to $K(\gamma)$. This proves the First Claim.

Second Claim: *If m is any positive common multiple of the exponents $m(1), \ldots, m(n)$, then there is a cyclic normal extension sequence from $K(R_m)$ to E.* For $i = 0, \ldots, n-1$ let

$$K_i = K(R_m \cup \{\alpha_i, \ldots, \alpha_{n-1}\})$$

and let $K_n = K(R_m)$. For $i = 0, \ldots, n-1$ we have

$$K_i = K_{i+1}(\alpha_i) \quad \text{and} \quad K_i \supseteq E_i$$

Apply Proposition 9 to $K_i : K_{i+1}$ and consider the sequence

$$(K_0, \ldots, K_{n-1}, K_n)$$

from $K(R_m) = K_n$ to $E = K_0$. This proves the Second Claim.

To conclude the proof, for an appropriate m, concatenate any cyclic normal extension sequence

$$(K(R_m), \ldots, K)$$

and a cyclic normal extension sequence

$$(E, \ldots, K(R_m))$$

to form a combined sequence

$$(E, \ldots, K(R_m), \ldots, K)$$

and behold. □

Let $F : K$ be a field extension and let $p \in K[X]$. The field F is called an (*absolute*) *splitting field* of p over K if for every further extension $E : F$, F is the relative splitting field of p in the extension $E : K$. Equivalently, $F \supseteq K$ is an absolute splitting field of $p \in K[X]$ if and only if all irreducible factors of p in $F[X]$ are linear, and

$$F = K(\alpha_1, \ldots, \alpha_n)$$

where the α_i are the roots of p in F.

An inductive application of the Imaginary Root Theorem yields:

Splitting Field Theorem. *Every polynomial p over a field K has an absolute splitting field F over K.*

If E, F are fields and φ any isomorphism from E to F, then an isomorphism $\overline{\varphi}$ from the ring $E[X]$ to $F[X]$ is given, for p written as $c_0 + c_1 X + \cdots + c_n X^n$ in $E[X]$, by

$$\overline{\varphi}(p) = \varphi(c_0) + \varphi(c_1)X + \cdots + \varphi(c_n)X^n$$

We shall write p_φ for $\overline{\varphi}(p)$. Associating to every coset $C \in E[X]/(p)$ its image set $\overline{\varphi}[C]$ establishes an isomorphism from $E[X]/(p)$ to $F[X]/(p_\varphi)$. Combining with Proposition 7 we get:

Proposition 10 *If $E' : E$ and $F' : F$ are field extensions, $\varphi : E \to F$ is an isomorphism, $\alpha \in E'$ is a root of an irreducible polynomial*

$p \in E[X]$, and $\beta \in F'$ is any root of p_φ, then φ can be extended to a unique isomorphism
$$\tau : E(\alpha) \to F(\beta)$$
that maps α to β.

Corollary. *Let p be a polynomial over a field K, and let*
$$E \supseteq K, \quad F \supseteq K$$
be two absolute splitting fields of p over K.

(i) *There is an isomorphism $\varphi : E \to F$ whose restriction to K is the identity.*
(ii) *The Galois groups $\mathcal{G}(E : K)$ and $\mathcal{G}(F : K)$ are isomorphic.*

Proof. (i) is obtained by recursive application of the proposition.

The group isomorphism claimed by (ii) is constructed by taking the field isomorphism $\varphi : E \to F$ and associating to every σ in the group $\mathcal{G}(E : K)$ the element $\varphi \cdot \sigma \cdot \varphi^{-1}$ of $\mathcal{G}(F : K)$. $\square$

Thus any two absolute splitting fields E, F over K of a polynomial $p \in K[X]$ are isomorphic, and if $E : K$ is a radical extension, then so is $F : K$. We can therefore call a polynomial $p \in K[X]$ *solvable by radicals* over K if an absolute splitting field E of p over K defines a radical extension $E : K$. There is always an absolute splitting field and it does not matter which one we take.

Consider, for example,
$$p = 7X^2 + 3X + \tfrac{1}{4}$$
in $\mathbb{Q}[X]$. In the subfield $\mathbb{Q}(\sqrt{2})$ of $\mathbb{R}$ generated by $\sqrt{2}$ it has the roots
$$(-3 + \sqrt{2})/(2 \cdot 7) \quad \text{and} \quad (-3 - \sqrt{2})/(2 \cdot 7)$$
The polynomial p, although having no rational root (why?), does have two roots in the radical extension $\mathbb{Q}(\sqrt{2}) : \mathbb{Q}$. It follows that $\mathbb{Q}(\sqrt{2})$ is a splitting field of p over $\mathbb{Q}$, and p is solvable by radicals over the rationals. Indeed every degree 2 polynomial
$$aX^2 + bX + c$$

over $\mathbb{Q}$ is so solvable: both

$$(-b+\delta)/2a \quad \text{and} \quad (-b-\delta)/2a$$

are roots in $\mathbb{C}$ where

$$\delta = \begin{cases} \sqrt{b^2-4ac} & \text{if } b^2 > 4ac \\ 0 & \text{if } b^2 = 4ac \\ i\sqrt{4ac-b^2} & \text{if } 4ac > b^2 \end{cases}$$

These roots belong to the subfield

$$E = \mathbb{Q}(i, \sqrt{|b^2-4ac|})$$

The field E is a splitting field of $aX^2 + bX + c$ over $\mathbb{Q}$ and $E : \mathbb{Q}$ is a radical extension.

Let E be a field of fractions of

$$\mathbb{Q}[X_1, X_2, X_3, X_4, X_5]$$

i.e., of the ring of polynomials in five indeterminates over $\mathbb{Q}$. In the ring $E[X]$ let p be the product of the five polynomials

$$X - X_i, \quad i = 1, \ldots, 5$$

Let $c_0, c_1, \ldots, c_5$ be the coefficients in E of this $p \in E[X]$,

$$p = c_0 + c_1 X + \cdots + c_5 X^5$$

For each $i = 0, 1, \ldots, 5$ we have

$$c_i = \sum_{\substack{J \subseteq \{1,\ldots,5\} \\ \operatorname{Card} J = 5-i}} (-1)^{5-i} \prod_{j \in J} X_j$$

In particular $c_0 = -X_1 X_2 X_3 X_4 X_5$ and $c_5 = 1$. (The reader may wish to calculate the other coefficients too.) Let K be the subfield of E generated by

$$c_0, c_1, \ldots, c_5$$

Obviously $\mathbb{Q}$ is a proper subfield of K, $p \in K[X]$, and E is an absolute splitting field of p over K. We intend to show that $\mathcal{G}(E : K)$ has a subgroup isomorphic to the symmetric group Σ_5. First, observe that if h is any automorphism of an integral domain D and Q is a field of fractions of D, then h can be extended to a unique automorphism

$\bar{h}$ of Q, defined by

$$\bar{h}(a/b) = h(a)/h(b) \quad \text{for all} \quad a, b \in D, \quad b \neq 0$$

Second, observe that for each permutation $\sigma \in \Sigma_5$, there is an automorphism h_σ of $\mathbb{Q}[X_1, \ldots, X_5]$ given by

$$h_\sigma\left(\sum_{i_1, \ldots, i_5} c(i_1, \ldots, i_5) X_1^{i_1} X_2^{i_2} X_3^{i_3} X_4^{i_4} X_5^{i_5} \right)$$

$$= \sum_{i_1, \ldots, i_5} c(i_1, \ldots, i_5) X_{\sigma(1)}^{i_1} X_{\sigma(2)}^{i_2} X_{\sigma(3)}^{i_3} X_{\sigma(4)}^{i_4} X_{\sigma(5)}^{i_5}$$

Associating to σ the extension $\bar{h}_\sigma$ of h_σ to the fraction field E yields an injective group homomorphism

$$\Sigma_5 \to \mathcal{G}(E : K)$$

Since Σ_5, which contains the alternating group A_5, is not a solvable group, $\mathcal{G}(E : K)$ is not solvable either. By the Solvability Theorem for Radical Extensions, the polynomial $p \in K[X]$ is not solvable by radicals over K.

Abel–Ruffini Theorem. *Not all polynomials of degree 5 are solvable by radicals.*

A general theory of solvability by radicals, in group-theoretical terms, was first developed by Evariste Galois, early in the last century. The group concept was invented by him. Galois had an excellent high-school education in Paris, failed the entrance examination at France's top engineering school, and died in a duel at the age of 21 the day after he committed his theory to writing.

EXERCISES

1. Let $E : F$ be a simple field extension with primitive element a. Show that a is the root of some polynomial over F if and only if the subring $F[a]$ of E generated by $F \cup \{a\}$ is the whole of E. Show that if $F[a]$ is not a field, then E is isomorphic to the field of rational fractions over F. Conclude that if $E : F$ is

GALOIS GROUPS AND IMAGINARY ROOTS   153

simple algebraic, then every element of E is the root of some polynomial over F.

2. Show that every function from a finite field to itself is polynomial.

3. Show that, over any given field, there is an infinity of irreducible polynomials.

4. Describe the Galois groups of $C : R$, $C : Q$, and $R : Q$. What is the fixed field of $\text{Aut}\, C$?

5. Show that every field E has a subfield K such that $\text{Aut}\, E = \mathcal{G}(E : K)$.

6. Let H be a normal subgroup of a group G. Verify that G/H is cyclic if and only if there is an element $a \in G$ such that $H \cup \{a\}$ generates the entire group G.

7. Let $h : G \to H$ be a surjective group homomorphism. Show that G is a solvable group if and only if both H and $\text{Ker}\, h$ are solvable.

8. Verify that if a field element a is a radical over a subfield, then so is $-a$ and $1/a$. Show that the exponents of a, together with zero and the negatives of these exponents, form a subgroup of Z.

9. Is $1 + \sqrt{2}$ a radical over Q? What about $1 + i$? Can you generalize?

10. In the ring Z, show that $a^{p-1} \equiv 1 \bmod pZ$ if p is a positive prime that does not divide the integer a.

11. Show that every element of the subring $Z[i]$ of C generated by i can be written in a unique fashion as $a + bi$ with $a, b \in Z$. Show that $Z[i]$ is Euclidean with norm $a^2 + b^2$ for nonzero elements. What are the units? See how far you can go in describing this ring of "Gaussian integers."

12. Consider in $Z[i]$ the equivalence relation $\equiv$ in which $a + bi \equiv c + di$ means that $a = qc$ and $b = qd$ for some positive rational q. Show that $\equiv$ is a multiplicative semigroup congruence in which zero forms a class by itself and all other classes are infinite. Call the infinite classes *Gaussian angles* and denote the quotient semigroup operation by $\oplus$. Verify that $\oplus$ defines a commutative group structure on the set of Gaussian angles. Verify

that $a + bi$ and $c + di$ have (belong to) the same angle α if and only if

$$a/\sqrt{a^2 + b^2} = c/\sqrt{c^2 + d^2}$$

$$b/\sqrt{a^2 + b^2} = d/\sqrt{c^2 + d^2}$$

Define $\sin \alpha$ as $b/\sqrt{a^2 + b^2}$ and $\cos \alpha$ as $a/\sqrt{a^2 + b^2}$. Prove

$$(\sin \alpha)^2 + (\cos \alpha)^2 = 1$$

$$\sin(\alpha \oplus \beta) = \sin \alpha \cos \beta + \cos \alpha \sin \beta$$

$$\cos(\alpha \oplus \beta) = \cos \alpha \cos \beta - \sin \alpha \sin \beta$$

BIBLIOGRAPHY

Jiři ADÁMEK, *Foundations of Coding: Theory and Applications of Error-Correcting Codes with an Introduction to Information Theory*. John Wiley & Sons 1991. This textbook explains how basic ideas from the theory of groups, rings, fields, and vector spaces can be used in noise-resistant or secret data transmission.

André BLANCHARD, *Les corps non commutatifs*. Presses Universitaires de France 1972. A pocket-size introduction to non-commutative fields, for the graduate student.

Richard A. DEAN, *Classical Abstract Algebra*. Harper & Row 1990. Contains much accessible material on Galois theory and covers special topics such as the classification of finite fields, compass and straightedge constructions, and the "fundamental theorem of algebra" stating that every complex polynomial has a complex root.

D. J. H. GARLING, *A Course in Galois Theory*. Cambridge University Press 1991. An advanced undergraduate textbook, short but self-contained.

Nathan JACOBSON, *Basic Algebra I*. Freeman 1985. Chapter 4 is an exposition of Galois theory oriented toward the issues of equation solvability and geometric constructions. Chapter 6 is devoted to polynomials over the real field.

Gregory KARPILOVSKY, *Topics in Field Theory*. North-Holland 1989. A contemporary monograph on field extensions, for researchers and graduate students.

Serge LANG, *Algebra*. Addison-Wesley 1971. A third of this text is devoted to the theory of fields, including algebraic extensions, Galois theory, and transcendental extensions.

Rudolf LIDL and Harald NIEDERREITER, *Finite Fields*. Addison-Wesley 1983. This comprehensive volume includes, beyond the complete structural classification of finite fields, a detailed treatment of polynomial factorization, a chapter on bijective polynomial functions, and combinatorial applications.

M. POHST and H. ZASSENHAUS, *Algorithmic Algebraic Number Theory*. Cambridge University Press 1989. Investigates questions of effective computation in number fields.

G. SZÁSZ, *Introduction to Lattice Theory*. Academic Press 1963. The concept of Galois connection between ordered sets will be of interest to the reader already familiar with rudimentary Galois theory. This concept is presented in Section 28. The reader need not know lattice theory in order to read this section; only some elementary definitions are used.

André WARUSFEL, *Structures algèbriques finies*. Hachette 1971. For the reader already comfortable with algebra, this volume offers a refreshing retrospective on finite groups, finite rings, and finite fields.

CHAPTER VI

VECTOR SPACES

1. BASES

For a commutative group $(V,+)$ the set $\text{End}\,V$ of all group endomorphisms can be given a binary algebraic structure in at least two ways. With composition $(\text{End}\,V, \circ)$ is a monoid. With addition of endomorphisms defined for $\sigma, \tau \in \text{End}\,V$ by

$$(\sigma + \tau)(x) = \sigma(x) + \tau(x) \quad \text{for all} \quad x \in V$$

$(\text{End}\,V, +)$ is obviously a commutative group. Suppose we have a field $(F, +, \cdot)$ and a function $p : F \to \text{End}\,V$ that happens to be a monoid homomorphism both

$$(F, \cdot) \to (\text{End}\,V, \circ) \quad \text{and} \quad (F, +) \to (\text{End}\,V, +)$$

Then $(V, +, p)$ is called a *vector space* over the field F. This is one short and unnatural way of defining vector spaces formally. The elements of V are called *vectors*, those of F *scalars*. For a scalar α and a vector v, let us simply write $\alpha \cdot v$ or αv instead of $[p(\alpha)](v)$, and call this element a *vector space product*. The original group operation "+" on V is called *vector addition* or *sum*. Let us write $\bar{0}$ for the neutral *null vector* in the group V. The additive inverse of any

$v \in V$ is denoted by $-v$. The fact that p is a function with codomain $\operatorname{End} V$ means that for all $\alpha \in F, v, w \in V$

$$\alpha(v + w) = \alpha v + \alpha w \qquad \text{(left distributivity)}$$

The homomorphism properties of p mean that for all $\alpha, \beta \in F, v \in V$

$$(\alpha + \beta)v = \alpha v + \beta v \qquad \text{(right distributivity)}$$

$$(\alpha \beta)v = \alpha(\beta v) \qquad \text{(mixed associativity)}$$

and we also have

$$0v = \bar{0}, \qquad 1v = v, \qquad (-1)v = -v,$$

$$\alpha \bar{0} = \bar{0}, \qquad \alpha(-v) = (-\alpha)v = -(\alpha v)$$

In practice we simply refer to "the vector space V" when F and p are understood. A *scalar multiple* of a vector v is any vector of the form αv where α is a scalar. Obviously the null vector is a scalar multiple of every vector, but the only scalar multiple of $\bar{0}$ is $\bar{0}$ itself. Every vector is a scalar multiple of itself.

Examples. (1) For any set S, $\mathcal{P}(S)$ is a vector space over the two-element field Z_2: vector addition is the symmetric difference

$$A + B = (A \setminus B) \cup (B \setminus A)$$

and the vector space product is given by

$$0 \cdot A = \emptyset, \qquad 1 \cdot A = A$$

The empty set is the null vector. (2) For any set S and any field F, the set F^S is a vector space over F: the vectors are functions, vector addition is given by

$$(f + g)(x) = f(x) + g(x)$$

and the vector space product by

$$(\alpha g)(x) = \alpha \cdot g(x)$$

The special case where S is a natural number interval $[1, n]$ may be familiar to the reader.

A subset U of a vector space V is called a *subspace* if it is a subgroup of $(V, +)$ and $\alpha v \in U$ for every scalar α and $v \in U$. Subspaces constitute an algebraic closure system on V. The smallest, *trivial subspace* consists of the null vector alone. The set of scalar multiples of any given vector is a subspace.

Let $M \subseteq V$ be any set of vectors, and let $(\alpha_i v_i : i \in I)$ be a finite family of scalar multiples of elements $v_i \in M$. The sum

$$\sum_{i \in I} \alpha_i v_i$$

is called a *linear combination* of elements of M.

For $n = 0$ we have the null vector as a linear combination, for $n = 1$ we have scalar multiples. It is easy to see that the subspace $\overline{M}$ generated by a set M of vectors consists precisely of all linear combinations of elements of M.

Example. Let M be the set of all singleton subsets of a given set S. The subspace $\overline{M}$ of $\mathcal{P}(S)$ consists of all finite subsets of S.

A set M of vectors is called (*linearly*) *independent* if no $v \in M$ belongs to the subspace generated by $M \setminus \{v\}$, i.e., if no element of M is a linear combination of the other elements. It follows that an independent set cannot contain the null vector, but for any nonnull vector v the singleton $\{v\}$ is linearly independent. If M consists of two distinct nonnull vectors v, w, then M is independent if and only if w is not a scalar multiple of v. Clearly every subset of an independent set is independent.

Combinatorial Example. Let R be a binary relation on a set S, $R \subseteq S^2$, and consider the vector space $\mathcal{P}(S)$. If R is a path, then the vector set

$$\{\{a, b\} : (a, b) \in R\}$$

is independent. If R is a cycle, then it is not independent. More about this in graph and matroid theory.

A set C of vectors is called (*linearly*) *dependent* if it is not independent. A dependent set C is called a *circuit* if it is minimal (i.e., every proper subset of C is independent).

Example. If R is a relational cycle on a set S, as in the previous example, then
$$\{\{a,b\} : (a,b) \in R\}$$
is a circuit in $\mathcal{P}(S)$.

Since the subspace closure system is algebraic, a set M of vectors is independent if and only if every finite subset of M is independent. It follows that every circuit is finite and that a set M is dependent if and only if it contains a circuit. It is also easy to see that a finite set M is independent if and only if a linear combination
$$\sum_{u \in M} \alpha_u \cdot u$$
is null only when each α_u is zero. It follows that for every circuit C there are nonzero scalars α_u, $u \in C$, such that
$$\sum_{u \in C} \alpha_u \cdot u = \overline{0}$$

Basis Characterization Theorem. *For any set B of vectors in a space V the following are equivalent:*

(i) *B is a maximal independent set,*
(ii) *B is a minimal generating set for the entire space V,*
(iii) *B is independent and generates V.*

Proof. Assume (i). Let $\overline{B}$ be the subspace of V generated by B. If $\overline{B} \neq V$, then for any $v \in V \setminus \overline{B}$ the set $B \cup \{v\}$ would be independent: therefore $\overline{B} = V$. The independence of B also implies that for every proper subset A of B, the members of $B \setminus A$ are not in $\overline{A}$. This proves (ii).

Assume (ii). If B were dependent, some $v \in B$ would lie in $\overline{B \setminus \{v\}}$. But then $B \subseteq \overline{B \setminus \{v\}}$ and therefore
$$V = \overline{B} \subseteq \overline{B \setminus \{v\}}$$
which is impossible by assumption. Thus B must be independent and (iii) is proved.

Assume (iii). If B were properly included in some larger independent set C, the vectors in $C \setminus B$ could not be in $\overline{B} = V$: this forces condition (i). □

A set B satisfying the conditions of the above theorem is called a *basis* of V.

Basis Existence Theorem. *Every vector space has a basis. Each independent set of vectors is contained in some basis.*

Proof. Apply Zorn's Lemma to the set of all independent sets (or those containing a given independent set) ordered by inclusion. □

Steinitz Exchange Theorem. *Let V be a vector space, $M \subseteq V$ an independent set, N a proper subset of M, and $v \in V \setminus M$ a vector such that $N \cup \{v\}$ is independent. Then for some vector $w \in M \setminus N$ the set $(M \setminus \{w\}) \cup \{v\}$ is independent.*

Remark. A simpler, more familiar statement is obtained for $N = \emptyset$. The reader is urged to reformulate without reference to N in this case. However, we shall prove and use the result in full.

Proof. If $M \cup \{v\}$ is independent, we are home. Otherwise let

$$C \subseteq M \cup \{v\}$$

be a circuit. Obviously v must belong to C, C cannot consist of v alone, and

$$C \not\subseteq N \cup \{v\}$$

Let

$$w \in C \cap (M \setminus N)$$

There are nonzero scalars α_u, $u \in C$, such that $\sum \alpha_u \cdot u = \overline{0}$. We claim that

$$M' = (M \setminus \{w\}) \cup \{v\}$$

is independent. Were this not so, M' would contain a circuit C'. Obviously $v \in C'$, $w \notin C'$. There are nonzero scalars β_u, $u \in C'$, with

$$\sum_{u \in C'} \beta_u \cdot u = \overline{0}$$

Let us define
$$\gamma_u = \begin{cases} \alpha_u/\alpha_v & \text{if } u \in C \setminus C' \\ -\beta_u/\beta_v & \text{if } u \in C' \setminus C \\ \alpha_u/\alpha_v - \beta_u/\beta_v & \text{if } u \in C \cap C' \end{cases}$$

We have
$$\sum_{u \in C \cup C'} \gamma_u \cdot u = \left(\sum_{u \in C} \alpha_u \cdot u\right) \Big/ \alpha_v - \left(\sum_{u \in C'} \beta_u \cdot u\right) \Big/ \beta_v = \bar{0} - \bar{0} = \bar{0}$$

Observe now that $\gamma_v = (\alpha_v/\alpha_v) - (\beta_v/\beta_v) = 0$ but $\gamma_w = \alpha_w/\alpha_v$ is not zero. Since
$$\sum_{u \in (C \cup C') \setminus \{v\}} \gamma_u \cdot u = \bar{0}$$

with not every γ_u zero, $(C \cup C') \setminus \{v\} \subseteq M$ is linearly dependent, which is absurd. □

Basis Equipotence Theorem. *Any two bases of a given vector space V are equipotent. Indeed any two maximal independent subsets of any set W of vectors in V are equipotent.*

Proof. We need only to prove the stronger second statement. Let $W \subseteq V$ and suppose that two maximal independent subsets A and B of W have different cardinalities, say $\operatorname{Card} A < \operatorname{Card} B$. We shall derive a contradiction by constructing a bijection $A \setminus B \to B \setminus A$.

An *exchange function* is defined as a bijection f from a subset S of $A \setminus B$ to some subset $\operatorname{Im} f \subseteq B \setminus A$ such that
$$(B \setminus \operatorname{Im} f) \cup S$$
is independent. The set $\mathcal{E}$ of exchange functions is nonempty (why?), and it can be ordered by restriction: $f \leq g$ means that the domain set S of f is a subset of the domain of g and $g|S = f$. The ordered set $(\mathcal{E}, \leq)$ satisfies the hypothesis of Zorn's Lemma: thus there is a maximal exchange function f. We claim that the domain S of f is $A \setminus B$ and $\operatorname{Im} f = B \setminus A$. It is easy to see that
$$S \neq A \setminus B \quad \text{if and only if} \quad \operatorname{Im} f \neq B \setminus A$$
For if we had $S = A \setminus B$ but $\operatorname{Im} f \neq B \setminus A$, then
$$(B \setminus \operatorname{Im} f) \cup S = A \cup [(B \setminus A) \setminus \operatorname{Im} f]$$

would be an independent set larger than A, which is absurd. On the other hand, if we had $\mathrm{Im}\, f = B \setminus A$ but $S \neq A \setminus B$, then the assumption $\mathrm{Card}\, A < \mathrm{Card}\, B$ would be contradicted. Therefore the only case we must be concerned with is $S \neq A \setminus B$, $\mathrm{Im}\, f \neq B \setminus A$. Take any
$$v \in (A \setminus B) \setminus S$$
and apply the Steinitz Exchange Theorem with
$$M = (B \setminus \mathrm{Im}\, f) \cup S = S \cup (A \cap B) \cup [(B \setminus A) \setminus \mathrm{Im}\, f]$$
$$N = S \cup (A \cap B)$$

This yields a vector
$$w \in M \setminus N = (B \setminus A) \setminus \mathrm{Im}\, f$$
such that
$$(M \setminus \{w\}) \cup \{v\}$$
is independent. Define now an extension g of f to $S' = S \cup \{v\}$ by $g(v) = w$. The set
$$(B \setminus \mathrm{Im}\, g) \cup S' = [B \setminus (\mathrm{Im}\, f \cup \{w\})] \cup (S \cup \{v\}) = (M \setminus \{w\}) \cup \{v\}$$
is independent and $f < g$ in the ordered set $\mathcal{E}$: a contradiction.

The only remaining possibility is $S = A \setminus B$, $\mathrm{Im}\, f = B \setminus A$, as claimed. □

The cardinality of a basis of a vector space V is called its *dimension*, and it is denoted by $\dim V$. The second statement of the Basis Existence Theorem implies that if U is a subspace of a vector space V, then $\dim U \leq \dim V$. If V has finite dimension and $U \neq V$, then $\dim U < \dim V$.

We conclude this section with an application of elementary linear algebra to vector spaces V over the real number field $\mathbb{R}$. A set C of vectors in V is called *convex* if $v, w \in C$, $0 \leq \alpha \leq 1$ imply
$$\alpha v + (1 - \alpha) w \in C$$

Examples. The entire vector space V is convex. Every singleton is convex. The empty set is convex. Every subspace of V is convex.

The set of convex sets constitutes an algebraic closure system on V. The closure of a set M of vectors is called the *convex hull* of M.

The reader should verify that this convex hull consists of all linear combinations
$$\alpha_0 v_0 + \cdots + \alpha_{m-1} v_{m-1}$$
of elements $v_i \in M$ with all coefficients α_i nonnegative and such that
$$\alpha_0 + \cdots + \alpha_{m-1} = 1$$
(This latter requirement forces $m \geq 1$.) Assume now that V has finite dimension n, and let $v_0, \ldots, v_{n+1}$ be any $n+2$ vectors, all distinct. The $n+1$ vectors
$$(v_0 - v_{n+1}), \ldots, (v_n - v_{n+1})$$
cannot be linearly independent. We have
$$\alpha_0(v_0 - v_{n+1}) + \cdots + \alpha_n(v_n - v_{n+1}) = \bar{0}$$
for some scalars $\alpha_0, \ldots, \alpha_n$, not all of which are zero. Define
$$\alpha_{n+1} = -(\alpha_0 + \cdots + \alpha_n)$$
Obviously
$$\alpha_0 v_0 + \cdots + \alpha_n v_n + \alpha_{n+1} v_{n+1} = \bar{0}$$
Let I be the set of indices i such that $\alpha_i > 0$, and let J be the complementary set of indices, $J = (n+2) \setminus I$. Both I and J are nonempty; $\sigma = \sum_{i \in I} \alpha_i$ is positive and equals $\sum_{j \in J} -\alpha_j$. The vector
$$\sum_{i \in I} (\alpha_i / \sigma) v_i = \sum_{j \in J} (-\alpha_j / \sigma) v_j$$
belongs to the convex hull of the set $\{v_i : i \in I\}$ as well as to that of $\{v_j : j \in J\}$. We have proved:

Radon's Theorem. *In an n-dimensional vector space over $\mathbb{R}$, $n \in \omega$, each set of $n+2$ vectors is the union of two disjoint subsets whose convex hulls intersect.*

This theorem can be used to determine the Helly number of the convex closure system on a finite dimensional vector space over $\mathbb{R}$. The corresponding result is Helly's Theorem, obtained by Eduard Helly just before World War I broke out and published by him, as well as by Radon and König, during the temporary peacetime of the 1920s. Helly-type theorems have been forthcoming ever since and also some previously known results have been recognized as Helly theorems, describing remarkable combinatorial aspects of various mathematical structures. Some are included in this volume.

Helly's Theorem. *Let V be an n-dimensional vector space over $\mathbb{R}$, $n \in \omega$. The Helly number of the convex closure system on V is $n + 1$.*

Proof. Let us show that if $\mathcal{F}$ is a finite nonempty set of convex sets such that for every nonempty subset $\mathcal{S}$ of $\mathcal{F}$ with $\text{Card}\,\mathcal{S} \leq n+1$ the intersection $\cap \mathcal{S}$ is not empty, then $\cap \mathcal{F}$ is not empty. Suppose this is not true and let $\mathcal{F}$ be a misbehaving set of convexes with the least possible number of members. We shall derive a contradiction. Certainly
$$\text{Card}\,\mathcal{F} \geq n+2$$
For each $K \in \mathcal{F}$, $\cap(\mathcal{F}\setminus\{K\})$ is not empty: let
$$c(K) \in \cap(\mathcal{F}\setminus\{K\})$$
For distinct K, K' the sets
$$\cap(\mathcal{F}\setminus\{K\}) \quad \text{and} \quad \cap(\mathcal{F}\setminus\{K'\})$$
are disjoint because $\cap \mathcal{F}$ is supposed to be empty. Thus $c(K) \neq c(K')$ if $K \neq K'$. It follows from Radon's Theorem that $\mathcal{F}$ is the union of two disjoint subsets $\mathcal{G}$ and $\mathcal{H}$ such that the convex hulls of
$$\{c(K) : K \in \mathcal{G}\} \quad \text{and} \quad \{c(K) : K \in \mathcal{H}\}$$
intersect. But any common element of these two convex hulls must belong to $\cap \mathcal{F}$, which is then nonempty, contrary to assumption. This shows that the Helly number sought is at most $n+1$.

On the other hand, if
$$B = \{v_0, \ldots, v_{n-1}\}$$
is a vector space basis, then consider the n subspaces H_i generated by the sets
$$B\setminus\{v_i\}, \quad i = 0, \ldots, n-1$$
Let H be the convex hull of B. Obviously
$$H \cap H_0 \cap \cdots \cap H_{n-1} = \emptyset$$

166 VECTOR SPACES

but the intersection of every n or less members of

$$\mathcal{F} = \{H, H_0, \ldots, H_{n-1}\}$$

is nonempty. Thus the Helly number cannot be less than $n + 1$. □

Historical Note. The theory of n-dimensional vector spaces, including but not limited to the three-dimensional geometry of everyday perception, was first developed by Hermann Grassmann. In the 1844 and 1862 editions of "Ausdehnungslehre" the view is explicitly put forward that mathematics is not a science of quantities. Beyond binary operations on vectors, Grassmann explores algebraic properties of the lattice of subspaces. In parallel with, but apparently independent of, contemporary Boolean algebra, this foreshadows, by some 50 years, the emergence of formal lattice theory and universal algebra.

EXERCISES

1. On a field F let $\leq$ be a linear order compatible with the additive group structure of F and such that the set of field elements greater than 0 forms a multiplicative subgroup of F^*. Develop a "convex closure" concept for vector spaces over F, and prove a Helly-type theorem. Give an example other than $F = \mathbb{R}$.

2. Show that no finite convex subset of a vector space over $\mathbb{R}$ can have more than one element.

3. Consider $\mathbb{R}$ as a one-dimensional vector space over itself. Show that a subset is convex in this vector space if and only if it is order convex in the standard real order. What is the Helly number of this convex closure system?

4. Referring to the formal definition $(V, +, p)$ of a vector space over a field F, does $p : F \to \operatorname{End} V$ have to be injective? Assume $\dim V > 0$.

5. Give examples of vector spaces V where every additive subgroup of V is a vector subspace and give counterexamples.

6. Design a program to perform the following tasks in the vector space $\mathcal{P}(S)$ over $\mathbb{Z}_2$ for finite $S \subset \omega$:

(a) Decide if a given set of vectors is independent.
(b) Express any given vector as a linear combination of members of a given basis.
(c) Find a basis containing a given independent set.
(d) Given $M \subseteq \mathcal{P}(S)$ and an independent $I \subseteq M$, find an independent set G with $I \subseteq G \subseteq M$ and having as many elements as possible.

2. LINEAR MAPS AND EQUATIONS

Let V and W be vector spaces over the same field F. A mapping $h : V \to W$ is called *linear* if it is a group homomorphism from $(V, +)$ to $(W, +)$ and if

$$\text{for all} \quad \alpha \in F, \quad x \in V, \quad h(\alpha \cdot x) = \alpha \cdot h(x)$$

Proposition 1 *The composition of two linear mappings $h : V \to W$ and $g : W \to Z$ is linear from V to Z. The identity mapping on each vector space V is linear. The inverse of a bijective linear map is linear.*

It is easy to see that the kernel and the image of a linear map are not only subgroups but also subspaces of the domain and codomain vector spaces, respectively. Let U be a subspace of a vector space V over a field F. As U is a (normal) subgroup of V, we have a quotient group structure $(V/U, +)$. Further, a vector space structure is defined on V/U by

$$\alpha \cdot C = \{\alpha v : v \in C\}$$

for every nonzero scalar α and coset C of U. This is called *quotient space*. Clearly the canonical surjection

$$V \to V/U$$

is linear and its kernel is U. Thus every vector subspace is the kernel of some linear map.

A bijective linear map $h : V \to W$ between vector spaces is called an *isomorphism* from V to W, and if such an h exists, then V and W are called *isomorphic*. Any isomorphism $h : V \to V$ is called an *automorphism* of V. Under composition the set of all automorphisms of V is a group; it is denoted by $\text{Aut}V$. Obviously it is a subgroup

of $\Sigma(V)$. As for the (not necessarily bijective) linear maps $V \to V$, these are called *endomorphisms* and they form a monoid under composition. The constant function $V \to V$ mapping everything to the null vector is an endomorphism. For each scalar α the map $h_\alpha : V \to V$ given by
$$h_\alpha(x) = \alpha \cdot x$$
is an endomorphism, and it is an automorphism if $\alpha \neq 0$. Not all endomorphisms and not even all automorphisms are of this form: the reader can find a counterexample for the two-dimensional vector space $\mathcal{P}(2)$ over $\mathbb{Z}_2$.

Linear Map Dimension Theorem. *Let $h : V \to W$ be a linear map between vector spaces. We have*
$$\dim V = \dim \operatorname{Ker} h + \dim \operatorname{Im} h$$
If $\dim V$ is finite and $h : V \to V$ is an endomorphism, then h is injective if and only if it is surjective onto V.

Proof. Let B_1 be any basis of $\operatorname{Ker} h$, C any basis of $\operatorname{Im} h$. For every $c \in C$, choose a $b_c \in V$ with $h(b_c) = c$. Then
$$B_2 = \{b_c : c \in C\}$$
is independent in V and it has the same cardinality as C. The union $B_1 \cup B_2$ is a basis of V. □

Coordinatization Theorem. *A subset B of a vector space V over a field F is a basis if and only if for every vector $v \in V$ there is a unique finite family of nonzero scalars $(\alpha_u : u \in I)$ indexed by a subset I of B such that*
$$v = \sum_{u \in I} \alpha_u \cdot u$$

The set $I \subseteq B$ described by the above theorem is called the *support* of v in B. For every basis element $u \in B$ we define the uth *coordinate function* c_u (with respect to B) from V to F by
$$c_u(v) = \begin{cases} \alpha_u & \text{if } u \in I, \\ 0 & \text{otherwise} \end{cases}$$

Consider now F as a vector space over itself: vector addition is field addition and vector space product is field product. Then every coordinate function $V \to F$ is linear. A linear function $V \to F$ is usually called a *linear form* on V. The set of all linear forms on V is a subspace of F^V. Note that the null vector of this space is the zero-valued constant function $V \to F$. It is referred to as the *null linear form*.

Dual Basis Theorem. *Let V be a finite dimensional vector space over a field F. If B is any basis of V, then the coordinate functions with respect to B constitute a basis of the space of all linear forms on V.*

Proof. Let f be any linear form. For every $v \in V$ we have

$$f(v) = f\left(\sum_{u \in B} c_u(v) \cdot u\right) = \sum_{u \in B} f(u) \cdot c_u(v)$$

This shows that $B^* = \{c_u : u \in B\}$ generates all linear forms in F^V. To show the independence of the set B^*, assume that for some family $(k_u : u \in B)$ of scalars, the function

$$\sum_{u \in B} k_u \cdot c_u$$

is constant zero. For every $w \in B$ we have

$$0 = \sum_{u \in B} k_u \cdot c_u(w) = k_w \cdot c_w(w) + \sum_{u \neq w} k_u \cdot c_u(w)$$

and since $c_u(w)$ is zero for $u \neq w$,

$$0 = k_w \cdot c_w(w) = k_w \cdot 1 = k_w \qquad \square$$

An important consequence of the Coordinatization Theorem is that any two spaces over the same field that have the same dimension are isomorphic. In particular any n-dimensional vector space V (n positive integer) over a field F is isomorphic to the space $F^{[1,n]}$: the isomorphism is established by taking any bijection f from $[1, n] = \{1, \ldots, n\}$ to a basis B of V and associating to every vector $x \in F^{[1,n]}$ the vector

$$\sum_{i \in [1,n]} x(i) \cdot f(i)$$

in V. Since isomorphism implies equipotence, V has $(\operatorname{Card} F)^n$ elements.

Application to Finite Fields. Let E be any finite field. The additive subgroup F of E generated by 1_E is easily seen to be closed also under the field's product operation. (Use distributivity.) F is a subring of E, isomorphic to a quotient ring of $\mathbb{Z}$. Since every subring of E is entire, F must be isomorphic to $\mathbb{Z}/p\mathbb{Z}$ for some prime number p. Thus F is a subfield of E and it has p elements. Consider now E as a vector space over F (the vector space product is the field product in E). As such, E has p^n elements, where n is the dimension of E. Thus the number of elements of any finite field is a prime power. (Indeed for any field extension $E : F$, E can be considered as a vector space over F. This has far-reaching consequences in Galois theory.)

Convention. Let F be any field and n a positive integer. Throughout this chapter we shall reserve the short notation F^n for the vector space $F^{[1,n]}$ (as opposed to its original use to denote the set of functions from the ordinal n to F). A typical vector

$$(v(i) : i \in [1,n])$$

in F^n shall also be written $(v_1, \ldots, v_n)$ like an n-tuple.

The *inner product* (or *dot product*) of any two vectors $v, w \in F^n$ is defined as the element

$$\sum_{1 \leq i \leq n} v(i) \cdot w(i)$$

of the field F, and it is denoted by $v \bullet w$. We have

$$v \bullet w = w \bullet v \quad \text{(commutativity)}$$

and if u is any third vector in F^n, then

$$u \bullet (v + w) = (u \bullet v) + (u \bullet w) \quad \text{(distributivity)}$$

Two vectors are called *orthogonal* if their dot product is zero. This concept is of pervasive importance in geometry and physics: the reader will appreciate how natural it is by taking $\mathbb{R}^2$ as a model for a flat physical surface and by using a sheet of paper to represent

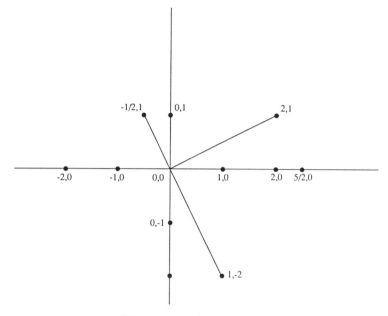

FIGURE 6.1 Orthogonality.

vectors in $\mathbb{R}^2$ as in Figure 6.1. Of course for any field F, (x,y) and $(y,-x)$ are always orthogonal in F^2.

Theorem of Pythagoras. *Let F be any field. If $v,w \in F^n$ are orthogonal and $u = v + w$, then $v \bullet v + w \bullet w = u \bullet u$.*

Proof. By distributivity

$$(v + w) \bullet (v + w) = (v \bullet v) + (w \bullet w) + (v \bullet w) + (w \bullet v) \qquad \square$$

Let m and n be two natural numbers. An $m \times n$ *matrix* over a field F is a family

$$a = (a_{ij} : 1 \leq i \leq m, \ 1 \leq j \leq n)$$

of elements of F indexed by the set of couples of integers (i,j) such that $1 \leq i \leq m$, $1 \leq j \leq n$. The field element a_{ij} is called the matrix's *entry at position i,j* (or in *row i, column j*). For each fixed i,

$$(a_{ij} : 1 \leq j \leq n)$$

is called the ith *row vector* of a (it is a vector in F^n) and it is denoted by $\vec{a_i}$. For each fixed j,

$$(a_{ij} : 1 \leq i \leq m)$$

belongs to F^m; it is called the jth *column vector* and it is denoted by $a_{[j]}$. This terminology corresponds to the usual tabular representation of matrices. For example, the 3×4 matrix a over $\mathbb{Q}$ given by $a_{ij} = i - j$ is represented as follows:

$$\begin{pmatrix} 0 & -1 & -2 & -3 \\ 1 & 0 & -1 & -2 \\ 2 & 1 & 0 & -1 \end{pmatrix}$$

We have here $\vec{a_2} = (1, 0, -1, -2)$ and $a_{[2]} = (-1, 0, 1)$.

For every $m \times n$ matrix a and $x \in F^n$, consider the family of dot products

$$(\vec{a_i} \bullet x : 1 \leq i \leq m)$$

Clearly this is an element of F^m. It is routine to verify that the mapping $h_a : F^n \to F^m$ given by

$$h_a(x) = (\vec{a_i} \bullet x : 1 \leq i \leq m)$$

is linear. Do we get every linear map $h : F^n \to F^m$ this way? Indeed we do: let a be the $m \times n$ matrix whose jth column $a_{[j]}$ is

$$h(\delta_{jk} : 1 \leq k \leq n)$$

where δ_{jk} is 0 or 1 according to whether $k \neq j$ or $k = j$. We have $h = h_a$. Also

$$h_a = h_b \quad \text{implies} \quad a = b$$

for the jth column vector of a cannot be anything else but

$$h_a(\delta_{jk} : 1 \leq k \leq n)$$

and the jth column vector of b is

$$h_b(\delta_{jk} : 1 \leq k \leq n)$$

Thus the map associating to each $m \times n$ matrix a over F the linear function $h_a : F^n \to F^m$ is a bijection. (It has many other interesting properties!) The matrix a and the linear function h_a are said to *correspond* to each other.

The theory of systems of linear equations can be developed in the language of matrices and corresponding linear functions. What are, for example, the real numbers u, y, z such that

$$5u + 2y - z = 3$$
$$u/3 - y/2 = 4$$

These are obviously given by the vectors $(u, y, z) \in \mathbb{R}^3$ such that

$$h_a(u, y, z) = (3, 4)$$

for the linear map $h_a : \mathbb{R}^3 \to \mathbb{R}^2$ corresponding to the matrix

$$a = \begin{pmatrix} 5 & 2 & -1 \\ \frac{1}{3} & -\frac{1}{2} & 0 \end{pmatrix}$$

The general question is then formulated as follows: given an $m \times n$ matrix a over a field F and its corresponding linear map $h_a : F^n \to F^m$ and given a vector $w \in F^m$, what are the vectors x in F^n such that

$$h_a(x) = w$$

It may happen that there are several solution vectors x, or that none exists, or that there is a unique solution. Trivial examples are given over the field $\mathbb{Q}$, for $m = n = 1$, by the 1×1 matrices whose sole entry is 0 or 1, and the vector $w \in \mathbb{Q}^1$ where $w(1)$ is 0 or 1: please verify all four combinations. In general, the existence of a solution means that

$$w \in \operatorname{Im} h_a$$

In this case *let x_0 be any solution, $h_a(x_0) = w$. The complete set of all solutions is*

$$\{x_0 + v : v \in \operatorname{Ker} h_a\}$$

This is the coset of $\operatorname{Ker} h_a$ that contains the vector x_0. In the special case of $m = n$ (*square matrices*) the Linear Map Dimension Theorem implies that *for any $m \times m$ matrix a the following conditions are equivalent*:

(i) *whatever is the choice of $w \in F^m$, $h_a(x) = w$ has at least one solution $x \in F^m$,*

(i) *whatever is the choice of $w \in F^m$, $h_a(x) = w$ has at most one solution x,*

(iii) *for the null vector $\bar{0} \in F^m$, $h_a(x) = \bar{0}$ has only the obvious solution $x = \bar{0}$,*

(iv) $h_a \in \operatorname{Aut} F^m$.

In the case where $m < n$ ("more unknowns than equations") the Linear Map Dimension Theorem implies that $h_a : F^n \to F^m$ cannot be injective, whether it is surjective or not. If $m < n$, then for any $w \in F^m$ either $h_a(x) = w$ has no solution x or it has several distinct solutions.

Finally in the case $m > n$ ("more equations than unknowns") $h_a : F^n \to F^m$ cannot be surjective, whether it is injective or not. If $m > n$, then for some $w \in F^m$ there is no solution $x \in F^n$ satisfying $h_a(x) = w$.

Several algorithmic procedures are known to solve systems of linear equations. The one we shall outline here is usually referred to as *Gaussian elimination*. It consists of reducing the problem of finding the solution vectors $x \in F^n$ of the system $h_a(x) = w$ to a similar problem for a simpler matrix b. First note that if all the n-column vectors of a are null vectors, then $h_a(x) = w$ has a solution if and only if w is a null vector, and in this case every $x \in F^n$ is a solution of the system. If a has $t \geq 1$ nonnull column vectors, then take any nonzero entry a_{ij}. For each row vector $\vec{a_k}$, $k \neq i$, consider the vector

$$b_k = \vec{a_k} - (a_{kj}/a_{ij}) \cdot \vec{a_i}$$

and the scalar

$$z_k = w_k - (a_{kj}/a_{ij}) \cdot w_i$$

Form the $(m-1) \times n$ matrix b whose row vectors are

$$b_1, \ldots, b_{i-1}, b_{i+1}, \ldots, b_m$$

and form the vector

$$z = (z_1, \ldots, z_{i-1}, z_{i+1}, \ldots, z_m) \in F^{m-1}$$

Then $x \in F^n$ is a solution of $h_a(x) = w$ if and only if

$$x \text{ is a solution of } h_b(x) = z \quad \text{and} \quad \vec{a_i} \bullet x = w_i$$

What did we gain? The matrix b has at most $t-1$ nonnull column vectors because

(i) if the lth column of a is null, then so is the lth column of b, $1 \leq l \leq n$,
(ii) the jth column of a is nonnull, but the jth column of b is null.

The nullity of the jth column of b means in particular that in every solution x of $h_b(x) = z$ we can replace x_j by any other scalar x'_j: the vector x' so obtained is again a solution of $h_b(x') = z$. On the other hand, $\vec{a_i} \bullet x = w_i$ means that only the scalar

$$x'_j = \left(w_i - \sum_{l \neq j} a_{il} x_l \right) \bigg/ a_{ij}$$

will do in such a replacement: every solution of $h_a(x) = w$ is derived from some solution of $h_b(x) = z$, and every solution of the latter simpler system determines a solution of $h_a(x) = w$. Now if b has only null columns (in particular it may be the empty matrix), then the solutions of $h_b(x) = z$ are obvious; otherwise we repeat the elimination process for a nonnull column of b. The reader is urged to write a computer program that for every input matrix a and vector w over the field $\mathbb{Q}$ either produces a solution vector x of the linear system $h_a(x) = w$ or determines its unsolvability.

EXERCISES

1. Show that every finite field F has infinitely many different simple algebraic extensions $E : F$. Is the extension field E finite in every such case?

2. Design a program to perform the following tasks in the vector spaces $\mathbb{Q}^n$ and $\mathbb{Z}_p^n$:
   (a) Decide if a given finite set of vectors is independent.
   (b) Express any given vector as a linear combination in a given basis.
   (c) Find a basis containing a given independent set.

(d) Given a finite set of vectors M and an independent $I \subseteq M$, find a maximal independent set G with $I \subseteq G \subseteq M$.

3. Verify that if V, W are vector spaces over the same field, B a basis of V and $f : B \to W$ any map, then there is a unique linear $h : V \to W$ extending f. For the cases where $V = W$ is $\mathbb{Q}^n$ or $\mathbb{Z}_p^n$, write a program to compute the matrix of h from any given B and f.

4. (a) Let F be a field, n a positive integer. If a and b are $n \times n$ square matrices with h_a and h_b as corresponding endomorphisms $F^n \to F^n$, then define $a \cdot b$ as the matrix corresponding to $h_a \circ h_b$. Show that this defines a monoid structure on the set of $n \times n$ matrices over F.

   (b) For $F = \mathbb{Q}$ or $\mathbb{Z}_p$, write a program to compute matrix products as defined above.

5. Show that $h : \mathbb{R}^2 \to \mathbb{C}$ given by $h(x, y) = x + yi$ is a vector space isomorphism (vector spaces over the real field $\mathbb{R}$). For $z \in \mathbb{C}$ what can you say about the inner product $h^{-1}(z) \bullet h^{-1}(z)$?

6. What simple connections can you see between linear polynomials, linear maps, and subspaces of dimension one?

3. AFFINE AND PROJECTIVE GEOMETRY

By a *geometry* we mean an inclusion-ordered closure system. The geometry of subspaces of a vector space V is called *projective geometry* on V: the term "projective" will be justified later. The set of cosets of the various subspaces, viewed as additive subgroups of V, together with the empty set $\emptyset$, constitutes another geometry, called the *affine geometry* of V. Thus the affine geometry includes the projective geometry. We shall be interested here mainly in these two kinds of geometries. Our all too general notion of geometry is not universally standard, but it is a convenient common denominator for the particular geometries that we wish to study. In any geometry $(\mathcal{G}, \subseteq)$ the closed sets (members of $\mathcal{G}$) are called *flats*, the smallest flat $\cap \mathcal{G}$ is called the *null flat* (it may or may not be empty), the flats covering the null flat in the order $(\mathcal{G}, \subseteq)$ are called *points*, any flat covering a point is called a *line*, and any flat covering a line is called a *plane*. (This terminology finds its justification in the affine geometry of $\mathbb{R}^3$.)

The largest flat in any geometry $(\mathcal{G},\subseteq)$ is $\cup \mathcal{G}$, it is called *space*, and any flat covered by it is called a *hyperplane*. For any three-dimensional vector space V, planes and hyperplanes coincide in the affine geometry of V. This coincidence fails in higher dimensions. Observe that every singleton subset of a vector space V is a point of the affine geometry, but the points of the projective geometry on V are the one-dimensional subspaces of V, i.e., the affine lines containing the null vector. As for the affine planes containing the null vector, these are precisely the projective lines. Generally, an affine flat belongs to the projective geometry if and only if it contains the null vector $\bar{0}$. Note that the null flat of the affine geometry is the empty set $\emptyset$, while the null flat of the projective geometry is the singleton $\{\bar{0}\}$. The projective geometry on V is denoted by Pro$V$.

Power Set Examples. Let S be any set. Consider $\mathcal{P}(S)$ as a vector space over $\mathbb{Z}_2$. Then the points of the projective geometry on $\mathcal{P}(S)$ are the pairs $\{\emptyset, X\}$ with $\emptyset \subset X \subseteq S$, while the lines are the sets
$$\{\emptyset, X, Y, Z\}$$
where X, Y, Z are any three distinct nonempty subsets of S and $X + Y = Z$. As for the affine geometry on $\mathcal{P}(S)$, its points are the singletons $\{X\}$ with $X \subseteq S$, while the lines are the pairs $\{X, Y\}$ where X and Y are any two distinct subsets of S. The affine planes are of the form
$$\{A, B, C, D\}$$
where A, B, C, D are distinct subsets of S and $A + B = C + D$. How would you describe affine hyperplanes?

Two geometries $\mathcal{G}$ on a set S and $\mathcal{H}$ on a set T are called *similar* if the orders $(\mathcal{G},\subseteq)$ and $(\mathcal{H},\subseteq)$ are isomorphic. Note that this does not require the equipotence of $S = \cup\mathcal{G}$ and $T = \cup\mathcal{H}$. (Counterexample?) If $\mathcal{I} \subseteq \mathcal{H}$ is any geometry contained in $\mathcal{H}$ and h is an order isomorphism from $(\mathcal{G},\subseteq)$ to $(\mathcal{I},\subseteq)$, then h is called an *embedding* of $(\mathcal{G},\subseteq)$ into $(\mathcal{H},\subseteq)$. In this case $(\mathcal{G},\subseteq)$ and $(\mathcal{I},\subseteq)$ are similar.

In a vector space V, for every $v \in V$, the set $\mathcal{A}_v$ of affine flats containing v constitutes a geometry on V. For the null vector $\bar{0}$, $\mathcal{A}_{\bar{0}}$ coincides with the projective geometry on V, and every other $\mathcal{A}_v$ is indeed similar to $\mathcal{A}_{\hat{0}}$: associate to every flat $U \in \mathcal{A}_{\bar{0}}$ the coset of U containing v. Thus the affine geometry includes many similar copies of the projective geometry.

We know that the additive group of any vector space V acts on V itself by translation, and it also acts on $\mathcal{P}(V)$ by translation. For every affine flat A and vector v the *translate*

$$v + A = \{v + x : x \in A\}$$

is again an affine flat. If U is a subspace, then $v + U$ is the coset of U containing v. The nonempty affine flats are precisely the translates of the various subspaces of V. The translates of any given subspace form a partition of V. The additive group of V acts on the set $\mathcal{A} \subseteq \mathcal{P}(V)$ of all affine flats by translation, and $\mathcal{A}$ is partitioned into translation orbits. Two flats in the same orbit are called *parallel*. Since the flats parallel to any given nonempty flat A are the cosets of some subspace U, these parallel flats partition V: every $v \subset V$ is contained in a unique flat parallel to A. For $V = \mathbb{R}^2$ and A any line, this is a widely known result in plane geometry.

Parallelism in a Power Set. Consider again the affine geometry of the vector space $\mathcal{P}(S)$ over $\mathbb{Z}_2$, where S is any set. Two distinct affine lines $\{A, B\}$ and $\{X, Y\}$ are parallel if and only if $A + X = B + Y$, i.e., precisely when their union is a plane.

Proposition 2 *If A is any flat and H is any hyperplane of an affine geometry, then either A is contained in some hyperplane parallel to H or A intersects all hyperplanes parallel to H.*

Proof. Suppose first that A and H are subspaces of the vector space V underlying the affine geometry. Then either $A \subseteq H$ or the subspace generated by $A \cup H$ is the full space V. In the latter case every $t \in V$ can be written as

$$t = u + w \quad \text{with some} \quad u \in A, \quad w \in H$$

If v is any element of a hyperplane H' parallel to H, i.e., of a translate $t + H$, then

$$v = t + z \quad \text{for some} \quad z \in H$$

and we have

$$v = u + w + z$$

with $u \in A$, $w \in H$. Clearly u and v are congruent modulo H, and thus $u \in A \cap H'$. This shows that if $A \not\subseteq H$, then A intersects every hyperplane H' parallel to H.

In the general case when A and H are not necessarily subspaces of V, take subspaces A_o and H_o parallel to A and H, respectively. If $A_o \subseteq H_o$, then A is contained in some parallel translate of H_o and therefore of H. If $A_o \not\subseteq H_o$, then A_o intersects every hyperplane parallel to H_o, i.e., to H, and the parallel translate A must then also intersect every such hyperplane. □

Corollary. *Any two nonparallel hyperplanes of an affine geometry intersect.*

In the affine geometry of a finite dimensional vector space V, let us now give a precise algebraic meaning to the concept of "projection." Let Aff V denote the affine geometry of V, and for any flat A in Aff V, let Aff A denote the geometry

$$\mathcal{P}(A) \cap \text{Aff } V$$

on A: this is obviously a geometry and we call it the *affine geometry of the flat A*. For any two flats A, B in Aff V we denote by $\overline{AB}$ the closure of $A \cup B$ in Aff V: this is the smallest flat containing both A and B. If $A \subseteq B$, then we say that A *lies* in B. The *dimension of a nonempty affine flat* A is defined as the dimension of the unique subspace A_o of V parallel to A; it is denoted by dim A. The geometries Aff A and Aff A_o are similar. The reader is urged to use inner vision to represent what follows, in the special case $V = \mathbb{R}^3$, and to use paper-assisted vision in the case $V = \mathbb{R}^2$. Affine points and lines are well illustrated by points and lines drawn on paper. Let us fix a point O in Aff V, and let H be a hyperplane not containing O. If L is any affine line containing O, then $L \not\subseteq H$ and either $L \cap H = \emptyset$ or $L \cap H$ is an affine point. We have $L \cap H = \emptyset$ precisely when L lies in the hyperplane H_O parallel to H and containing O. If X is any affine point not lying in H_O, then $\overline{OX}$ is an affine line and

$$\overline{OX} \cap H$$

is again an affine point, called the *projection of the point X (from O to H)*. In general the *projection of a set* $S \subseteq V \setminus H_O$ is defined as the union of the projections of all the points X contained in S. For any

affine flat
$$A \subseteq V \setminus H_O$$

the flat $\overline{OA}$ coincides with the union of all lines $\overline{OX}$ for points X lying in A. The projection of A is

$$\overline{OA} \cap H$$

For $A \neq \emptyset$ we have

$$\dim(\overline{OA} \cap H) = \dim \overline{OA} - 1 = \dim A$$

The projection of $\emptyset$ is $\emptyset$. The projection of a d-dimensional flat is a d-dimensional flat. The map associating to each flat $A \subseteq V \setminus H_O$ its projection is surjective to Aff H, but it is generally not injective.

Let us now consider *projections from the origin*: O is the singleton of the null vector, and H is a hyperplane not containing the null vector. In this case two flats

$$A, B \subseteq V \setminus H_O$$

have the same projection P from O to H if and only if they generate the same subspace of V. This subspace then coincides with

$$\overline{OA} = \overline{OB} = \overline{OP}$$

These considerations lead us to consider the *projective completion map* π from Aff H to the projective geometry Pro V on V, given by

$$\pi(A) = \overline{OA}$$

It is an embedding of Aff H into Pro V because

$$A = \overline{OA} \cap H \quad \text{for every} \quad A \in \text{Aff } H$$

It is not surjective onto Pro V: its image is

$$\text{Pro } V \setminus \text{Pro } H_O$$

The members of the subspace projective geometry Pro H_O are called *flats at infinity*. Every $A \in$ Aff H is parallel in Aff V to a unique flat at infinity, called the *direction* of A. Any two (or more) parallel flats in Aff H have intersecting projective completions: the intersection is their common direction, a flat at infinity.

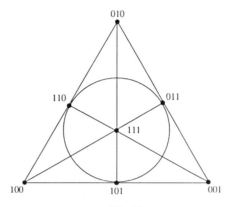

FIGURE 6.2 The Fano plane.

In particular, let L and D be two distinct parallel lines of Aff H. The intersection of their projective completions is

$$\overline{OL} \cap \overline{OD}$$

This is a line in Aff V but a point (not a line) in Pro V, and in Pro H_O as well, because it covers the nonempty null flat in these projective geometries. The familiar optical illusion that parallel lines meet at a point at infinity has now a precise mathematical meaning, and this has nothing to do with far-away distances.

Points at infinity have nothing to do with infinite cardinalities either. Consider the three-dimensional vector space $V = \mathbb{Z}_2^3$ over the two-element field $\mathbb{Z}_2$. This space is isomorphic to the space $\mathcal{P}(3)$ over $\mathbb{Z}_2$ and it has eight elements. For each vector v, $v + v$ is the null vector

$$\overline{0} = (0, 0, 0)$$

and thus $\{\overline{0}, v\}$ is a subspace of dimension 1 for each $v \neq \overline{0}$: there are seven of these. Each two-dimensional subspace consists of the vectors orthogonal to a fixed $v \neq \overline{0}$. There are seven of these subspaces, each consisting of $\overline{0}$ plus three nonnull vectors: the reader is urged to identify these in the ideogram of Figure 6.2. Pro $\mathbb{Z}_2^3$ is usually called the *Fano projective plane*; it has seven points and seven lines. Verify that the parallel affine lines

$$100, 001 \quad \text{and} \quad 111, 010$$

have a common direction, generated by 101.

EXERCISES

1. Prove a Helly-type theorem for the affine closure system of a finite dimensional vector space V over any field F. For $F = \mathbb{Q}$ or $\mathbb{Z}_p$, design a computer program that for any finite set of affine flats locates a common point if the flats intersect. Show that every flat in Aff V is of the form $h^{-1}(y)$ where $y \in V$ and $h : V \to V$ is linear. [To represent the data, use the fact that a flat $h^{-1}(y)$ is fully determined by y and the matrix of h.]

2. Let V be a finite dimensional vector space over a field F. Show that $A \subseteq V$ is an affine flat if and only if every linear combination

$$\alpha_0 v_0 + \cdots + \alpha_{n-1} v_{n-1}$$

of vectors $v_i \in A$ with $\alpha_0 + \cdots + \alpha_n = 1$ belongs to A.

3. Verify that Pro V is an upper section of (Aff $V, \subseteq$).

4. Verify that the inclusion-ordered set of flats of any geometry is a complete lattice. Characterize lub's in projective and affine geometries.

5. Show that if V is any vector space, then the smallest geometry on V containing all hyperplanes of Aff V is the whole of Aff V. What about Pro V?

6. Show that two distinct affine lines are parallel if and only if their union generates an affine plane.

7. Let V be a three-dimensional vector space over a field with n elements. In Pro V, show that
   (a) any two distinct points lie on a unique common line,
   (b) any two distinct lines contain a unique common point,
   (c) V is a plane,
   (d) every line contains $n + 1$ points, and every point is contained in $n + 1$ lines,
   (e) there are $n^2 + n + 1$ points and the same cardinal number of lines.

4. HYPERPLANES IN LINEAR PROGRAMMING

Consider the vector space $\mathbb{R}^n$ over the real numbers, where n is a fixed positive integer. Let $(f_i : i \in I)$ be a finite nonempty family of nonnull linear forms $\mathbb{R}^n \to \mathbb{R}$, referred to as *constraint functions*, and let $(b_i : i \in I)$ be a family of real numbers indexed by the same set I. Let g be a nonnull linear form $\mathbb{R}^n \to \mathbb{R}$; it shall be called the *objective function*. Linear programming is the theory and computational practice of finding vectors $x = (x_1, \ldots, x_n)$ in $\mathbb{R}^n$ subject to the constraints

$$f_i(x) \leq b_i \qquad \text{for all} \quad i \in I$$

and with the value $g(x)$ as large as possible. In one typical industrial application each b_i represents the maximum available quantity of some limited resource, such as a particular raw material. Each $x_1, \ldots, x_n$ represents a quantified production level of a particular product that incorporates or otherwise consumes the various resources considered. If the amount of the ith resource used by the jth product is a_{ij}, then

$$a_{i1}x_1 + \cdots + a_{in}x_n = f_i(x)$$

cannot exceed b_i. Similarly, if the profit contributed by each unit of the jth product is c_j, then the aggregate profit

$$c_1 x_1 + \cdots + c_n x_n = g(x)$$

is to be maximized. A variety of seemingly different logistical problems can be formulated in the language of linear programming. Here let us just note that

- minimization of a cost objective function g can be modeled by maximizing $-g$,
- minimum production quotas and the "natural" nonnegativity of quantities to be produced are modeled by $-x_j \leq -\text{quota}_j$ and $-x_j \leq 0$,
- generally it is unnecessary to work with constraints of the form $f_i(x) \geq b_i$ for these are equivalent to $-f_i(x) \leq -b_i$.

Now back to mathematics. Formally a *linear programming problem* (*LP problem*) is defined as the couple $(g, \mathcal{C})$ where

$$\mathcal{C} = \{(f_i, b_i) : i \in I\}$$

Each (f_i, b_i) is called a *constraint*. A *feasible solution* is a vector x in $\mathbb{R}^n$ with
$$f_i(x) \leq b_i \quad \text{for all} \quad i \in I$$
and the LP problem is called *feasible* if such an x exists. If for every other feasible solution x' we have
$$g(x') \leq g(x)$$
then x is called an *optimal solution*. For every $i \in I$ the set
$$H_i = \{x \in \mathbb{R}^n : f_i(x) = b_i\}$$
is a hyperplane in $\text{Aff}\,\mathbb{R}^n$—it is called the *hyperplane associated with the constraint* (f_i, b_i). The latter is a *tight constraint* for a feasible solution x if $x \in H_i$, and H_i is then called a *constraining hyperplane* of x. A feasible solution x that has a tight constraint is called a *constrained solution*. For any $J \subseteq I$, the intersection
$$\bigcap_{i \in J} H_i$$
is obviously an affine flat; if it is not void, then it is called a *boundary flat*. For each constrained solution x, there is a smallest boundary flat containing x, namely the intersection of all of its constraining hyperplanes: call this the *minimal boundary flat* of x.

We shall need the auxiliary notion of ordered lines. If L is any line in $\text{Aff}\,\mathbb{R}^n$ and f a linear form $\mathbb{R}^n \to \mathbb{R}$ that is not constant on L, then an order $\leq_f$ is defined on L by
$$x \leq_f y \quad \text{if and only if} \quad f(x) \leq f(y)$$
The restriction $f|L$ is an order isomorphism from $(L, \leq_f)$ to $\mathbb{R}$. If g is another such linear form, then $\leq_g$ coincides either with $\leq_f$ or with the dual of $\leq_f$. These orders on L are called *natural orderings*. If we fix any two distinct vectors $x, y \in L$, take any basis of $\mathbb{R}^n$ that includes
$$u = y - x$$
and let f be the uth coordinate function, then
$$\leq_f \quad \text{and} \quad \leq_{-f}$$
are the only two natural, mutually dual orders on L. We have $x \leq_f y$ if $f(u)$ is positive.

Let x be any feasible solution with the largest possible number of tight constraints. Let B be the minimal boundary flat of x if x is constrained; otherwise let $B = \mathbb{R}^n$. Obviously if a constraint (f_i, b_i) is tight for x, then f_i is constant on B. We claim that f_i is constant on B for all nontight constraints as well. Nonconstancy of such an f_i on B means that the hyperplane H_i associated with (f_i, b_i) intersects B. Let $S \subseteq I$ be the set of indices i such that (f_i, b_i) is not tight for x but f_i is nonconstant on B. We shall derive a contradiction from $S \neq \emptyset$. Take any $j \in S$ and take an element z of $H_j \cap B$. Let L be the line containing x and z. Of course $L \subseteq B$. Take one natural ordering $\leq$ of L. The set

$$P = \{i \in S : H_i \cap L \neq \emptyset\}$$

is not empty. For every $i \in P$, there is a $y_i \in L$ with

$$f_i(y_i) = b_i$$

and the set

$$\{t \in L : f_i(t) \leq b_i\}$$

is either the lower section $(\leftarrow, y_i]$ in $(L, \leq)$ or the upper section $[y_i, \rightarrow)$. Let

$$Y = \{y_i : i \in P\}$$

If $[x, \rightarrow) \cap Y \neq \emptyset$, then let y be its minimum in the chosen order on L, else let y be the maximum of $(\leftarrow, x] \cap Y$. Every vector in the interval of $(L, \leq)$ generated by $\{x, y\}$ is a feasible solution, and y has all the tight constraints that x has, plus at least one more: contradiction with the definition of x. As claimed, each nontight (and tight) constraint function f_i must be constant on B. Since $x \in B$ is feasible, all vectors in B are feasible. Of course B is the minimal boundary flat of all of its elements.

Suppose in addition that x is an optimal solution. We claim that the objective function g too is constant on the minimal boundary flat B of x. Consider the "optimal hyperplane"

$$G = \{t \in \mathbb{R}^n : g(t) = g(x)\}$$

If $B \not\subseteq G$, then B would intersect every hyperplane parallel to G, i.e., $t \in B$ would exist with arbitrarily high values $g(t)$. All $t \in B$ being feasible, this contradicts the optimality of x. Thus $B \subseteq G$ and g is constant with value $g(x)$ on B, i.e., all vectors in B are optimal:

If an LP problem has a feasible solution, then every vector in some minimal boundary flat is a feasible solution. If it has an optimal solution then every vector in some minimal boundary flat is an optimal solution.

This means that since boundary flats can be described by simultaneous linear equations $f_i(x) = b_i$, where the (f_i, b_i) are some of the constraints, the feasibility of an LP problem can be determined by Gaussian elimination. Furthermore, if we know that an optimal solution exists, one can be found by taking an arbitrary vector in each minimal boundary flat: the one with the highest objective function value is an optimal solution. But how can we detect a situation where feasible solutions exist but none is optimal?

Either the objective function g is constant on each minimal boundary flat or some minimal boundary flat intersects every class of the equivalence induced on R^n by g. (These classes are parallel hyperplanes.) Whichever is the case, take an arbitrary representative vector in each minimal boundary flat: we have a finite set S of representatives. Take a real number b greater than all $g(x)$, $x \in S$. Add to the original LP problem the new constraint $(-g, -b)$: this new LP problem is feasible if and only if the original one is feasible without optimal solution.

Identifying all minimal boundary flats and a representative in each by Gaussian elimination is a rather obvious computational procedure, sure but rather slow. Better procedures have been known ever since linear programming was introduced to model problems of military logistics at the end of World War II, and progress is still being made both in theoretical understanding and computational efficiency. However, the reader should realize that some very basic principles of affine geometry are responsible for the fundamental fact that the solution of LP problems is amenable to algorithmic computation.

EXERCISES

1. Verify that the set of feasible solutions of an LP problem $(g, (f_i, b_i) : i \in I)$ is convex in R^n. Show that if for every $J \subseteq I$ with $\operatorname{Card} J \leq n + 1$ the LP problem $(g, (f_i, b_i) : i \in J)$ has a feasible solution, then the original problem has a feasible solution as well.

2. On a field F let $\leq$ be a linear order compatible with the additive group structure of F and such that the set of $x > 0$ forms a multiplicative subgroup of F^*. Develop a rudimentary theory of linear programming without real numbers. For $F = \mathbb{Q}$, design a computer program to solve LP problems.

3. Formulate PERT/CPM in LP language.

4. Can you describe some connection between convexity in $\mathbb{R}^n$ and natural order convexity on affine lines?

5. TIME AND SPEED IN SPECIAL RELATIVITY

Here is an axiomatic view of the where-and-when physics of the Einstein railroad.

The *spacetime universe* is defined as the set (vector space) $\mathbb{R}^2$: its elements are called *worldpoints*, or *events*. Choose any positive real number c: it may be called *lightspeed*. In $\mathbb{R}^2$ consider the two one-dimensional subspaces L and L^- generated respectively by $(c,1)$ and $(-c,1)$: any line of the affine geometry Aff $\mathbb{R}^2$ parallel to either one of L or L^- is called a *photon*. We define *optical causality* as the binary relation O on the spacetime universe in which $(x,t)O(x',t')$ means that $t \leq t'$ and some photon contains both worldpoints (x,t) and (x',t'). Optical causality is reflexive, antisymmetric, and acyclic, but it is not transitive. However, its restriction to any line D of Aff $\mathbb{R}^2$ is an order: a chain coinciding with one of the two natural orders of D if D is a photon and an antichain otherwise. Indeed a subset $D \subseteq \mathbb{R}^2$ is a photon if and only if D is a maximal set of worldpoints on which optical causality is a total order.

Causality is the order relation C defined on the spacetime universe as the transitive closure of optical causality. *Material causality* is the binary relation

$$M = (C \setminus O) \cup I$$

where I is the identity relation on $\mathbb{R}^2$. With respect to the basis

$$\{(c,1), (-c,1)\}$$

of $\mathbb{R}^2$, every worldpoint w has a unique expression

$$w = \alpha \cdot (c,1) + \beta \cdot (-c,1)$$

In the order C the section $[(00), \rightarrow)$ consists of those worldpoints w whose coefficients α, β are nonnegative. For $u, v \in \mathbb{R}^2$,

$$uCv \quad \text{if and only if} \quad v - u \in [(00), \rightarrow)$$

Let now S be the set of worldpoints consisting of the origin (00) and of those w with both coefficients α, β positive. It is easy to see that

$$(00)Mv \quad \text{if and only if} \quad v \in S$$

and in general

$$uMv \quad \text{if and only if} \quad v - u \in S$$

It follows that material causality is an order relation.

Every photon is an order-convex chain in the causality order C. Indeed a subset $D \subseteq \mathbb{R}^2$ is a photon if and only if D is a maximal order-convex chain in the causality order. For any two distinct events u, v we have uMv if and only if uCv and the segment $[u, v]$ is not a chain in the causality order C. Thus material causality can be defined directly from the causality order C, without explicit reference to optical causality. Finally, optical causality can also be derived from material causality: for u, v distinct, uOv holds if and only if $u \| v$ in M but

$$[u, \rightarrow) \supseteq [v, \rightarrow) \setminus \{v\}$$

where the upper sections refer to the material causality order M. The reader can now conclude that the three relations O, C, and M have the same automorphisms,

$$\operatorname{Aut} O = \operatorname{Aut} C = \operatorname{Aut} M$$

Any bijection $r : \mathbb{R}^2 \to \mathbb{R}^2$ is called a *reference system*. For any worldpoint (event) w, if

$$r(w) = (x, t)$$

then w is said to have, or to *occur at*, *location* x and *time* t in the reference system r. The *trace of the reference system* r is the set of worldpoints having location 0 in r. Throughout this section we shall use T to denote the set

$$\{(0, t) : t \in \mathbb{R}\}$$

The trace of any reference system r is thus the set $r^{-1}[T]$.

An *inertial line* is a line E in Aff $\mathbb{R}^2$ such that not all (x,t) in E have the same value t. In this case for every $t \in \mathbb{R}$ there is a unique $x(t) \in \mathbb{R}$ such that
$$(x(t), t) \in E$$
The difference $x(1) - x(0)$ is called the *slope* of E. For real numbers $t_1 < t_2$, the quotient $[x(t_2) - x(t_1)]/(t_2 - t_1)$ equals the slope. Photons are precisely the inertial lines with slope c or $-c$. Given a reference system r, a subset D of $\mathbb{R}^2$ whose image $r[D]$ is an inertial line is called an *inertial motion* in r. The slope of $r[D]$ is then called the *velocity* of D in r. The absolute value of the velocity is called *speed*. The trace of any reference system r is an inertial motion of speed and velocity 0 in the system r itself.

For any given $u \in \mathbb{R}^2$ and positive real numbers a, b we can define a map $r : \mathbb{R}^2 \to \mathbb{R}^2$ as follows. Recall that every worldpoint w has a unique expression
$$w = \alpha \cdot (c, 1) + \beta \cdot (-c, 1)$$
with $\alpha, \beta \in \mathbb{R}$. Define the map r by
$$r(w) = u + a \cdot \alpha \cdot (c, 1) + b \cdot \beta \cdot (-c, 1)$$
Obviously r is a reference system and it is an automorphism of each of the relations O, C, and M. We call r an *optical reference system*. Observe that a subset D of $\mathbb{R}^2$ is an affine line if and only if $r[D]$ is one. In particular, the trace of r is an affine line. Also, in any optical reference system r, photons are precisely the inertial motions of speed c.

Each affine line D in $\mathbb{R}^2$ is either a chain in the material causality order M or it is a material causality antichain: it is a chain if and only if it is an inertial line with slope strictly between $-c$ and c. In this case the material causality order coincides with one of the two natural orders on D.

Proposition 3 *For any affine line D in $\mathbb{R}^2$ the following conditions are equivalent:*

(i) *D is a chain in the order of material causality,*
(ii) *D is the trace of some optical reference system,*
(iii) *in some optical reference system, D is an inertial motion of speed less than c,*

(iv) *in all optical reference systems, D is an inertial motion of speed less than c.*

Proof. The equivalence of (i), (ii), and (iv) is clear, in view of the foregoing remarks.

Condition (ii) means that for some optical reference system r,

$$r[D] = \{(0,t) : t \in R\}$$

In this case obviously D is an inertial motion of velocity 0 in r, implying (iii).

Assume (i). We shall prove (ii). Choose two distinct elements u, v of D such that uMv. We have $(0,0)M(v-u)$ and

$$v - u = a \cdot (c,1) + b \cdot (-c,1)$$

for some positive real numbers a, b. Let $p : \mathbb{R}^2 \to \mathbb{R}^2$ be the map defined for each worldpoint

$$w = \alpha \cdot (c,1) + \beta \cdot (-c,1)$$

by

$$p(w) = (\alpha/a) \cdot (c,1) + (\beta/b) \cdot (-c,1)$$

Note that p is linear. Let $r : \mathbb{R}^2 \to \mathbb{R}^2$ be defined by

$$r(w) = -p(u) + p(w)$$

Then r is an optical reference system and its trace is D. □

If a line D satisfies the conditions of Proposition 3, then D is called a *material motion*. See Figure 6.3. (Here the zero-slope line T is drawn vertically.) Of course a material motion D may have different velocities in different optical reference systems.

Optical reference systems form a permutation group on $\mathbb{R}^2$ that we take the liberty of calling the *Lorentz group* $\mathcal{L}_2$. This group $\mathcal{L}_2$ is a subgroup of $\operatorname{Aut} O$. Within $\mathcal{L}_2$ those reference systems whose trace is the line

$$T = \{(0,t) : t \in R\}$$

i.e., $r[T] = T$, constitute a subgroup S_T of $\mathcal{L}_2$. The reference systems r in S_T have the rare property that $(c,1)$ and $(-c,1)$ occur at the

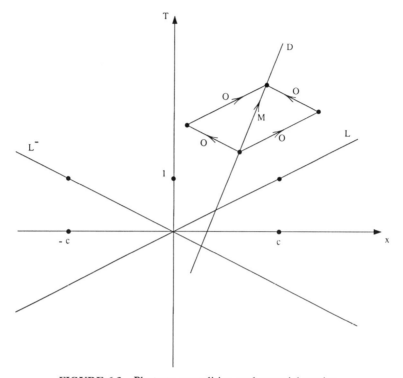

FIGURE 6.3 Photons, causalities, and material motion.

same time in r. For every $t \in \mathbb{R}$, the *time shift* $s_t : \mathbb{R}^2 \to \mathbb{R}^2$ given by

$$s_t(w) = (0,t) + w$$

is in S_T. For every positive real q, the *dilatation* d_q given by

$$d_q(w) = q \cdot w$$

is in S_T. Note that $r, s \in \mathcal{L}_2$ have the same trace if and only if

$$rs^{-1} \in S_T$$

Thus there are many different optical reference systems that have the same trace: for every $r \in \mathcal{L}_2$ and $g \in S_T$, the composition gr and r have the same trace.

Let $r \in \mathcal{L}_2$. Then $h : \mathbb{R}^2 \to \mathbb{R}^2$, defined by

$$h(w) = r(w) - r(0,0) \qquad \text{for all} \quad w \in \mathbb{R}^2$$

is linear, $h \in \mathcal{L}_2$, and

$$h(c,1) = (ca,a) \quad \text{and} \quad h(-c,1) = (-cb,b)$$

for some positive real numbers a, b. If m is the 2×2 matrix

$$\begin{pmatrix} (a+b)/2 & c(a-b)/2 \\ (a-b)/2c & (a+b)/2 \end{pmatrix}$$

and if h_m is the corresponding linear map $\mathbb{R}^2 \to \mathbb{R}^2$, then it is easy to verify that

$$h_m(c,1) = h(c,1) \quad \text{and} \quad h_m(-c,1) = h(-c,1)$$

and consequently $h_m = h$. (The reader can probably find out how m was actually discovered.) Let us abbreviate

$$(a+b)/2 = k, \quad (a-b)/2 = l$$

The matrix m can be rewritten as

$$\begin{pmatrix} k & lc \\ l/c & k \end{pmatrix}$$

Observe that $k > 0$ and $k^2 > l^2$. It is then easy to verify that

$$h(lc/(l^2 - k^2), k/(k^2 - l^2)) = (0,1)$$

which implies that the common slope of $h^{-1}[T]$ and $r^{-1}[T]$ is

$$[lc/(l^2 - k^2)]/[k/(k^2 - l^2)] = -lc/k$$

Let ρ denote this slope.

Consider now any material motion D and let δ be its slope. The subspace E of $\mathbb{R}^2$ generated by $(\delta, 1)$ is parallel to D, and the images

$$h[E], \quad h[D], \quad r[E], \quad r[D]$$

are all parallel. Their common slope is easily determined from the vector

$$h(\delta, 1) \in h[E]$$

This vector is obviously

$$(k\delta + lc, l\delta/c + k) = (k(\delta - \rho), k - k\delta\rho/c^2)$$

and thus the common slope of $h[E]$, $r[D]$, etc., is $(\delta - \rho)/(1 - \delta\rho/c^2)$:

Proposition 4 *Let r be an optical reference system and D a material motion with slope δ. If the trace of r has slope ρ, then the image line $r[D]$ has slope*

$$(\delta - \rho)/(1 - \delta\rho/c^2)$$

Let now r and s be two optical reference systems and D any material motion. Then $rs^{-1} \in \mathcal{L}_2$ and $s[D]$ is an inertial line. The slope of the trace of rs^{-1} is by definition the slope of

$$(sr^{-1})[T] = s[r^{-1}[T]]$$

It coincides with the velocity of the trace of r in the system s: let us write v_{rs} to denote this relative velocity. The slope of $s[D]$ is by definition the velocity of D in the system s; let us denote it by v_s. Proposition 4, applied to the optical reference system rs^{-1} and the inertial line $s[D]$, says that

$$(rs^{-1})[s[D]] = r[D]$$

has slope

$$(v_s - v_{rs})/(1 - v_s v_{rs}/c^2)$$

But this slope is the velocity of D in r—let us denote it by v_r:

Velocity Composition Theorem. *Let r and s be two optical reference systems and D any material motion. If the velocities of D in r and s are v_r and v_s, respectively, and if v_{rs} is the velocity of the trace of r in the system s, then*

$$v_r = (v_s - v_{rs})/(1 - v_s v_{rs}/c^2)$$

Experimental Remark. If standing by the rail track you see a train D going at $v_s = 100$ km/hr and you see another, slow train following it at $v_{rs} = 60$ km/hr, then the formula shows that a physicist on the slow train will measure the velocity v_r of D as somewhat more than 40 km/hr. Accurate instruments to confirm such facts are only available since late in the last century, and to most poorly instrumented passengers of the slow train, the relative velocity of the fast train still seems to be not more than but equal to 40 km/hr. To them the Earth seems to be flat too, and these false perceptions cause admittedly very little inconvenience.

EXERCISES

1. Is $\mathcal{L}_2$ a proper subgroup of $\mathrm{Aut}\,O$?
2. What can you say about the structure of the Lorentz group?
3. Can you develop a rudimentary theory of velocities without real numbers, using $\mathbb{Q}$ instead of $\mathbb{R}$?

BIBLIOGRAPHY

Emil ARTIN, *Geometric Algebra*. John Wiley & Sons 1988. A short and enlightening volume on algebra and geometry—affine, projective, and more.

Reinhold BAER, *Linear Algebra and Projective Geometry*. Academic Press 1952. A classical reading on the "essential structural identity of projective geometry and linear algebra." Includes the proof of some well-known, yet not easily found fundamental results.

Garrett BIRKHOFF and Saunders MAC LANE, *Algebra*. Macmillan 1979. In this classical undergraduate text the discussion of vector spaces over fields is preceded by that of modules over rings and followed by multilinear algebra, which provides the proper setting for determinants.

Vašek CHVÁTAL, *Linear Programming*. Freeman 1983. Covers many aspects of the subject and presupposes only a good mastery of high school mathematics. Key concepts are introduced by examples followed by formal definitions. Simplex algorithms are a central subject, and the more recent ellipsoid method is covered in an appendix. Applications range from Polly's diet problem to industrial optimization.

Ludwig DANZER, Branko GRÜNBAUM, and Victor KLEE, Helly's theorem and its relatives. *Proc. Symp. Pure Math. (Amer. Math. Soc.)* 7, 1963, pp. 101–118. A classical survey article on the combinatorial aspects of convexity.

Richard A. DEAN, *Classical Abstract Algebra*. Harper & Row 1990. Section 10.1 offers a concise and clear review of the basic facts about vector spaces.

Robert GOLDBLATT, *Orthogonality and Spacetime Geometry*. Springer 1987. An essentially combinatorial orthogonality relation is shown to be at the basis of special relativity.

P. R. HALMOS, *Finite Dimensional Vector Spaces*. Springer 1974. Bases, linear maps, and matrices, and an entire chapter devoted to orthogonality. Also included are some multilinear algebra and determinants and a short chapter on analysis. The book is accessible at the undergraduate level.

A. HEYTING, *Axiomatic Projective Geometry*. North-Holland 1980. The theorems of Desargues and Pappos are derived from simple combinatorial postulates rather than using linear algebra. Logical implications among variants of these theorems are investigated. The last chapter is devoted to the concept of order.

Daniel R. HUGHES and Fred C. PIPER, *Projective Planes*. Springer 1973. The combinatorial theory of the point-line containment relation in projective planes is properly set in an algebraic framework that generalizes fields and vector spaces. A detailed graduate text with few formal prerequisites.

Havard KARLOFF, *Linear Programming*. Birkhäuser 1991. Designed for advanced undergraduate or graduate students, this short volume includes a discussion of the simplex and ellipsoid methods and of Karmarkar's algorithm.

Serge LANG, *Linear Algebra*. Addison-Wesley 1971. A geometrically motivated introduction to vector spaces, accessible to undergraduates.

CHAPTER VII

GRAPHS

1. TREES AND MEDIAN GRAPHS

Let V be any set and let E be any set of two-element subsets of V. Then $G = (V, E)$ is called a *graph*. The elements of V are called *vertices*, those of E *edges*. Any two vertices forming an edge are called *adjacent*.

Example. Let $V = \{1, 2, 3, 4\}$, $E = \{\{1, 2\}, \{2, 3\}, \{3, 4\}, \{4, 1\}\}$. The graph $G = (V, E)$ is "graphically" displayed in Figure 7.1.

If E consists of all non-singleton pairs of vertices, then (V, E) is called a *complete* graph. If $V = \emptyset$, then $E = \emptyset$ and (V, E) is called the *empty graph*. If V is finite, then (V, E) is called a *finite graph*.

For any graph the binary *adjacency relation* R on V is defined by xRy if and only if x and y are adjacent. This is an irreflexive and symmetric relation. Every irreflexive and symmetric binary relation R on a set V is the adjacency relation of a graph, namely whose edges are the pairs $\{x, y\}$ with xRy. There is some truth to the view that the theory of graphs coincides with the theory of binary relations.

Paths and cycles were defined for binary relations in Chapter II. Corresponding concepts in graph theory are of crucial importance. If P is a relation-theoretical path in the adjacency relation of a graph

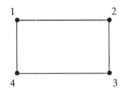

FIGURE 7.1 A graph on four vertices with four edges.

$G = (V, E)$, from vertex a to b, then the edge set

$$\{\{x,y\} : (x,y) \in P\}$$

is called a *path in the graph G from a to b*, or simply *between a and b*. Similarly, if C is a cycle in the adjacency relation, then

$$\{\{x,y\} : (x,y) \in C\}$$

is a *cycle in the graph G*.

Historical Comment. A graph-theoretical argument was first used by Euler in 1735, in his "Solutio problematis ad geometriam situs pertinentis." It was an argument involving paths and cycles.

The number of edges in a path, or a cycle, is called the *length* of the path, or the cycle. The *vertices on the path*, or *on the cycle*, are those belonging to the edges of the path, or cycle. If between two given vertices a and b a path exists (which is by no means guaranteed), then those with the least possible length are called *geodesics* between a and b. The *distance* of a and b, denoted by $d(a,b)$, is then defined as the length of such a geodesic, or 0 if $a = b$.

A graph is *connected* if between any two distinct vertices there is at least one path. The distance is then always defined, and for any $a, b, c \in V$ we have the *triangle inequality*

$$d(a,b) + d(b,c) \geq d(a,c)$$

(It is a simple but worthwhile exercise to verify this rigorously, say using induction.) The *vertex interval* between a and c, denoted by $I(a,c)$, is the set

$$\{b \in V : d(a,b) + d(b,c) = d(a,c)\}$$

We say that $K \subseteq V$ is a *convex set of vertices* if $a, c \in K$ implies $I(a,c) \subseteq K$. Convex vertex sets constitute an algebraic closure system on V.

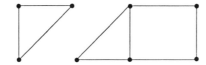

FIGURE 7.2 A graph with two connected components.

The graph $G' = (V', E')$ is a *subgraph* of $G = (V, E)$ if

$$V' \subseteq V \quad \text{and} \quad E' \subseteq E$$

If $G' \neq G$, then G' is a *proper subgraph* of G. If $V' = V$, then G' is called a *spanning subgraph* of G. If

$$E' = \{\{x, y\} \in E : x, y \in V'\}$$

then G' is called a *full* or *induced subgraph* of G (induced by the set V'). For any graph G, the preorder closure of the adjacency relation is an equivalence relation on the set of vertices. The subgraph of G induced by each equivalence class is connected, and it is called a *connected component* of G. The graph of Figure 7.2 has two connected components. (As a simple exercise in graphical observation, the reader may wish to verify also that this graph has 8 vertices, 9 edges, $6 + 6$ paths of length 2, $6 + 4$ paths of length 3, 7 paths of length 4, no paths longer than that, 4 cycles, and maximal vertex distance 2.)

A connected graph without a cycle is called a *tree*. A graph of any kind without a cycle is called a *forest*. A graph is a forest if and only if its connected components are trees. Let a be any vertex of a graph G: the graph G is a tree if and only if for every vertex $b \neq a$ there is a unique path from a to b. In a tree, a set of vertices is convex if and only if it induces a subgraph that is connected, i.e., a subgraph that is again a tree. Given any path P in a tree, from a to b, the vertices on P are precisely the members of the interval $I(a, b)$.

Observe that the edges of a graph $G = (V, E)$ are particular vectors in the space $\mathcal{P}(V)$ over $\mathbb{Z}_2$. The cycles of G are precisely those circuits of $\mathcal{P}(V)$ that consist of graph edges. Therefore G is a forest if and only if its edge set E is linearly independent in the vector space $\mathcal{P}(V)$. In any case, let $\mathcal{F}(G)$ be the set of forest subgraphs of G. This set is naturally ordered by letting $F \leq H$ mean, for $F, H \in \mathcal{F}(G)$, that F is a subgraph of H. By the second statement of the Basis Equipotence Theorem (Chapter VI) all maximal forests of

G have the same cardinal number of edges. Assume now that G is connected. A maximal forest $F \in \mathcal{F}(G)$ is obviously a spanning subgraph. Could F fail to be connected? If so, then there are paths in G between vertices lying in different connected components of F. A shortest possible path of this kind can consist of no more than a single edge. By adding this edge to F we would obtain a larger forest F', with F as a proper subgraph of F', which is impossible. The reader can easily conclude that in any connected graph, maximal cycle-free subgraphs and spanning tree subgraphs coincide.

An *orientation* of a graph $G = (V, E)$ is an irreflexive and antisymmetric relation D on V such that

$$E = \{\{x, y\} : xDy\}$$

Orientations that are strongly acyclic are of interest because these generate, as covering relations, discrete orders on the vertex set V. Of course the covering relation of any order on a set V determines a graph on V with edge set

$$E = \{\{x, y\} : x \text{ is covered by } y\}$$

This graph (V, E) is called the *covering graph* of the order in question. Some graphs cannot arise as covering graphs because they possess no strongly acyclic orientation at all: consider any graph having a cycle of length 3. On the other hand, all orientations of a graph G are strongly acyclic if and only if all orientations are acyclic, and this is the case if and only if G is a forest.

Among all the orientations of a tree $T = (V, E)$ let us consider those that, as covering relations of a discrete order on V, correspond to an order with a minimum $u \in V$. For each vertex u of T there is exactly one such orientation D_u: we have $x D_u y$ for an edge $\{x, y\}$ if and only if

$$d(u, x) + 1 = d(u, y)$$

(Observe first that no two adjacent vertices of a tree can have the same distance from a third vertex.) The orientation D_u is referred to as a *directed tree* with *basepoint* u. The corresponding order (of which D_u is the covering relation) is called the *basepoint order* $\leq_u$. If P is any path in T between vertices a and b, then the vertex interval $I(a, b)$ consists precisely of the vertices lying on P. Further, $I(u, a)$

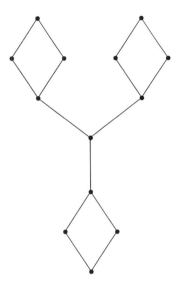

FIGURE 7.3 A median graph.

coincides with the order interval

$$[u,a] = (\leftarrow, a] \quad \text{in} \quad \leq_u$$

and $I(u,b)$ is $[u,b] = (\leftarrow, b]$ in this basepoint order. Observe that any order segment in $\leq_u$ is a finite chain. It follows that $I(u,a) \cap I(u,b)$ is a finite chain: let m be its maximum. Then $m \in I(a,b)$, indeed

$$I(u,a) \cap I(u,b) \cap I(a,b) = \{m\}$$

The reader can draw two conclusions from this. First, every basepoint order on a tree is a lower semilattice. Second, if u, a, b are any vertices of a tree T, then the vertex intervals $I(u,a)$, $I(u,b)$, and $I(a,b)$ have a unique common point $m(u,a,b)$. If T is any connected graph, not necessarily a tree, that enjoys this second property, then T is called a *median graph* and $m(u,a,b)$ is called the *median* of the vertices u, a, b. Median graphs that are not trees do exist. For example, let $(\mathcal{P}(S), \subseteq)$ be the lattice of all subsets of any finite set S. Then the covering graph of this lattice is median. For subsets X, Y, Z of S,

$$m(X,Y,Z) = (X \cap Y) \cup (X \cap Z) \cup (Y \cap Z)$$

Another example is displayed in Figure 7.3.

Helly Theorem for Median Graphs. *On the vertex set of any median graph, the closure system formed by all convex sets has Helly number at most 2.*

Proof. Thanks to the Lemma for Helly Number 2 (Chapter II) we need only to show that in a median graph, if three convex vertex sets A, B, C intersect pairwise, then $A \cap B \cap C \neq \emptyset$. Let $a \in B \cap C$, $b \in A \cap C$, $c \in A \cap B$. Then

$$I(a,b) \subseteq C, \quad I(a,c) \subseteq B, \quad I(b,c) \subseteq A$$

Since the intersection of these three vertex intervals is nonvoid and contained in $A \cap B \cap C$, this latter intersection cannot be void either. □

EXERCISES

Develop computer programs to perform the following tasks for finite graphs:

1. Find the connected components.
2. Find a shortest path between any two distinct vertices of a connected graph.
3. Construct a spanning tree in a connected graph.
4. Decide if a given graph is median and, if so, find the median of any three vertices.
5. Decide if a given graph is a forest.
6. Find a maximum length cycle in a given graph that is not a forest.
7. Determine if the edges of a given graph can be partitioned into disjoint cycles and exhibit such a partition if it exists.
8. Determine if a given graph is a covering graph (of some order).

2. GAMES

Games with two opponents, a finite number of conceivable situations, alternating (rather than simultaneous) decision making by the players, and a finite number of moves available to each can be quite well described by trees. Chess is such a game, and the model can be adapted as well to multiplayer chance-dependent games such as poker. Games of speculative gain, pathfinding, and strategic conquest have been played for ages in various social settings. Their symbolism has fascinated even the true professionals of military and economic pursuit, oriental kings, and gold-diggers of the West. The twentieth century mathematical theory of games too was developed in a military and economic context—witness the book *The Theory of Games and Economic Behaviour* by Morgenstern and von Neumann. Here we limit our exposition to what is usually called deterministic two-person zero-sum games in extensive form.

Let $T = (V, E)$ be a tree on a finite set of vertices, called *positions*, and let D be a basepoint orientation of T, where the basepoint u is called *initial position*. The couples (x, y) in D are called *moves*. The maximal vertices of the basepoint order are called *terminal positions*. Let V_1 and V_2 be two complementary subsets of V, $V_1 \cup V_2 = V$, $V_1 \cap V_2 = \emptyset$. For $i = 1, 2$, let

$$D_i = \{(x, y) \in D : x \in V_i\}$$

Clearly $D_1 \cup D_2 = D$, $D_1 \cap D_2 = \emptyset$. Positions in V_i and moves in D_i are said to belong to *player i*. Let R be a linearly ordered set, called the set of *outcomes*. (For example, the outcomes may be real numbers with the standard order, each such outcome representing a payment due to player 1 by player 2. However, we shall not need any arithmetic property of the outcomes.) Let p be a function associating to each terminal position v an outcome $p(v)$: we call this a *payoff function*. A *game* is formally defined as

$$(D, V_1, V_2, R, p)$$

Note that the underlying *game tree* T is implicit, but it is fully determined by these data. A *scenario* is defined as a relation-theoretical path in the directed tree D from the initial position u to some terminal position v: then $p(v)$ is called the *outcome* of the scenario. [For a trivial game with a single position u, the empty set is defined to be a scenario with outcome $p(u)$.] For given i (1 or 2) a relation σ_i contained in D_i such that for every nonterminal $x \in V_i$ there is a unique $y \in V$ with $(x, y) \in \sigma_i$ is called a *strategy of player i*. A scenario S is

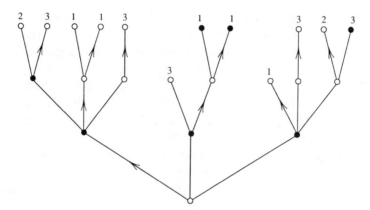

FIGURE 7.4 A strategic equilibrium.

said to be *compatible* with σ_i if

$$S \cap D_i \subseteq \sigma_i$$

It is an excellent exercise in graph theory to verify that given strategies σ_1 and σ_2 of the two players, there is a unique scenario $S(\sigma_1 \sigma_2)$ compatible with σ_1 and σ_2. The outcome of this scenario is called the *result* of the *strategic couple* $(\sigma_1 \sigma_2)$, and it is denoted by $r(\sigma_1 \sigma_2)$. It is easy to see that if S is any scenario compatible with a strategy σ_i of player i, then there is a strategy σ_j of the other player j with $S = S(\sigma_1 \sigma_2)$. A strategic *equilibrium* is a strategic couple $(\sigma_1 \sigma_2)$ such that for every other strategy σ'_1 of player 1,

$$r(\sigma'_1 \sigma_2) \leq r(\sigma_1 \sigma_2)$$

and for every other strategy σ'_2 of player 2,

$$r(\sigma_1 \sigma_2) \leq r(\sigma_1 \sigma'_2)$$

The interpretation is that if player 1 wants to maximize the outcome and player 2 wants to minimize it, then by choosing σ_i, player i can guarantee an outcome no worse for himself than $r(\sigma_1 \sigma_2)$, regardless of what the other player does. Figure 7.4 illustrates a strategic equilibrium with the result 1. There are $1 + 7 + 7$ positions belonging to player 1 ("white"), including the initial position. There are $3 + 1 + 3$ positions belonging to player 2 ("black"). There are $5 + 3 + 4$ terminal positions. There are $3 + 8 + 5 + 5$ possible moves and 3 possible outcomes. Moves marked with an arrow belong to a strategy of the equilibrium. At this game, it is easy to be a grandmaster.

Equilibrium Theorem. *Every game has a strategic equilibrium.*

Proof. Consider a game $\mathcal{G} = (D, V_1, V_2, R, p)$ with the notation introduced earlier. For every position v in $V = V_1 \cup V_2$ define the *residual game*
$$\mathcal{G}_v = (D', V_1', V_2', R', p')$$
as follows. The position set V' of $\mathcal{G}_v$ is the section $[v, \rightarrow)$ in the basepoint order of V (based on the initial position u). Let D' be the restriction of D to V': this is precisely the basepoint orientation, with basepoint v, of the subgraph T' of the game tree T induced by V'. (The basepoint order $\leq_v$ on V' also coincides with the restriction to V' of the basepoint order $\leq_u$ of V.) Let
$$V_1' = V_1 \cap V', \qquad V_2' = V_2 \cap V', \qquad R' = R$$
and let p' be the restriction of p to the terminal positions in V'. This fully defines $\mathcal{G}_v$. Clearly $\mathcal{G}_u = \mathcal{G}$.

We shall prove, by induction, that every residual game of $\mathcal{G}$ has an equilibrium, and this will apply to $\mathcal{G}_u = \mathcal{G}$. Suppose the claim is false: let $\mathcal{G}_v$ be a residual game without an equilibrium and having as few positions as possible. Trivially, v cannot be a terminal position of $\mathcal{G}$. Let S be the set of those positions that cover v in the basepoint order $\leq_u$,
$$S = \{w \in V : (v, w) \in D\}$$
For every $w \in S$, $\mathcal{G}_w$ has an equilibrium since it has fewer positions than $\mathcal{G}_v$: let r_w be the result of such an equilibrium (σ_1^w, σ_2^w). Observe that if y is another element of S, then $\sigma_i^w \cap \sigma_i^y = \emptyset$ for $i = 1, 2$, because the games $\mathcal{G}_w$ and $\mathcal{G}_y$ have no moves or positions in common. Now v belongs either to player 1 or to player 2. In the first case choose $y \in S$ with the largest possible value r_y and let
$$\sigma_1 = \{(v, y)\} \cup \bigcup_{w \in S} \sigma_1^w, \qquad \sigma_2 = \bigcup_{w \in S} \sigma_2^w$$
In the second case choose $y \in S$ with the smallest possible value r_y and let
$$\sigma_1 = \bigcup_{w \in S} \sigma_1^w, \qquad \sigma_2 = \{(v, y)\} \cup \bigcup_{w \in S} \sigma_2^w$$

In both cases, $(\sigma_1 \sigma_2)$ is an equilibrium of $\mathcal{G}_v$, refuting the assumption that none exists. □

A game may well have several equilibria. However, if $(\sigma_1 \sigma_2)$ and $(\tau_1 \tau_2)$ are two equilibria, then

$$r(\sigma_1 \sigma_2) \leq r(\sigma_1 \tau_2) \leq r(\tau_1 \tau_2) \leq r(\tau_1 \sigma_2) \leq r(\sigma_1 \sigma_2)$$

and therefore all strategic equilibria yield the same result.

This does apply to chess. Players 1 and 2 are called *white* and *black*. A position is a map from a subset of $2^{8 \cdot 8}$ (the "chessboard") to a finite set of pieces, together with a repetition indicator of first, second, or third occurrence. (This ensures finiteness. In particular all chessboard maps on the third occurrence are terminal.) There are three outcomes in R: (1) *black wins*, (2) *draw*, and (3) *white wins*, ordered as follows:

$$(1) < (2) < (3)$$

Game theory tells us that of the three possible outcomes of the chess game, one and only one can be secured by a strategic equilibrium. Which one then? The answer can be obtained by straightforward computation, and the knowledge of equilibrium strategies would be highly prized, by serious chess-players at least. Yet the solution is not at hand; present-day computers are still not powerful enough to carry out the computation. Conceptually straightforward even though technically beyond reach: is that what the mathematically inclined Edgar Poe and the esthetically minded Hardy called, in turn, the frivolity of chess, trivial and unimportant mathematics, complexity mistaken for what is profound? Or did they miss the point and something else remains to be understood?

EXERCISES

1. Simplify the rules of chess to the point where you can settle the question of whether either black or white has a decisive initial advantage. You may reduce the number of pieces and the number of squares, impose maximum length-of-game rules, etc.

2. Can you invent a reasonably interesting chesslike game in which neither black nor white has an initial advantage?

3. CHROMATIC POLYNOMIALS

Let $G = (V, E)$ and $G' = (V', E')$ be two graphs. A function h from V to V' such that $h[e] \in E'$ for every $e \in E$ is called a graph *homomorphism* from G to G'. In shorthand we write $h : G \to G'$.

Proposition 1 *The composition of two graph homomorphisms*

$$h : G \to G' \quad \text{and} \quad g : G' \to G''$$

is a homomorphism from G to G''. For every graph $G = (V, E)$ the identity mapping on V is a homomorphism from G to itself.

If h is bijective and its inverse is a homomorphism as well, then h is called an *isomorphism*. The graphs G and G' are *isomorphic* if such an isomorphism exists.

Let us denote by K_V the complete graph on a vertex set V. If n is the cardinality of V, then K_V is isomorphic to K_n. Any homomorphism from a graph G to K_n is called an *n-coloring* of G (or *coloring with n colors*). The term comes from thinking about assigning "colors" to the vertices in such a way that no two adjacent vertices get the same color. If $m \leq n$, then every m-coloring is also an n-coloring. If G has n vertices, then obviously it has an n-coloring. The least cardinal m such that G has an m-coloring is called the *chromatic number* of G. Only the graph with empty vertex set has a 0-coloring, and only graphs with no edges have a 1-coloring.

Proposition 2 *A graph G has a 2-coloring if and only if G has no cycle of odd length.*

Proof. Let E_c be the set of edges in an odd-length cycle, say of length $n = 2k + 1$. Let $V_c = \cup E_c$. Any 2-coloring of G, restricted to V_c, would yield a 2-coloring of the subgraph $G_c = (V_c, E_c)$. The graph G_c is isomorphic to the graph Z_n on vertex set $\mathbb{Z}_n$ whose edges are the pairs $\{z, z+1\}$, $z \in \mathbb{Z}_n$. Also, observe that K_2 is isomorphic to the similarly defined graph Z_2 on $\mathbb{Z}_2$. Every graph homomorphism $h : Z_n \to Z_2$ must obey the rule

$$h(z+1) = h(z) + 1 \quad \text{for all} \quad z \in \mathbb{Z}_n$$

We leave it to the reader to verify that no such function h can exist. Thus a graph with a 2-coloring can have no odd cycles.

Conversely, assume a graph $G = (V, E)$ has no odd-length cycle. Choose one vertex in each connected component of G. For each $x \in V$ there is precisely one such chosen vertex $c(x)$ that is in the same component as x. Define the 2-coloring $h : V \to \mathbb{Z}_2$ by letting $h(x)$ be the congruence class mod $2\mathbb{Z}$ of the distance $d(x, c(x))$. □

Usually a graph with chromatic number 2 is called *bipartite*: its vertex set is the union of two disjoint sets V_1, V_2 such that no two vertices within the same part V_i are adjacent.

The *chromatic function* of a finite graph G is the function P_G associating with each natural number n the number of n-colorings of G. Thus

$$P_G : \omega \to \omega$$

If $G = (V, E)$ has m vertices and no edges, then every map $h : V \to n$ is an n-coloring. Therefore we have, for such edgeless graphs,

$$P_G(n) = n^m \quad \text{for all} \quad n \in \omega$$

This suggests that the behavior of P_G in general may be described by some polynomial depending on G.

For any finite graph $G = (V, E)$ and $e \in E$ we can form two new graphs from G. The *edge-deleted graph* $G - e$ has the same vertices as G, and its edges are those of G except that e is removed. The *edge-contracted graph* $G \cdot e$ has vertex set $(V \backslash e) \cup \{\bar{e}\}$ where $\bar{e}$ is any element not belonging to $V \backslash \{e\}$, and the edge set

$$\{d \in E : d \cap e = \emptyset\} \cup \{\{v, \bar{e}\} : v \in V \backslash e, \{v, x\} \in E \text{ for some } x \in e\}$$

The set H of n-colorings of G is a proper subset of the set H_e of n-colorings of $G - e$. We claim that there is a bijection from $H_e \backslash H$ to the set of n-colorings of $G \cdot e$. Every $h \in H_e \backslash H$ has the same value on the two vertices x, y belonging to e. The desired bijection is established by associating with each $h \in H_e \backslash H$ the map $g : (V \backslash e) \cup \{\bar{e}\} \to n$ such that

$$g|(V \backslash e) = h|(V \backslash e) \quad \text{and} \quad g(\bar{e}) = h(x) = h(y)$$

The maps g so obtained are precisely the n-colorings of $G \cdot e$. The equipotence of $H_e \backslash H$ with the set of n-colorings of $G \cdot e$ means that we have the *edge recursion formula*

$$P_G(n) = P_{G-e}(n) - P_{G \cdot e}(n)$$

Chromatic Polynomial Theorem. *The chromatic function of a finite graph $G = (V, E)$ is the restriction to ω of a unique polynomial function on the ring $\mathbb{Z}$, defined by a unique polynomial $p \in \mathbb{Z}[X]$ of degree $n = \text{Card}\, V$.*

Proof. By induction on the number of edges. For $E = \emptyset$ we have $p = X^n$. If $E \neq \emptyset$, let $e \in E$. Assume that the result is true for the graphs $G - e$ and $G \cdot e$, both of which have less edges than G. Let the polynomials p_d and p_c of degree n and $n - 1$ correspond to the chromatic functions of $G - e$ and $G \cdot e$. Then by the edge recursion formula the polynomial $p = p_d - p_c$ does the job for the chromatic function of G.

Uniqueness follows from the observation that if q were another such polynomial, then $p - q$ in $\mathbb{Z}[X] \subseteq \mathbb{Q}[X]$ would have an infinity of roots (namely all the natural numbers) and therefore $p - q$ would coincide with the zero polynomial, i.e., $p = q$. □

The polynomial $p \in \mathbb{Z}[X]$ defined by the above theorem is called the *chromatic polynomial* of G. The edge recursion formula for the chromatic function can be restated in terms of polynomials:

Recursion Formula (For Chromatic Polynomials). *Let $p(G)$ denote the chromatic polynomial of a finite graph G. If e is any edge of G, then we have*

$$p(G) = p(G - e) - p(G \cdot e)$$

Since no nonempty graph has a 0-coloring, 0 is a root of the chromatic polynomial, and X divides in $\mathbb{Z}[X]$ the chromatic polynomial $p(G)$ of every nonempty graph G. Assume in addition that G has k connected components $G_1, \ldots, G_k$. The reader should verify that $p(G)$ is equal to the product of the various $p(G_i)$,

$$p(G) = p(G_1) \cdots p(G_k)$$

It follows that for any graph with k components, X^k divides the chromatic polynomial $p(G)$. Using the Recursion Formula and the expression of the chromatic polynomial as the product of the chromatic polynomials of its components, one easily proves the following result.

Proposition 3 *The chromatic polynomial of a forest F with k tree components and m edges is $X^k(X-1)^m$.*

This immediately yields many nonisomorphic graphs with the same chromatic polynomial.

EXERCISES

1. Design computer programs for finite graphs,
    (a) to find out if a graph is bipartite,
    (b) to determine the chromatic number of a graph,
    (c) to calculate the coefficients of the chromatic polynomial.
2. (a) Verify that the set of isomorphisms from a graph $G = (V, E)$ to itself is a permutation group on V. Let us denote it by $\operatorname{Aut} G$. Does every permutation group arise this way?
    (b) Let H be a cyclic group. Construct a graph G such that $\operatorname{Aut} G$ is isomorphic to H. What if H is not cyclic?
3. Call a finite graph $G = (V, E)$ *planar* if there is an injection $p : V \to \mathbb{R}^2$ such that for $e, d \in E$ we have the following relationship among convex hulls in $\mathbb{R}^2$:
$$\overline{p[e]} \cap \overline{p[d]} = \overline{p[e \cap d]}$$
(The bar symbol on top of a set denotes its hull.) What can you prove or conjecture about planar graphs and their chromatic numbers?

BIBLIOGRAPHY

Claude BERGE, *Théorie générale des jeux à n personnes*. Gauthier-Villars 1957. An accessible exposition of n-person game theory, written in the spirit of the theory of graphs.

Claude BERGE, *Graphes et hypergraphes*. Dunod 1973. A classic text on graphs and the more general hypergraph concept. Hypergraphs differ from graphs in that edges can have cardinality different from 2 (some authors call them simply set systems).

Norman L. BIGGS, *Algebraic Graph Theory*. Cambridge University Press 1974. In addition to chromatic and Tutte polynomials, this research-level tract discusses several applications of linear algebra to graph theory. A good third of the volume is devoted to graph automorphisms.

J. A. BONDY and U. S. R. MURTY, *Graph Theory with Applications*. North-Holland 1979. This self-contained introduction to graph theory has a well-rounded chapter on planar graphs, including a limpid proof of Kuratowski's characterization theorem. Trees and vertex colorings are dealt with in dedicated chapters. There is an informal but rigorous discussion of key algorithms and a marked concern for concepts relevant to operations research.

Fred BUCKLEY and Frank HARARY, *Distance in Graphs*. Addison-Wesley 1990. An expanded and updated account of the distance-related topics covered in Harary's *Graph Theory*. Chapter 11 is devoted to algorithms.

Herbert FLEISCHNER, *Eulerian Graphs and Related Topics, Part 1*. North-Holland, Volume 1, 1990, Volume 2, 1991. A comprehensive research monograph on paths and related concepts. Complete with historical and contemporary research references.

Alan GIBBONS, *Algorithmic Graph Theory*. Cambridge University Press 1989. Path-related problems, planarity, matchings, and colorings are treated constructively. The last chapter is devoted to inherent complexity.

Frank HARARY, *Graph Theory*. Addison-Wesley 1972. For over two decades, many students and researchers have been introduced to the subject by this book.

Paul C. KAINEN and Thomas L. SAATY, *The Four-Color Problem: Assaults and Conquest*. Dover Publications 1986. Part I states this long-open problem and reviews the solution claimed by Appel and Haken. Part II is devoted to graph theory and algebra connected to the four-color problem.

H. M. MULDER, *The Interval Function of a Graph*. Amsterdam Mathematical Centre Tracts 132, 1980. A research monograph on median graphs and related concepts.

Oystein ORE, *The Four-Color Problem*. Academic Press 1967. Graph theory presented in light of a simple combinatorial conjecture that has defied solution for over a century.

R. C. READ and W. T. TUTTE, Chromatic polynomials. In *Graph Theory*, Volume 3, pp. 15–42, Academic Press 1988. An accessible expository article that reviews elementary facts and discusses several still unsolved research problems as well.

W. T. TUTTE, *Graph Theory*. Addison-Wesley 1984. An advanced, yet self-contained text on connectivity, reconstruction from subgraphs, connections with matroids and linear algebra, chromatic and other polynomials, combinatorial maps, bridges, planarity, and other concepts.

CHAPTER VIII

LATTICES

1. COMPLEMENTS AND DISTRIBUTIVITY

A lattice is by definition an ordered set $(L, \leq)$ in which every pair of elements $\{x, y\}$ possesses a least upper bound (lub) as well as a greatest lower bound (glb). Two commutative semigroup operations $\vee$ (*join*) and $\wedge$ (*meet*) are then defined on L by

$$x \vee y = \text{lub}\{x, y\}, \qquad x \wedge y = \text{glb}\{x, y\}$$

The semigroups $(L, \vee)$ and $(L, \wedge)$ are called the *join semigroup* and the *meet semigroup* of L, respectively. Join and meet are linked by the *absorption* laws

$$x \vee (x \wedge y) = x, \qquad x \wedge (x \vee y) = x$$

A purely algebraic approach to lattices is possible as well. Let $(L, \vee)$ and $(L, \wedge)$ be two commutative semigroups on the same set L, nothing to do a priori with order relations. Assume that they are linked by the two absorption laws as above. The binary relation

$$R = \{(x, y) \in L^2 : x \vee y = y\}$$

is then an order on L in which every pair of elements has a lub as well as a glb. In fact $\text{lub}\{x, y\}$ happens to coincide with $x \vee y$ in the semigroup $(L, \vee)$ and $\text{glb}\{x, y\}$ coincides with $x \wedge y$ in $(L, \wedge)$.

One may therefore think of lattices either as orders on which two algebraic operations are defined or as algebraic structures with a convenient order relation. Our terminology corresponds to the first conception.

A subset S of a lattice L that is closed both under join and meet is called a *sublattice*. The restriction of the order $\leq$ to S is then obviously a lattice. Note that this restriction $(S,\leq)$ may well be a lattice without S being a sublattice: in the lattice $(\mathcal{P}(3),\subseteq)$ of all subsets of $3 = \{0,1,2\}$ consider

$$S = \{\emptyset, \{0\}, \{1\}, \{0,1,2\}\}$$

The sublattices of L constitute an algebraic closure system on L. In contrast, the subsets S of L such that $(S,\leq)$ is a lattice do not form a closure system: consider S in $\mathcal{P}(3)$ as in the preceding example, let

$$T = \{\emptyset, \{0\}, \{1\}, \{0,1\}\}$$

and note that $(S \cap T, \subseteq)$ is not a lattice, even though both $(S, \subseteq)$ and $(T, \subseteq)$ are lattices. Every convex segment $[x,y]$ of a lattice is a sublattice—the converse obviously need not be true.

A lattice $(L,\leq)$ is *bounded* if it has both a maximum and a minimum. Equivalently, it is enough to require that L have both a lub and a glb. Every nonempty finite lattice is bounded. The min and the max coincide only in the *trivial* one-element lattice. In any lattice L, every segment $[x,y]$ constitutes a bounded sublattice. Note that a sublattice S of $(L,\leq)$ may possess upper and lower bounds in L without $(S,\leq)$ being a bounded lattice: in $(\mathbb{R},\leq)$ consider

$$S = \{x \in \mathbb{R} : 0 < x < 1\}$$

We therefore reserve the term *bounded sublattice* to mean a sublattice that possesses a maximum and a minimum. Any two elements x,y of a lattice generate a finite, bounded sublattice, namely

$$\{x, y, x \wedge y, x \vee y\}$$

Let x be any element of a bounded lattice $(L,\leq)$ with minimum u and maximum w. Any $y \in L$ such that

$$x \wedge y = u, \qquad x \vee y = w$$

is called a *complement* of x in L. Complements need not always exist: look at any chain with at least three elements. For a more

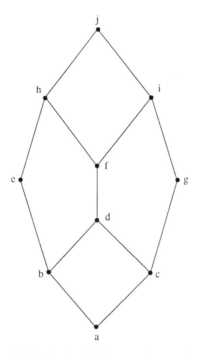

FIGURE 8.1 A bounded lattice in which some elements have no complement.

complex case, look at the lattice of Figure 8.1. In the bounded lattice $(\mathcal{P}(A), \subseteq)$, on the other hand, where A is any set, every $X \in \mathcal{P}(A)$ has $A \setminus X$ as its unique complement. If $B \subseteq C \subseteq A$, then every X in $[B, C]$, i.e., every set X such that $B \subseteq X \subseteq C$, has

$$(C \setminus X) \cup B$$

as a complement in $[B, C]$. In any lattice L, if $x, y \in L$, then x and y are complements of each other in the sublattice $\{x, y, x \wedge y, x \vee y\}$.

Unique Complementation Theorem. *For any lattice L the following conditions are equivalent:*

(i) *in every bounded sublattice S, each element has at most one complement in S,*
(ii) *for all $x, y, z \in L$, $x \wedge (y \vee z) = (x \wedge y) \vee (x \wedge z)$,*
(iii) *for all $x, y, z \in L$, $x \vee (y \wedge z) = (x \vee y) \wedge (x \vee z)$.*

216 LATTICES

Proof. Observe first that (ii) implies
$$(x \vee y) \wedge (x \vee z) = [(x \vee y) \wedge x] \vee [(x \vee y) \wedge z] = x \vee [(x \vee y) \wedge z]$$
$$= x \vee [(x \wedge z) \vee (y \wedge z)] = x \vee (y \wedge z)$$
i.e., (ii) implies (iii). Similarly (iii) can be shown to imply (ii), yielding the equivalence of (ii) and (iii).

Next, observe that if in some bounded sublattice S, with minimum u and maximum w, an element x has two distinct complements y and z, then using (ii), we get
$$y = y \wedge w = y \wedge (x \vee z) = (y \wedge x) \vee (y \wedge z) = u \vee (y \wedge z) = y \wedge z$$
that is, $y \leq z$. In the same way we can derive $z \leq y$ too, forcing the equality $y = z$. Therefore (ii) implies (i).

Conversely, assume (i). First, let us show that (iii) holds in the special case $x \leq z$. Were this not true, we would have
$$x \vee (y \wedge z) < (x \vee y) \wedge (x \vee z) = (x \vee y) \wedge z$$
Let us abbreviate $x \vee (y \wedge z) = a$, $(x \vee y) \wedge z = b$. Then
$$y \vee a = y \vee b = x \vee y$$
and
$$y \wedge a = y \wedge b = y \wedge z$$
This means that in the sublattice
$$\{a, b, y, y \wedge z, x \vee y\}$$
the element y has two complements, contradicting (i). Thus a must coincide with b, i.e.,
$$x \vee (y \wedge z) = (x \vee y) \wedge z \quad \text{for all} \quad x \leq z$$
This will be referred to as the *modular identity*.

Second, let us show that (i) implies the *median equality*
$$(x \wedge y) \vee (x \wedge z) \vee (y \wedge z) = (x \vee y) \wedge (x \vee z) \wedge (y \vee z)$$
for all lattice elements x, y, z. Were this not true, for some x, y, z,
$$d = (x \wedge y) \vee (x \wedge z) \vee (y \wedge z)$$
would be distinct from
$$e = (x \vee y) \wedge (x \vee z) \wedge (y \vee z)$$

As in any case $d \leq e$ (because $x \wedge y$, $x \wedge z$, and $y \wedge z$ are all less than or equal to e), we must have $d < e$. Let

$$a = d \vee (x \wedge e), \qquad b = d \vee (y \wedge e), \qquad c = d \vee (z \wedge e)$$

We have

$$a \vee b = d \vee (x \wedge e) \vee (y \wedge e) = d \vee (x \wedge (y \vee z)) \vee [y \wedge (x \vee z)]$$

By the modular identity and $x \wedge (y \vee z) \leq x \vee z$, this is equal to

$$d \vee \{[(x \wedge (y \vee z)) \vee y] \wedge (x \vee z)\}$$

and, in view of $y \leq y \vee z$ within the square brackets, equal to

$$d \vee \{[(y \vee x) \wedge (y \vee z)] \wedge (x \vee z)\} = d \vee e = e$$

Thus $a \vee b = e$. Similarly $a \vee c = e$ and $b \vee c = e$. As for meets, observe first that the modular identity implies, in view of $d \leq e$, that

$$a = (d \vee x) \wedge e, \qquad b = (d \vee y) \wedge e, \qquad c = (d \vee z) \wedge e$$

Now one can prove

$$a \wedge b = a \wedge c = b \wedge c = d$$

in a way similar to the proof of

$$a \vee b = a \vee c = b \vee c = e$$

(The role of $\wedge$ is taken by $\vee$.) In conclusion, the uniqueness of complements guaranteed by (i) would be violated in the sublattice $\{a,b,c,d,e\}$. Thus (i) must imply the median equality as claimed.

To conclude the proof that (i) implies (ii), consider $a = x \wedge (y \vee z)$ for arbitrary lattice elements x, y, z. Trivially

$$a = x \wedge (x \vee y) \wedge (x \vee z) \wedge (y \vee z)$$

By the median equality

$$a = x \wedge [(x \wedge y) \vee (x \wedge z) \vee (y \wedge z)]$$

and by the modular identity, in view of $(x \wedge y) \vee (x \wedge z) \leq x$, we obtain

$$[(x \wedge y) \vee (x \wedge z)] \vee [(y \wedge z) \wedge x] = a$$

But this latter expression of a is obviously equal to just the term $(x \wedge y) \vee (x \wedge z)$, proving

$$x \wedge (y \vee z) = a = (x \wedge y) \vee (x \wedge z)$$

as intended. □

A lattice satisfying the conditions of the above theorem is called *distributive*. The equivalence of (ii) and (iii) is one of the earliest published results of abstract lattice theory, in Schröder's *Algebra der Logik*, Teubner-Verlag, Leipzig, 1890.

Examples. (1) Every chain is a distributive lattice. (2) The lattice of subsets of any set is distributive, by Proposition 14 of Chapter I. (3) Let A be any set, Card $A \geq 2$. Let $\emptyset \subset B \subset A$ and $C = A \backslash B$. Let

$$S_A = \{A, \emptyset\}, \qquad S_B = \{B, \emptyset\}, \qquad S_C = \{C, \emptyset\},$$

$$S_0 = \{\emptyset\}, \qquad S = \{A, B, C, \emptyset\}$$

Each of these five sets is a subgroup of the symmetric difference group $(\mathcal{P}(A), +)$. In the lattice of all subgroups,

$$\{S_A, S_B, S_C, S_0, S\}$$

is a sublattice, and it is not distributive. On the other hand, the lattice of subgroups of $\mathbb{Z}$ is distributive, and so is the lattice of subgroups of any cyclic group.

One obvious consequence of the definition of distributivity is that every sublattice of a distributive lattice is distributive. It seems rather vacuous to restate this as follows: a lattice L is distributive if and only if every sublattice of L is distributive. But could this statement be strengthened so that the distributivity of some particular sublattices of L would already suffice to infer the distributivity of L? For one thing, if all the bounded sublattices of L are distributive, that is sufficient. Birkhoff's Forbidden Sublattice Criterion goes much further in this direction.

Forbidden Sublattice Criterion (For Distributive Lattices). *A lattice L is distributive if and only if it has no five-element sublattice*

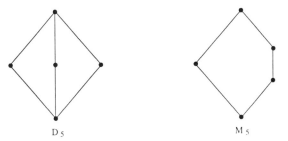

FIGURE 8.2 Nondistributive lattices.

isomorphic to either one of the lattices D_5 or M_5 illustrated in Figure 8.2.

Proof. It is easily seen that D_5 and M_5 are nondistributive. Conversely, the construction of M_5 or D_5 using nondistributivity is essentially contained in the proof of the Unique Complementation Theorem. First, if the modular law

$$x \vee (y \wedge z) = (x \vee y) \wedge z$$

fails to hold for some $x \leq z$, then

$$\{y, y \wedge z, x \vee y, x \vee (y \wedge z), (x \vee y) \wedge z\}$$

is a sublattice of type M_5. Second, if the modular law does hold, but the median law

$$(x \wedge y) \vee (x \wedge z) \vee (y \wedge z) = (x \vee y) \wedge (x \vee z) \wedge (y \vee z)$$

fails for some lattice elements x, y, z, then the five elements

$$(x \wedge y) \vee (x \wedge z) \vee (y \wedge z) = d$$

$$(x \vee y) \wedge (x \vee z) \wedge (y \vee z) = e$$

$$a = d \vee (x \wedge e) = (d \vee x) \wedge e$$

$$b = d \vee (y \wedge e) = (d \vee y) \wedge e$$

$$c = d \vee (z \wedge e) = (d \vee z) \wedge e$$

form a sublattice of type D_5. (If neither the modular nor the median law fails, then the lattice is distributive according to the proof of the Unique Complementation Theorem.) □

One more noteworthy result can be easily extracted from the proof of the Unique Complementation Theorem. In that proof the median equality

$$(x \wedge y) \vee (x \wedge z) \vee (y \wedge z) = (x \vee y) \wedge (x \vee z) \wedge (y \vee z)$$

was established for all elements x, y, z of any distributive lattice. Should, on the other hand, a lattice fail to be distributive, then in a five-element forbidden sublattice, whether of type D_5 or M_5, the reader can surely find elements x, y, z violating the median law:

Median Equality Theorem. *A lattice L is distributive if and only if for every x, y, z in L we have*

$$(x \wedge y) \vee (x \wedge z) \vee (y \wedge z) = (x \vee y) \wedge (x \vee z) \wedge (y \vee z).$$

Covering Lemma. *Let x and y be elements of a distributive lattice.*

(i) *If x covers $x \wedge y$, then $x \vee y$ covers y.*
(ii) *If $x \vee y$ covers y, then x covers $x \wedge y$.*

Proof. (i) Assume that x covers $x \wedge y$. If y is not covered by the join $x \vee y$, then let

$$y < z < x \vee y$$

and look at

$$\{x \wedge y, x, y, z, x \vee y\}$$

It is a forbidden sublattice of the M_5 type. Thus y must be covered by $x \vee y$.

The proof of (ii) is similar. □

Remark. Why not state this result in "if and only if" form? The reader will see in the next section that the development of lattice theory benefits from splitting these *lower* (i) and *upper* (ii) *covering conditions*.

Let $a \leq b$ in a discrete order. The *relative height* of b above a, denoted by $h(a, b)$, is defined for $a < b$ as the length of the shortest path from a to b in the covering relation, or $h(a, b) = 0$ if $a = b$. Equivalently, if C is a maximal chain in $[a, b]$ having as few elements

as possible, then
$$h(a,b) = \operatorname{Card} C - 1$$
A discrete order is said to satisfy the *Jordan–Dedekind chain condition* if for all $a < b$, every path from a to b in the covering relation has length $h(a,b)$ and not more. Equivalently, this means that within each segment $[a,b]$ all maximal chains have the same cardinality, namely $h(a,b) + 1$. For example, the lattice D_5 in Figure 8.2 satisfies the Jordan–Dedekind condition, but M_5 does not.

Proposition 1 *Every discrete distributive lattice satisfies the Jordan–Dedekind chain condition.*

Proof. Suppose that for some $a < b$, some maximal chain K of $[a,b]$ has more than $h(a,b) + 1$ elements. Among all such misbehaving couples (a,b) take one with smallest possible $h(a,b)$. However, the segment $[a,b]$ has a maximal chain C with
$$\operatorname{Card} C = h(a,b) + 1 < \operatorname{Card} K$$
By the minimal choice of $h(a,b)$, $C \cap K = \{a,b\}$. Let $c \in C$ cover a, and let $k \in K$ cover a. Then $a = c \wedge k$, and according to the Covering Lemma $c \vee k$ covers both c and k. We have
$$h(c,b) = h(a,b) - 1, \qquad h(c \vee k, b) = h(a,b) - 2,$$
$$h(k,b) = h(a,b) - 1$$
Therefore all maximal chains of $[k,b]$ have cardinality $h(k,b) + 1$, in particular
$$\operatorname{Card}(K \setminus \{a\}) = h(k,b) + 1$$
which forces $\operatorname{Card} K = h(a,b) + 1$. □

Corollary. *In a discrete distributive lattice, if $a \leq b \leq c$, then*
$$h(a,b) + h(b,c) = h(a,c)$$

Let us turn our attention from covering relations to covering graphs. More precisely, we wish to analyze the behavior of geodesics in the covering graph G of a discrete lattice L. This graph is certainly connected. (We can concatenate paths from x to $x \vee y$ and from $x \vee y$ to y.) The distance $d(x,y)$ is at most

$$h(x, x \vee y) + h(y, x \vee y)$$

and similarly

$$d(x,y) \leq h(x \wedge y, x) + h(x \wedge y, y)$$

If P is an x-to-y geodesic path in G, with vertices

$$x = x_0, x_1, \ldots, x_n = y, \qquad \text{where} \quad n = d(x,y)$$

and if $\{x_i, x_{i+1}\}$ are the edges of P for $0 \leq i \leq n-1$, then define the *step differential* $s_P(x, x_i)$ for $i = 0, \ldots, n$ as the integer

$$\text{Card}\{j \in i : x_j < x_{j+1}\} - \text{Card}\{j \in i : x_j > x_{j+1}\}$$

Define the *x-based elevation* of P as the sum $\sum_{i=0}^{n} s_P(x, x_i)$, denoted by $H_x(P)$. Obviously

$$-d(x,y)^2 \leq H_x(P) \leq d(x,y)^2$$

For $0 < i < n$, call the vertex x_i on P *locally high* if $x_{i-1} < x_i$ and $x_i > x_{i+1}$ in L. Call x_i *locally low* if $x_{i-1} > x_i$ and $x_i < x_{i+1}$. We say that P is a *high geodesic* if it has no locally low vertices. Clearly P is high if and only if for some $0 \leq j \leq n$,

$$x_0 \leq \cdots \leq x_j \qquad \text{and} \qquad x_j \geq \cdots \geq x_n$$

We say that P is a *low geodesic* if it has no locally high vertex. Obviously P is low if and only if, for some $0 \leq j \leq n$,

$$x_0 \geq \cdots \geq x_j \qquad \text{and} \qquad x_j \leq \cdots \leq x_n$$

There is always a maximum in the set of vertices of a high geodesic, and there is always a minimum among the vertices on a low geodesic. Observe that a geodesic P in G is simultaneously high and low precisely when its vertices form a chain in L.

In a distributive lattice L, an x-to-y geodesic P with maximum x-based elevation must be high. (The reason is that if x_i were a locally low vertex on P, $x_{i-1} > x_i$, $x_i < x_{i+1}$, then $x_i = x_{i-1} \wedge x_{i+1}$ would be covered by both x_{i-1} and x_{i+1}. Thus $x_i' = x_{i-1} \vee x_{i+1}$ would cover both x_{i-1} and x_{i+1}, and we could form a path P' of even higher elevation if we replaced in P the vertex x_i with x_i' and the edges $\{x_{i-1}, x_i\}$, $\{x_i, x_{i+1}\}$ with $\{x_{i-1}, x_i'\}$, $\{x_i', x_{i+1}\}$.) Thus a high geodesic does exist between any two elements x, y of a distributive lattice. Assume that P is any such high geodesic, say of length n. We claim that

$$x_j = \max\{x_i : 0 \leq i \leq n\}$$

coincides with $x \vee y$. Obviously $x \vee y \leq x_j$, and therefore, by the Corollary of Proposition 1,

$$h(x, x \vee y) + h(x \vee y, x_j) = h(x, x_j) = j$$
$$h(y, x \vee y) + h(x \vee y, x_j) = h(y, x_j) = n - j$$

which add up to

$$h(x, x \vee y) + h(y, x \vee y) + 2h(x \vee y, x_j) = n = d(x, y)$$

This forces $h(x \vee y, x_j) = 0$, $x_j = x \vee y$ as claimed, and

$$d(x, y) = h(x, x \vee y) + h(y, x \vee y)$$

Similarly we can show, using a low geodesic, that

$$d(x, y) = h(x \wedge y, x) + h(x \wedge y, y)$$

Combined with

$$h(x \wedge y, x \vee y) = h(x \wedge y, x) + h(x, x \vee y) = h(x \wedge y, y) + h(y, x \vee y)$$

this further implies

$$h(x, x \vee y) = h(x \wedge y, y) \quad \text{and} \quad h(y, x \vee y) = h(x \wedge y, x)$$

and therefore

$$d(x, y) = h(x \wedge y, x) + h(x, x \vee y) = h(x \wedge y, x \vee y)$$

for all elements x, y of a discrete distributive lattice. In particular

$$d(x, y) = h(x, y) \quad \text{if} \quad x \leq y$$

Proposition 2 *Let L be a discrete distributive lattice with covering graph G. Then a set C of lattice elements forms an order-convex sublattice of L if and only if C is a convex set of vertices in the covering graph G.*

Proof. Suppose C is an order-convex sublattice. Let $x, y \in C$. We need to show that every vertex of every x-to-y geodesic P in G belongs to C. Clearly it suffices to show that every vertex $x = x_0, x_1, \ldots, x_n = y$ on P belongs to the segment $[x \wedge y, x \vee y]$ of L. Assuming this assertion false, let P be an x-to-y geodesic with maximum elevation $H_x(P)$ among those that allegedly have some vertices outside this segment. Clearly P is not a high geodesic, for then its top vertex x_j would coincide with $x \vee y$, forcing all vertices on P to be in

$$[x, x \vee y] \cup [y, x \vee y] \subseteq [x \wedge y, x \vee y]$$

Thus P has a locally low vertex x_i, covered by x_{i-1} and x_{i+1},

$$x_i = x_{i-1} \wedge x_{i+1}$$

A higher elevation x-to-y geodesic P' can then be constructed, as in an earlier argument, on the same vertices as P except that x_i is exchanged for

$$x_i' = x_{i-1} \vee x_{i+1}$$

Every vertex of this P' must belong to $[x \wedge y, x \vee y]$ because of the maximal-elevation choice of P. But from

$$x_i < x_i' \leq x \vee y, \qquad x \wedge y \leq x_{i-1}, \qquad x \wedge y \leq x_{i+1}$$

it follows that $x_i = x_{i-1} \wedge x_{i+1}$ is in $[x \wedge y, x \vee y]$ as well, a contradiction proving that all x-to-y geodesics are within this segment.

Conversely, suppose that $C \subseteq L$ is convex in the covering graph G. If $x \leq y$ are in C, then for any $x \leq z \leq y$ we have the equality $h(x, z) + h(z, y) = h(x, y)$, i.e.,

$$d(x, z) + d(z, y) = d(x, y)$$

in G, and therefore $z \in C$, proving the order convexity of C. To see that C is a sublattice of L, let $x, y \in C$. A highest elevation x-to-y geodesic contains $x \vee y$ as a vertex, and thus $x \vee y \in C$. □

Remark. Distributivity is not necessary for Proposition 2 to hold. Consider the five-element lattice D_5 of Figure 8.2.

Interval Theorem. *A discrete lattice L is distributive if and only if for all $x, y \in L$ the interval $I(x, y)$ of the covering graph coincides with the order segment $[x \wedge y, x \vee y]$.*

Proof. Assume distributivity. Let $z \in [x \wedge y, x \vee y]$, that is,
$$x \wedge y \leq z \leq x \vee y$$
This implies $x \vee z \leq x \vee y$, $y \vee z \leq x \vee y$, and $(x \vee z) \wedge (y \vee z) = z$. We have

$$\begin{aligned} d(x,y) &= h(x, x \vee y) + h(y, x \vee y) \\ &= h(x, x \vee z) + h(x \vee z, x \vee y) + h(y, y \vee z) + h(y \vee z, x \vee y) \\ &= h(x, x \vee z) + h(z, x \vee z) + h(z, y \vee z) + h(y, y \vee z) \\ &= d(x, z) + d(z, y) \end{aligned}$$

which means that $z \in I(x,y)$. Thus $[x \wedge y, x \vee y] \subseteq I(x,y)$. The reverse inclusion follows from Proposition 2.

Conversely, assume that $I(x,y) = [x \wedge y, x \vee y]$ for all $x, y \in L$. If L were not distributive, then in some segment of L an element a would have two distinct complements b, c:

$$a \vee b = a \vee c \quad \text{and} \quad a \wedge b = a \wedge c$$

We would have $b \in [a \wedge c, a \vee c]$, i.e., $b \in I(a,c)$, and thus, in the covering graph of L, $d(a,b) < d(a,c)$. But from $c \in [a \wedge b, a \vee b]$ we would derive the opposite inequality, $d(a,c) < d(a,b)$. This contradiction shows that L must be distributive. □

Median Diagram Theorem. *A discrete lattice L is distributive if and only if its covering graph is median.*

Proof. Let L be a discrete distributive lattice. Let $x, y, z \in L$. Consider the three intervals $I(x,y)$, $I(x,z)$, $I(y,z)$ in its covering graph. By the Interval Theorem we have, e.g., $I(x,y) = [x \wedge y, x \vee y]$, and therefore

$$I(x,y) \cap I(x,z) \cap I(y,z)$$

is precisely the segment

$$[(x \wedge y) \vee (x \wedge z) \vee (y \wedge z), (x \vee y) \wedge (x \vee z) \wedge (y \vee z)]$$

According to the Median Equality Theorem, this segment is a singleton.

Conversely, suppose that the covering graph G of L is median. Let us show that for $x, y \in L$, if C_x is any maximal chain in $[x, x \vee y]$ and C_y is any maximal chain in $[y, x \vee y]$, then $\text{Card}(C_x \cup C_y) - 1$ equals the distance $d(x, y)$ in G. Let us abbreviate $\text{Card}(C_x \cup C_y) - 1$ as $|C_x C_y|$. Obviously

$$C_x \cap C_y = \{x \vee y\}$$

and the inequality

$$|C_x C_y| \geq d(x, y)$$

holds in any case. Suppose equality is not always achieved, and choose x, y, C_x, C_y with strict inequality

$$|C_x C_y| > d(x, y)$$

and $|C_x C_y|$ as small as possible. Since a covering graph cannot have a three-vertex complete subgraph, we have $|C_x C_y| \geq 3$, and at least one of

$$h(x, x \vee y) \geq 2 \quad \text{or} \quad h(y, x \vee y) \geq 2$$

must hold. Assume $h(x, x \vee y) \geq 2$; in the alternative case the proof would be similar. Let s be the element of C_x covering x and let

$$K_s = C_x \setminus \{x\}, \qquad K_y = C_y$$

We have $s \vee y = x \vee y$, $|K_s K_y| = |C_x C_y| - 1$, and therefore $|K_s K_y| = d(s, y)$. Consequently

$$d(s, y) > d(x, y) - 1$$

Since x and s are adjacent in G, $d(s, y)$ is equal either to $d(x, y)$ or to $d(x, y) + 1$. The former alternative is impossible because

$$I(x, s) \cap I(x, y) \cap I(s, y) = \emptyset$$

would contradict the median hypothesis. Therefore $d(s, y)$ is equal to $d(x, y) + 1$. Let z be the element of K_s covering s and let

$$T_z = K_s \setminus \{s\}, \qquad T_y = K_y$$

We have $z \vee y = x \vee y$ and

$$|T_z T_y| = |K_s K_y| - 1 = |C_x C_y| - 2$$
$$|T_z T_y| = d(z, y) = d(s, y) - 1$$

Thus $d(x,y) = d(z,y) = d(s,y) - 1$. Also $d(x,z) = 2$ and the reader can easily verify that

$$I(x,z) = [x,z]$$

Clearly $x \notin I(z,y)$ and $z \notin I(x,y)$. Can any other element m of $I(x,z)$ belong to $I(z,y)$ or $I(x,y)$? Let

$$U_m = \{m\} \cup T_z, \quad U_y = T_y$$

The element m is covered by z, $m \vee y = x \vee y$, $|U_m U_y| = |K_s K_y|$, and therefore $d(m,y) = d(s,y)$. Obviously m cannot belong to $I(z,y)$ or $I(x,y)$. Thus

$$I(x,z) \cap I(x,y) \cap I(z,y) = \emptyset$$

which contradicts the median property. The absurdity of the assumption $|C_x C_y| > d(x,y)$ has been demonstrated. For all maximal chains C_x in $[x, x \vee y]$ and C_y in $[y, x \vee y]$ we have

$$\text{Card}(C_x \cup C_y) - 1 = d(x,y)$$

and thus $C_x \cup C_y \subseteq I(x,y)$.

Since every element of $[x, x \vee y]$ belongs to some maximal chain C_x and every element of $[y, x \vee y]$ belongs to some C_y, we must conclude that

$$[x, x \vee y] \cup [y, x \vee y] \subseteq I(x,y)$$

Similarly we can derive from the median property of the covering graph the inclusion

$$[x \wedge y, x] \cup [x \wedge y, y] \subseteq I(x,y)$$

How could L fail to be distributive if its covering graph is median? Then L would have a five-element forbidden sublattice F of the type D_5 or M_5 of Figure 8.2. If it is of type D_5, then for the three pairwise incomparable elements x, y, z of F we would have, as a consequence of the inclusion relations established above,

$$\{\max F, \min F\} \subseteq I(x,y) \cap I(x,z) \cap I(y,z)$$

which contradicts the singleton intersection property of median graphs. If F is of type M_5, then for the two comparable complements

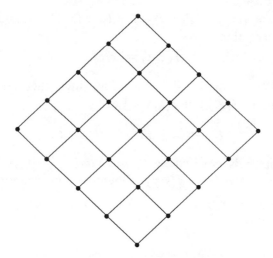

FIGURE 8.3 A distributive lattice.

$y < z$ of some x in F, we would have

$$z \in [y, x \vee y], \qquad y \in [x \wedge z, z]$$

and consequently $\{y, z\} \subseteq I(x,y) \cap I(x,z) \cap I(y,z)$, which is again absurd. Thus L must be distributive if its covering graph is median. □

It is perhaps using the Median Diagram Theorem that the distributivity of the lattice-looking lattice of Figure 8.3 can most insightfully be verified. Better, try the *infinite grid lattice* $\mathbb{Z}^n$ for any positive integer n, where the order is given by

$$(x_0, \ldots, x_{n-1}) \leq (y_0, \ldots, y_{n-1}) \qquad \text{if and only if}$$
$$x_0 \leq y_0, \ldots, x_{n-1} \leq y_{n-1}$$

EXERCISES

1. Design programs
   (a) to compute $h(a,b)$ for elements $a \leq b$ in a finite ordered set,
   (b) to determine if a finite lattice is distributive,
   (c) to find $a \vee b$ and $a \wedge b$ for given elements a, b of a finite distributive lattice,

(d) to determine if a given element of a finite distributive lattice possesses a complement.

2. Find all order-convex sublattices of $(\mathcal{P}(3), \subseteq)$.

3. What is the order dimension of the lattice of Figure 8.3? What is the order dimension of the infinite grid lattice $\mathbb{Z}^n$?

4. Let R and Q be two order relations on the same set A. Show that if lub's in R coincide with lub's in Q, then $R = Q$. Verify that the same holds for glb's.

5. Verify that if L is a lattice and $x_1, \ldots, x_n \in L$, then $x_1 \vee \cdots \vee x_n$ is a lub of the set $\{x_1, \ldots, x_n\}$; verify that a similar observation holds for meets.

6. Verify that the dual order of a lattice is a lattice and the dual of a distributive lattice is distributive.

7. Do equivalence relations form a sublattice of the inclusion-ordered lattice $(\mathcal{P}(A^2), \subseteq)$ of all relations on a set A? What about preorders, transitive relations, or symmetric relations?

8. Verify that the transitive relations on a given set form a lattice, of which the equivalence relations form a sublattice. Are these lattices distributive?

9. Show that the congruence relations on a groupoid G form a sublattice of the inclusion-ordered lattice of all equivalence relations on G.

10. Show that
    (a) if G is a group, then the subgroups of G form a sublattice of the lattice of subgroupoids of G,
    (b) the normal subgroups of a group form a sublattice of the lattice of subgroups,
    (c) the ideals of a ring form a sublattice of the lattice of subgroups of the ring's additive group,
    (d) the subspaces of a vector space form a sublattice of the lattice of subgroups of the vector addition group.

11. Show that if the lattice of subgroups of a group G is distributive, then the lattice of subgroups of any quotient group of G is distributive as well.

12. Let S be any set of subgroups of a group G that constitutes a closure system on G. Show that the cosets of the various members of S plus the empty set $\emptyset$ constitute another, larger closure system on G. Give examples.

13. Show that the lattice of ideals of a principal entire ring with identity is distributive.

14. On $\mathbb{R}^2$ consider the causality order of special relativity. Is it a (distributive) lattice? What about material causality?

15. Let L be a lattice and let S be the set of its sublattices. Verify that $(S, \subseteq)$ is a lattice. Is it a sublattice of $(\mathcal{P}(L), \subseteq)$? Describe the sublattice of S generated by the segments $[x, y]$ of L.

16. Show that a bijective map from a lattice to itself is an order automorphism if and only if it is an automorphism of the join semigroup of the lattice. (What about the meet semigroup?)

2. BOOLEAN ALGEBRA

A bounded distributive lattice L in which every element x has a complement x' is called a *Boolean lattice*.

Examples. For any set S, $(\mathcal{P}(S), \subseteq)$ is a Boolean lattice. The simplest interesting instance is $S = 2 = \{0, 1\}$. On the other hand, let $S = \omega$. The sublattice of $\mathcal{P}(\omega)$ consisting of the finite subsets of ω and their complements is also a Boolean lattice. The sublattice of $\mathcal{P}(\omega)$ consisting of all lower sections of $(\omega, \leq)$ is not Boolean.

Boolean lattices were indeed studied before the emergence of abstract lattice theory. In George Boole's *Investigation of the Laws of Thought* (1854) collections consisting of collections are endowed with order and algebraic structure. Lattices and universal algebras have since proven to be powerful tools of mathematical metamathematics.

Mathematical logic originates with the recognition that the truth or falsehood of certain composite sentences like

$$\text{``}7^8 + 8^9 \text{ is prime} \quad \text{and} \quad (\sqrt{2} + \sqrt{5+6+7})^2 \in \mathbb{Z}\text{''}$$
$$\text{``}7^8 + 8^9 \text{ is prime} \quad \text{or} \quad (\sqrt{2} + \sqrt{5+6+7})^2 \in \mathbb{Z}\text{''}$$

depends quite predictably on the respective truth or falsehood of the statements

$$\text{``}7^8 + 8^9 \text{ is prime''}$$
$$\text{``}(\sqrt{2} + \sqrt{5+6+7})^2 \in \mathbb{Z}\text{''}$$

Thus the logical connectives "and," "or," "only if," etc., can be thought of as functions $c : 2^2 \to 2$. The function c for the connective "and" is given by

$$c(00) = c(01) = c(10) = 0, \quad c(11) = 1$$

and the function k for "only if " by

$$k(10) = 0, \quad k(00) = k(01) = k(11) = 1$$

Expressed in the Boolean lattice $2 = \{0, 1\}$ we have

$$c(x, y) = x \wedge y \quad \text{and} \quad k(x, y) = x' \vee y$$

where x' denotes the complement of x. Connectives combining three sentences such as "if ... then ... or ... " are modeled by functions $c : 2^3 \to 2$, in this instance

$$c(x, y, z) = (x \wedge y' \wedge z')'$$

This motivates the following concept of Boolean functions.

Let n be a fixed natural number. A function $c : 2^n \to 2$ is called a *Boolean function* in n variables. The *characteristic set* of c is defined as

$$\{\mathbf{x} \in 2^n : c(\mathbf{x}) = 1\}$$

The set B_n of all Boolean functions in n variables is ordered by letting $c \leq k$ mean that the characteristic set of c is a subset of the characteristic set of k. Associating to each $c \in B_n$ its charateristic set establishes an order isomorphism between $(B_n, \leq)$ and $(\mathcal{P}(2^n), \subseteq)$. Thus the ordered set $(B_n, \leq)$ of n-variable Boolean functions is a Boolean lattice. We shall return to this in a moment.

The following classical result generalizes statement (v) of Proposition 14, Chapter I. It is easy to prove and serves in turn as a useful tool for manipulating expressions in Boolean lattices.

De Morgan's Laws. *In any Boolean lattice, glb, lub, and complements are linked by the identities*

$$(x \wedge y)' = x' \vee y'$$

$$(x \vee y)' = x' \wedge y'$$

Let $(L, \leq)$ be a Boolean lattice, with x' denoting the complement of $x \in L$. A ring structure $(L, +, \cdot)$ is defined on L by

$$x + y = (x \wedge y') \vee (x' \wedge y)$$

$$x \cdot y = x \wedge y$$

In this *ring associated with the Boolean lattice* the idempotent law $x^2 = x$ is obviously satisfied by every element x. Obviously the lattice maximum serves as ring identity.

Conversely, let $(R, +, \cdot)$ be any *idempotent ring*, i.e.,

$$x^2 = x \qquad \text{for all} \quad x \in R$$

We are assuming that the ring has an identity element. In this ring, the divisibility preorder is an order: $a|b$ and $b|a$ means

$$b = qa, \qquad a = pb \qquad \text{for some} \quad q, p \in R$$

and hence

$$b = qa = qpb, \qquad a = pb = pqa$$

$$a = pqa = pq^2 a = qpqa = qpb = b$$

Let $\leq$ be the dual of the divisibility order on R. Now the ring zero 0_R is the smallest and 1_R is the greatest element. It is not difficult to verify that $(R, \leq)$ is a lattice in which

$$x \vee y = x + y + x \cdot y$$

$$x \wedge y = x \cdot y$$

for all elements x, y, and indeed a Boolean lattice, with the complement $x' = x + 1_R$. (Key: $x + x = 0_R$ follows from ring idempotence.) This lattice $(R, \leq)$ is called the *lattice associated with the idempotent ring* R.

If $\mathcal{R}(L)$ denotes the ring associated with a Boolean lattice L and $\mathcal{L}(R)$ denotes the lattice associated with an idempotent ring R, then

$$\mathcal{L}\mathcal{R}(L) = L \quad \text{and} \quad \mathcal{R}\mathcal{L}(R) = R$$

If L is isomorphic to L' as an order, then $\mathcal{R}(L)$ is isomorphic to $\mathcal{R}(L')$, and conversely. The ring associated with a power set lattice $(\mathcal{P}(S), \subseteq)$ is nothing else than the familiar symmetric difference ring $(\mathcal{P}(S), +, \cap)$.

In any lattice with a smallest element, the elements covering the minimum are called *atoms*. These will play an important role in this section and the next one.

Representation Theorem (Finite Case). *Every finite Boolean lattice is isomorphic to some power set lattice* $(\mathcal{P}(S), \subseteq)$.

Proof. Let S be the set of atoms of a finite Boolean lattice $(L, \leq)$. With every $A \subseteq S$ let us associate $\operatorname{lub} A$ in L. This defines a mapping from $\mathcal{P}(S)$ to L. We claim that it is surjective onto L. For $y \in L$ let

$$A_y = \{a \in S : a \leq y\}$$

$$z = \operatorname{lub} A_y$$

and let z' be the complement of z in L. The reader should verify that y and z are both complements of z' in L, and therefore $y = z$. This shows surjectivity as claimed.

Obviously if $B \subseteq A \subseteq S$, then $\operatorname{lub} B \leq \operatorname{lub} A$. We shall show that, conversely, $\operatorname{lub} B \leq \operatorname{lub} A$ for sets of atoms B, A implies $B \subseteq A$, and this will establish the desired isomorphism. If this is not true, let A be a set of atoms of smallest possible cardinality such that for some other set of atoms B,

$$\operatorname{lub} B \leq \operatorname{lub} A \quad \text{but} \quad B \not\subseteq A$$

The case of A being empty or a singleton is quickly ruled out. Let then A_1, A_2 be disjoint nonempty sets partitioning A, $A_1 \cup A_2 = A$.

Let $B_0 = B \setminus A$. By assumption this set is nonempty. Let

$$\text{Card } A_1 = m, \quad \text{Card } A_2 = n, \quad \text{Card } B_0 = t$$

Let

$$A_1 = \{a_{11}, \ldots, a_{1m}\}, \quad A_2 = \{a_{21}, \ldots, a_{2n}\}, \quad B_0 = \{b_1, \ldots, b_t\}$$

Using distributivity, we have

$$\text{lub } A_1 \wedge \text{lub } A_2 = (a_{11} \vee \cdots \vee a_{1m}) \wedge (a_{21} \vee \cdots \vee a_{2n})$$
$$= \text{lub}\{a_{1i} \wedge a_{2j} : 1 \leq i \leq m, \ 1 \leq j \leq n\}$$
$$= \min L$$

as well as

$$\text{lub } A_1 \wedge \text{lub}(A_2 \cup B_0) = \min L$$

and of course

$$\text{lub } A_1 \vee \text{lub } A_2 = \text{lub } A$$

$$\text{lub } A_1 \vee \text{lub}(A_2 \cup B_0) = \text{lub } A \vee \text{lub } B_0 = \text{lub } A$$

Thus $\text{lub } A_2$ and $\text{lub}(A_2 \cup B_0)$ are complements of $\text{lub } A_1$ in the segment sublattice $[\min L, \text{lub } A]$ and therefore

$$\text{lub}(A_2 \cup B_0) = \text{lub } A_2$$

Since $\text{Card } A_2 < \text{Card } A$, we must have $A_2 \cup B_0 \subseteq A_2$, which is absurd. Thus the map associating with each set of atoms its lub in L is an isomorphism between the lattices $(\mathcal{P}(S), \subseteq)$ and $(L, \leq)$. □

The surjectivity of the lub map used in this proof deserves a separate statement:

Corollary. *Every element x of a finite Boolean lattice is the least upper bound of the atoms $a \leq x$.*

In the Boolean lattice $(B_n, \leq)$ of all n-variable Boolean functions the atoms are those functions $f : 2^n \to 2$ whose characteristic set is a singleton $\{\mathbf{x}\}$, $\mathbf{x} \in 2^n$. For $i \in n$, let the Boolean function $x_i \in B_n$ be defined by

$$x_i(\mathbf{y}) = \mathbf{y}(i) \quad \text{for every} \quad \mathbf{y} \in 2^n$$

(Remember that **y** is a function from the cardinal n to 2.) Let $\overline{x_i}$ in B_n be defined by letting $\overline{x_i}(\mathbf{y}) = x_i(\mathbf{y})'$, the complement of $x_i(\mathbf{y})$ in $(2, \leq)$. Of course x_i and $\overline{x_i}$ are complements in B_n. Both these functions are called *literals*. A *simple conjunction* c in B_n is a glb of n distinct literals,

$$c = l_0 \wedge \cdots \wedge l_{n-1}$$

such that for each $0 \leq i \leq n-1$ the literal l_i is either x_i or $\overline{x_i}$. The reader can verify that simple conjunctions are precisely the atoms of B_n. In $(B_n, \leq)$ least upper bounds are usually called *disjunctions*. The Corollary of the Representation Theorem therefore implies that every Boolean function in n variables is the disjunction of a unique set of simple conjunctions.

Examples. Let c be the three-variable Boolean function "if–then–or" given earlier by $c(x, y, z) = (x \wedge y' \wedge z')'$. The complement of c in B_3 is the simple conjunction $x_0 \wedge \overline{x_1} \wedge \overline{x_2}$. The function c itself is not an atom in B_3; it is the disjunction of no less than seven simple conjunctions. The two-variable "either–or" function $(x_0 \wedge \overline{x_1}) \vee (\overline{x_0} \wedge x_1)$ is an example of a nonatomic member of B_2.

Remark on Notation. In a Boolean lattice, the meet $x \wedge y$ coincides with the product $x \cdot y$ in the associated ring. Several texts use the product notation, to the exclusion of "$\wedge$," especially in the context of Boolean functions. Thus $x_0 \wedge \overline{x_1} \wedge \overline{x_2}$ can also be denoted by $x_0 \cdot \overline{x_1} \cdot \overline{x_2}$ or simply $x_0 \overline{x_1} \overline{x_2}$.

Consider the covering graph H of a finite Boolean lattice. In view of the Representation Theorem, we assume that the lattice in question is a power set lattice $(\mathcal{P}(S), \subseteq)$. Then $A, B \in \mathcal{P}(S)$ are adjacent in H if and only if the symmetric difference $A + B$ is a singleton. It is elementary to conclude that every vertex of H is adjacent to the same number of vertices. Such graphs in general are called *regular*. It turns out that this property of the covering graph distinguishes Boolean lattices from other distributive lattices.

Regular Diagram Theorem. *A finite lattice is Boolean if and only if its covering graph is median and regular.*

Proof. In view of the Median Diagram Theorem and the preceding comments, only the sufficiency of the graphic conditions remains to be established. Let L be a finite lattice with median and regular covering graph. We already know that L is distributive; let us show that it must be Boolean as well.

Let d be the number of atoms of L. By regularity, the number of elements covering or covered by any element x of L is d.

We claim that the lub of the atoms of L, denoted by m, is the max of L. For every atom a let a' be the lub of all the other atoms of L. Then a and a' are complements in $[\min L, m]$. According to the Covering Lemma, m covers a'. Because of unique complementation, $a' \neq b'$ if $a \neq b$. Thus the number of elements covered by m is at least d. By the regularity of the covering graph, it cannot be more than d, and there is no more room for any element to cover m. Thus m is the max of L, as claimed.

For any $x \in L$, let x' be the lub of those atoms which are not less than or equal to x. Then x' is a complement of x in L. □

EXERCISES

1. How would a program decide whether a given finite lattice is Boolean?

2. Does the Representation Theorem hold for infinite Boolean lattices?

3. Show that every ordered set is order isomorphic to some subset (restriction) of some Boolean lattice.

4. Show that every Boolean lattice is order isomorphic to its dual.

5. Show that every convex segment $[a, b]$ of a Boolean lattice is again a Boolean lattice.

6. Let L be a Boolean lattice. Show that those Boolean sublattices of L that contain both max L and min L form an algebraic closure system on L.

7. Let c be any simple conjunction in n variables, $c \in B_n$. Show that the complement of c in B_n is the disjunction of $2^n - 1$ simple conjunctions. Can you generalize?

3. MODULAR AND GEOMETRIC LATTICES

Among distributive lattices we defined Boolean lattices by adding a condition, the existence of complements: this proved to be quite restrictive. We now proceed by generalization to the broader classes of modular and geometric lattices. Some of our arguments are already contained in the discussion of distributive lattices, which we shall invite the reader to review as warranted.

A lattice L is called *modular* if for every $x \leq z$ we have the *modular identity*

$$x \vee (y \wedge z) = (x \vee y) \wedge z$$

The concept arose naturally in the proof of the Unique Complementation Theorem for distributive lattices. All distributive lattices are modular. On the other hand, the five-element nondistributive lattice D_5 of Figure 8.2 is modular as well. Obviously, every sublattice of a modular lattice is modular.

Modular Characterization Theorem. *For any lattice L the following conditions are equivalent:*

(i) *L is modular,*
(ii) *both the lower and upper covering relations hold (i.e., x covers $x \wedge y$ if and only if $x \vee y$ covers y) in every discrete sublattice S of L,*
(iii) *L has no five-element sublattice isomorphic to M_5 of Figure 8.2,*
(iv) *no element of a bounded sublattice S has two distinct comparable complements in S.*

Proof. Assume (i). Let $\prec$ denote the covering relation of the lattice L. If (ii) failed, then either

$$x \wedge y \prec x \quad \text{and} \quad y < z < x \vee y$$

for some x, y, z in L or

$$y \prec x \vee y \quad \text{and} \quad x \wedge y < z < x$$

In the first case

$$y \vee (x \wedge z) = y \vee (x \wedge y) = y \quad \text{and} \quad (y \vee x) \wedge z = z$$

even though $y < z$, contradicting modularity. The absurdity of the second case is shown similarly.

Condition (ii) implies (iii) and (iii) implies (iv) obviously.

To close the circle, the argument to deduce (i) from (iv) is already contained in the proof of the Unique Complementation Theorem, where the modular identity was first introduced. □

Example from Group Theory. Let $\mathcal{N}$ be the set of normal subgroups of a group G. Ordered by inclusion, $\mathcal{N}$ is a closure system on G and therefore a lattice. For $A, B \in \mathcal{N}$ we have

$$A \wedge B = A \cap B \quad \text{and} \quad A \vee B = \{a \cdot b : a \in A, b \in B\}$$

We leave it as an exercise to prove that the lattice $(\mathcal{N}, \subseteq)$ is modular. If G is commutative, then every subgroup is normal, and therefore the lattice of subgroups of a commutative group is always modular.

Counterexample from Group Theory. Consider the symmetric group Σ_4 of all permutations of $\{1,2,3,4\}$. Consider the following five subgroups: the alternating group A_4 made up of all even permutations; the trivial subgroup O; the three-element subgroup B generated by the permutation (123); the two-element subgroup C generated by (12)(34); and the four-element subgroup D generated by $\{(12)(34), (13)(24)\}$. Then $\{A_4, O, B, C, D\}$ in a sublattice of type M_5 in the lattice of all subgroups of Σ_4: this lattice is therefore not modular.

Jordan–Dedekind Chain Theorem (For Modular Lattices). *A lattice is modular if and only if every discrete sublattice satisfies the Jordan–Dedekind chain condition.*

Proof. If a lattice is not modular, then the Jordan–Dedekind condition obviously fails in a five-element sublattice of type M_5.

Conversely, assume that L is a modular lattice. Let the reader review the proof of Proposition 1. Was distributivity fully used or would modularity have sufficed? □

We wish to investigate bounded discrete lattices. Foremost in mind we have the affine geometry $\text{Aff } V$ of a finite dimensional vector space V over a field F. The set of affine flats, ordered by inclusion, is a lattice because $\operatorname{Aff} V$ is a closure system. Clearly $\operatorname{Pro} V$ is a sublattice of $\operatorname{Aff} V$.

Let us explore first the sublattice $\operatorname{Pro} V$ of $(\operatorname{Aff} V, \subseteq)$. It consists of all vector subspaces of V. It is a bounded lattice thanks to the trivial subspace $\{\overline{0}\}$ and the whole space V. For distinct comparable subspaces $U \subset W$ we have

$$\dim U < \dim W \leq \dim V$$

which implies that every chain in every segment $[U, W]$ is finite and consequently $(\operatorname{Pro} V, \subseteq)$ is a discrete lattice. In general it is not distributive: let $\dim V = 2$ and consider the sublattice of $\operatorname{Pro} V$ generated by three distinct one-dimensional subspaces of V. What about modularity? Note that in the lattice $\operatorname{Pro} V$,

$$U \wedge W = U \cap W \quad \text{and} \quad U \vee W = \{u + w : u \in U, w \in W\}$$

Therefore $\operatorname{Pro} V$ is also a sublattice of the inclusion-ordered modular lattice of all subgroups of the additive group of V. Thus $\operatorname{Pro} V$ is a modular lattice as well.

Turning our attention to the full lattice $(\operatorname{Aff} V, \subseteq)$ we see that it is bounded by the empty flat $\emptyset$ and the full space V. For fixed t in V, the translation map that associates with each flat $A \in \operatorname{Aff} V$ the flat $t + A$ is an order automorphism of $(\operatorname{Aff} V, \subseteq)$. Therefore, if $n \geq 1$ distinct nonempty affine flats form a chain $A_1 \subset \cdots \subset A_n$, then for any $v \in A_1$ the translates

$$(-v + A_1) \subset \cdots \subset (-v + A_n)$$

form a chain in $\operatorname{Pro} V$, and thus n cannot exceed $\dim V + 1$. Taking into account the empty flat as well, it follows that no chain in $(\operatorname{Aff} V, \subseteq)$ has more than $\dim V + 2$ members. Consequently $\operatorname{Aff} V$ is a discrete lattice.

In contrast with its sublattice $\operatorname{Pro} V$, the lattice $\operatorname{Aff} V$ is generally not modular. Indeed, in a two-dimensional space V take two distinct parallel affine lines L_1, L_2. Then $V = L_1 \vee L_2$ covers L_2, but L_1 does not cover $L_1 \wedge L_2 = \emptyset$. Thus the upper covering condition is not satisfied, and $\operatorname{Aff} V$ is not modular.

However, the lower covering condition is always satisfied in $\operatorname{Aff} V$. To prove this, let A cover $A \wedge B = A \cap B$.

Case 1. $A \wedge B \neq \emptyset$. Choose any $v \in A \cap B$. Let $t = -v$. The meet
$$(t + A) \wedge (t + B)$$
of the translates is covered by $t + A$. These translates belong to the modular sublattice Pro$V$, and thus the join
$$(t + A) \vee (t + B)$$
covers $t + B$. Retranslating by v, we find that
$$(v + t + A) \vee (v + t + B) = A \vee B$$
covers $v + t + B = B$.

Case 2. $A \wedge B = \emptyset$. Then A is a point $\{w\}$. If $B = \emptyset$, then the join $A \vee B = \{w\}$ covers B. So let $B \neq \emptyset$ and choose any $v \in B$. Let $t = -v$. The translate $t + B$ is a vector subspace of V. The join
$$(t + A) \vee (t + B)$$
is the vector subspace of V generated by the set $(t + B) \cup \{t + w\}$. As $t + w \notin t + B$,
$$\dim[(t + A) \vee (t + B)] = \dim(t + B) + 1$$
and thus $(t + A) \vee (t + B)$ covers $t + B$. Retranslating by v, we find that $A \vee B$ covers B in this case too.

An additional remarkable property of the lattice Aff V is that every flat is the union of points, i.e., every lattice element is the lub of a set of atoms. Let us remember how useful this property was in describing the structure of finite Boolean algebras.

A *geometric lattice* is defined as a bounded discrete lattice L satisfying the lower covering condition and such that every $x \in L$ is the lub of the atoms $a \leq x$.

Examples. Geometric lattices include all affine and projective geometries of finite dimensional vector spaces, all finite Boolean lattices, and for any set S and natural number n, the inclusion-ordered set
$$\{A \in \mathcal{P}(S) : A = S \text{ or Card } A \leq n\}$$

Reviewing once more the proof of Proposition 1, the reader can see that the result hinges on the lower covering condition alone:

Jordan–Dedekind Chain Theorem (For Geometric Lattices). *The Jordan–Dedekind chain condition is satisfied in every geometric lattice.*

Corollary 1 *In any geometric lattice, if $a \leq b \leq c$, then*
$$h(a,b) + h(b,c) = h(a,c)$$

Corollary 2 *Every geometric lattice is a complete lattice.*

Proof. Let o be the minimum and m the maximum of the lattice. Let A be any set of lattice elements. Let F be a finite subset of A with the value $h(o, \text{glb } F)$ as small as possible. Let G be a finite subset of A with the value $h(\text{lub } G, m)$ as small as possible. Then $\text{glb } F$ is a glb of A, and $\text{lub } G$ is a lub of A. □

Remark. Clearly the Jordan–Dedekind condition is not satisfied in every sublattice of every geometric lattice. (Consider any nonmodular geometric lattice.) Thus geometric lattices can have nongeometric sublattices. This contrasts sharply with distributivity and modularity, as these properties are inherited by all sublattices. However, some sublattices do inherit the geometric property: every convex segment sublattice $[a,b]$ of a geometric lattice is again geometric. (Verification of this is an excellent exercise in the application of the lower covering condition.)

In any geometric lattice with minimum o, the relative height $h(o,a)$ of an element a over o will be called simply the *height* of a, and it will be denoted by $h(a)$. For $a \leq b$ we have $h(a,b) = h(b) - h(a)$.

Examples. (1) In a finite power set lattice $(\mathcal{P}(S), \subseteq)$ the height of $A \in \mathcal{P}(S)$ is Card A. (2) In the geometric lattice $(\text{Pro } V, \subseteq)$ of all subspaces of a finite dimensional vector space, the height of a subspace $A \in \text{Pro } V$ is dim A. Both in $(\text{Pro } V, \subseteq)$ and in $(\text{Aff } V, \subseteq)$ the lattice elements of heights 1, 2, and 3 are the points, lines, and planes of the projective or affine geometry, respectively.

EXERCISES

1. Is the lattice of ideals of a ring modular?

2. Is the lattice of equivalence relations on a set modular? Under what conditions is it geometric?

3. What is the simplest distributive nongeometric lattice you can think of?

4. Verify that the covering graph of a geometric lattice is bipartite.

5. Design a program to determine if a finite lattice is modular and if it is geometric.

6. Design a program to find, in any finite geometric lattice L, a set A of atoms with as few elements as possible having the property that $\operatorname{lub} A = \max L$.

BIBLIOGRAPHY

Raymond BALBES and Philip DWINGER, *Distributive Lattices*. University of Missouri Press 1974. Universal algebras and categories are introduced before lattices, as a conceptual support for the material to be presented. Separate chapters are devoted to various classes of distributive lattices whose role in logic parallels that of the Boolean class.

C. BENZAKEN, Post's closed systems and the weak chromatic number of hypergraphs. *Discrete Math.* 23, 1978, pp. 77–84. An intriguing connection between Boolean functions and a chromatic number concept.

Garrett BIRKHOFF, *Lattice Theory*. A.M.S. Colloquium Publications 1967. Entire chapters of this classical introduction to lattice theory concern themselves with monoids, groups, rings, fields, vector spaces, metric and topological spaces, and universal algebras.

Peter CRAWLEY and Robert P. DILWORTH, *Algebraic Theory of Lattices*. Prentice-Hall 1973. Even though it is self-contained, this advanced text is perhaps best suited for graduate students and researchers who have some familiarity with lattices and matroids.

George GRÄTZER, *General Lattice Theory*. Birkhäuser 1978. After presenting distributive, modular, and other classes of lattices, the concept of equational class is discussed in Chapter V. The reader may wish to compare this to the theory of equational classes of universal algebras.

Paul R. HALMOS, *Lectures on Boolean Algebras*. Van Nostrand 1963. This concise account of the subject is most profitably read with some background knowledge of topology.

Peter L. HAMMER and Sergiu RUDEANU, *Boolean Methods in Operations Research and Related Areas*. Springer 1968. A classic text on Boolean and pseudo-Boolean functions, on equations and inequalities stated in terms of these, and on applications to graphs and other models used in operations research. Mathematically rigorous and presupposes only some knowledge of elementary set theory.

Hans HERMES, *Einführung in die Verbandtheorie*. Springer 1967. This introduction to lattice theory covers modular, distributive, and Boolean lattices and devotes a separate chapter to connections with mathematical logic.

Nathan JACOBSON, *Basic Algebra I*. Freeman 1985. Chapter 8 opens a window on lattice theory in the context of classical algebra. Relatively short as this chapter is, it includes Möbius functions on arbitrary ordered sets and the fundamental theorem of projective geometry.

J. KUNTZMANN, *Fundamental Boolean Algebra*. Blackie and Son 1967. A detailed study of the algebra of Boolean functions. Accessible at the undergraduate level and even without any knowledge of general lattice theory.

Sergiu RUDEANU, *Boolean Functions and Equations*. North-Holland 1974. Presents the theory of Boolean functions in the spirit of lattice theory and universal algebra.

G. SZÁSZ, *Introduction to Lattice Theory*. Academic Press 1963. A self-contained, limpid and well-rounded presentation of the basic theory. Accessible to undergraduates and worthwhile for researchers.

CHAPTER IX

MATROIDS

1. LINEAR AND ABSTRACT INDEPENDENCE

In a vector space V, independence of a set of vectors was defined with reference to the subspace closure operator on V. With respect to any closure operator σ on any set M, we can say that a subset I of M is *independent* if no $x \in I$ belongs to the closure $\sigma(I\setminus\{x\})$. Obviously every subset of an independent set is independent. We say that σ is a *matroid closure operator* if

  (i) the closed sets form a Noetherian closure system,
  (ii) for every $A \subseteq M$ and $x, y \in M$,

$$\text{if} \quad x \notin \sigma(A) \quad \text{but} \quad x \in \sigma(A \cup \{y\}), \quad \text{then} \quad y \in \sigma(A \cup \{x\})$$

The corresponding closure system is called a *matroid closure system* on M. Recall that condition (i) means that the closure system is algebraic and every closed set is finitely generated.

Examples. (1) The subspace closure system on a vector space is a matroid closure system if and only if the vector space has finite dimension. The verification of (ii) in a vector space hinges on the observation that

$$\text{if} \quad x \in \sigma(A \cup \{y\}) \setminus \sigma(A), \quad \text{then} \quad x = v + \beta \cdot y$$

where v is a linear combination of elements of A and $\beta \neq 0$: this implies $y = \beta^{-1} \cdot x - \beta^{-1} \cdot v$. (2) For natural n, the *n-uniform* matroid closure on any set M is defined by

$$\sigma(A) = \begin{cases} A & \text{if } \operatorname{Card} A \leq n \\ M & \text{otherwise} \end{cases}$$

If σ is a closure operator on a set V, then for any $M \subseteq V$ the *relative closure* operator σ_M on M is defined by letting, for $A \subseteq M$,

$$\sigma_M(A) = \sigma(A) \cap M$$

If σ is a matroid closure operator on V, then the relative closure operator σ_M is a matroid closure operator on M. For example, let σ be the subspace closure operator on a finite dimensional vector space $F^{[1,n]}$, and let M be the set of row vectors of some $m \times n$ matrix. Then σ_M defines a matroid closure on the finite set M. This is where the terminology comes from. In the research literature, matroid closure systems on infinite sets are not always included in the matroid concept.

A *matroid* is formally defined as a couple (M, σ) where σ is a matroid closure operator on M. To simplify the notation, we often write M instead of (M, σ) and refer to the members of M as the *elements of the matroid*. A set C of elements is called *dependent* if it is not independent. If no proper subset of a dependent C is dependent, then C is called a *circuit*. This is consistent with vector space terminology. Further, if $x \in \sigma(A)$ for a set A of matroid elements, then let us say that x *depends* on A (or on the elements of A).

Example. Let $G = (V, E)$ be any finite graph. Consider $\mathcal{P}(V)$ as a vector space over $\mathbb{Z}_2$. The subspace closure operator σ is a matroid closure operator on $\mathcal{P}(V)$. We have $E \subseteq \mathcal{P}(V)$. Consider the relative matroid (E, σ_E) on the edge set of G. The circuits are precisely the cycles of G. Therefore (E, σ_E) shall be called the *cycle matroid* of the graph G.

Notation. Often we use the notation $\overline{A}$ instead of $\sigma(A)$ to denote the closure of a set A. With respect to a singleton $\{x\}$, we shall write $A + x$ for $A \cup \{x\}$ and $A - x$ for $A \setminus \{x\}$. The property characterizing matroid closure systems among the Noetherian systems is then the requirement that

$$x \notin \overline{A}, \quad x \in \overline{A+y} \quad \text{imply} \quad y \in \overline{A+x}$$

Proposition 1 *If C is a circuit of a matroid M, then each element x of C depends on $C - x$.*

Proof. Some element a of C obviously depends on $C - a$, because otherwise C would be independent. But could some other element x of C fail to depend on $C - x$? Since $C - x$ is independent, we have $a \notin \overline{C \setminus \{a, x\}}$ but $a \in \overline{C - a}$, i.e., $a \in \overline{(C \setminus \{a, x\}) + x}$ and therefore

$$x \in \overline{C \setminus \{a, x\} + a}, \quad x \in \overline{C - x} \qquad \square$$

The Noetherian property of matroid closure systems postulates a fortiori that all matroid closure systems are algebraic. As in vector spaces, this implies that a set A of matroid elements is independent if and only if every finite subset of A is independent. Consequently, every circuit is finite. A set A is dependent if and only if it contains a circuit. Further, the proof of the Basis Characterization Theorem carries over, verbatim, to the general matroid context. Thus for any set B of matroid elements the following are equivalent:

(i) B is a maximal independent set,
(ii) B is a minimal generating set for the entire matroid set M,
(iii) B is independent and generates M.

Consistently with linear algebra, such a set B is called a *basis* of the matroid (M, σ). For any subset S of M, a set $I \subseteq S$ is independent with respect to $\sigma = \sigma_M$ if and only if it is independent with respect to the relative operator σ_S. Thus the maximal independent subsets of S are precisely the bases of the relative matroid (S, σ_S). By a *basis* of S we mean a basis of (S, σ_S).

Example. In the cycle matroid of a finite graph $G = (V, E)$ a set $B \subseteq E$ is a basis if and only if B is the union of the edge sets of one spanning tree in each connected component of G.

The Noetherian property of matroid closure systems implies that all independent sets in a matroid are finite. It then becomes quite obvious that the Basis Existence Theorem for vector spaces remains valid in matroids: every matroid has a basis, and each independent set of matroid elements is contained in some basis.

Let G be a set of elements that generate the entire matroid M. If B is a maximal independent subset of G, then the closure of B contains G, and therefore B generates the entire matroid M. Thus every set of elements that generate the matroid M contains a basis. Even more generally, if $G \subseteq M$ generates M and $I \subseteq G$ is independent, then there is a basis B such that $I \subseteq B \subseteq G$.

Let S be any set of matroid elements, and let $x \in \overline{S} \setminus S$. Take any basis B of S. The finite set $B \cup \{x\}$ is dependent. Take any minimal dependent subset C of $B \cup \{x\}$. Obviously C is a circuit, and $C \setminus S = \{x\}$. The reader can conclude that in any matroid M, the closure of any set $S \subseteq M$ is given by

$$\overline{S} = S \cup \{x \in M : \text{for some circuit } C, \ C \setminus S = \{x\}\}$$

Proposition 2 *If A, B are distinct circuits of a matroid and v belongs to $A \cap B$, then there is a circuit C such that $C \subseteq (A \cup B) \setminus \{v\}$.*

Proof. Could the set $(A \cup B) - v$ be independent? Let $w \in A \setminus B$. Obviously w belongs to $(A \cup B) - v$. We claim that w depends on

$$(A \cup B) - v - w$$

Observe that in the inclusion

$$A - w \subseteq (A - w - v) \cup \overline{B - v}$$

the larger set has the same closure as

$$(A - w - v) \cup (B - v) = (A \cup B) - w - v.$$

Therefore $\overline{A - w}$ is contained in the closure of $(A \cup B) - w - v$, and since $w \in \overline{A - w}$, w does belong to the closure of $(A \cup B) - w - v$ as claimed. Thus $(A \cup B) - v$ is dependent and it must contain a circuit. $\square$

Matroid Exchange Theorem. *Let V be a matroid, $M \subseteq V$ an independent set, N a proper subset of M, and $v \in V \setminus M$ such that $N \cup \{v\}$ is independent. Then for some $w \in M \setminus N$ the set*

$$(M \setminus \{w\}) \cup \{v\}$$

is independent.

Proof. (An abstract and therefore simpler version of the Steinitz Exchange proof.)

The case where $M + v$ is independent needs no discussion. Let therefore $C \subseteq M + v$ be a circuit. We have $\{v\} \subset C$, $C \not\subseteq N + v$. Let
$$w \in C \cap (M \setminus N)$$
Let us show that $M' = M - w + v$ is independent. Were this not so, take a circuit $C' \subseteq M'$. Obviously, $v \in C'$, $w \notin C'$. Thus
$$C \neq C' \quad \text{and} \quad v \in C \cap C'$$
According to Proposition 2, $(C \cup C') - v$ would be dependent, which is absurd. □

Basis Equipotence Theorem (For Matroids). *Any two bases of a given matroid are equipotent.*

Proof. Review the proof of the vector space Basis Equipotence Theorem. (In fact the situation is simpler. Since all matroid independent sets are finite, so is the set $\mathcal{E}$ of exchange functions and Zorn's Lemma is not really needed.) □

A matroid closure system will also be referred to as a *matroid geometry*. This is consistent with the convention that any closure system can be called a geometry. The closed sets of a matroid can be referred to as flats. The cardinality of any basis of a matroid (M, σ) is called the *geometric dimension* of the matroid. For $S \subseteq M$ the geometric dimension of (S, σ_S) is denoted by $\operatorname{gd} S$. (The term *rank* used in many texts is synonymous.) If S is closed and $S \subset T \subseteq M$, then $\operatorname{gd} S < \operatorname{gd} T$.

Examples. (1) The projective geometry $\operatorname{Pro} V$ on any finite dimensional vector space V is a matroid geometry. For every flat W (subspace of V) we have $\operatorname{gd} W = \dim W$. (2) The inclusion-ordered power set $\mathcal{P}(S)$ of any finite set S is a matroid geometry. For every flat A (subset of S) we have $\operatorname{gd} A = \operatorname{Card} A$. (3) Let $G = (V, E)$ be a finite graph with n vertices and k connected components. In the cycle matroid we have $\operatorname{gd} E = n - k$. (Excellent exercise in graph theory.)

The inclusion-ordered set of flats of any matroid (M,σ) is a lattice, because it is a closure system. Let us show that this lattice is geometric. It is bounded by $\sigma(\emptyset) = \overline{\emptyset}$ and $\sigma(M) = \overline{M} = M$. It is discrete because every chain of flats comprises at most $1 + \operatorname{gd} M$ flats. Does it satisfy the lower covering condition? Observe first that a flat G covers a flat $F \subset G$ if and only if

$$\operatorname{gd} F + 1 = \operatorname{gd} G$$

Assume that F covers $F \wedge G = F \cap G$. Let B be a basis of $F \wedge G$, B_G a basis of G containing B, and b any element of $F \backslash G$. Then $B \cup \{b\}$ is independent, and since $\operatorname{gd} F = \operatorname{gd}(F \wedge G) + 1$, $B \cup \{b\}$ is a basis of F. If follows that $B_G \cup \{b\}$ generates $F \vee G$, and it is in fact a basis of $F \vee G$. Thus

$$\dim F \vee G = \dim G + 1$$

and $F \vee G$ covers G. This proves that the lattice of matroid flats satisfies the lower covering condition. Finally, observe that the atoms below any flat F are the flats A with $\operatorname{gd} A = 1$ contained in F. For $v \in F$ either the singleton $\{v\}$ is dependent, in which case the element v belongs to $\overline{\emptyset}$ and thus belongs to every flat, or $\{v\}$ is independent, in which case $\overline{\{v\}} \subseteq F$ and $\overline{\{v\}}$ is an atom. Obviously F is the least upper bound of all such atoms $\overline{\{v\}}$. This completes the proof that the lattice of matroid flats is geometric.

Matroid Lattice Theorem. *The set of flats of any matroid, ordered by inclusion, constitutes a geometric lattice. Every geometric lattice is order-isomorphic to the lattice of flats of some matroid.*

Proof. The first statement was proved above.

As for the second statement, let L be a geometric lattice, let M be the set of atoms of L, and define $\sigma : \mathcal{P}(M) \to \mathcal{P}(M)$ by

$$\sigma(A) = \{x \in M : x \leq \operatorname{lub} A\}$$

Obviously σ is a closure operator on M and we can write $\overline{A}$ for $\sigma(A)$. For any closed sets A, B we have

$$A \subseteq B \quad \text{if and only if} \quad \operatorname{lub} A \leq \operatorname{lub} B$$

For every $l \in L$, the set of atoms less than or equal to l is closed and the lub of this closed set is l. Thus associating with every closed set its lub defines an order isomorphism between the lattice $\mathcal{C}$ of closed sets and L. Therefore $(\mathcal{C}, \subseteq)$ is a geometric lattice as well. In particular every chain in $(\mathcal{C}, \subseteq)$ is finite. The closure system $\mathcal{C}$ is therefore Noetherian.

Next, observe that the atoms of $(\mathcal{C}, \subseteq)$ are precisely the singleton subsets of M. To conclude that we have a matroid closure system, let

$$A \subseteq M, \quad x, y \in M, \quad x \notin \overline{A} \quad \text{but} \quad x \in \overline{A+y}$$

Obviously $\overline{A} \subset \overline{A+x} \subseteq \overline{A+y}$. In the geometric lattice $(\mathcal{C}, \subseteq)$, the atom $\overline{\{y\}} = \{y\}$ covers

$$\overline{\{y\}} \wedge \overline{A} = \{y\} \wedge \overline{A} = \emptyset$$

and therefore $\overline{\{y\}} \vee \overline{A} = \overline{A+y}$ covers $\overline{A}$. This forces

$$\overline{A+x} = \overline{A+y} \quad \text{and} \quad y \in \overline{A+x}$$

proving that $\mathcal{C}$ is a matroid closure system. We already know that $(\mathcal{C}, \subseteq)$ is order isomorphic to L. □

Indeed a closure system constitutes a matroid geometry if and only if the inclusion-ordered set of closed sets is a geometric lattice. Thus both the projective and affine geometries on any finite dimensional vector space are matroid geometries.

In any matroid, the geometric dimension of a matroid flat is equal to its height in the geometric lattice of flats. A note of caution is in order regarding the geometric dimension concept in matroids arising from a vector space V. In the matroid geometry Pro$V$ we have gd $F = \dim F$ for every flat F (subspace of V). This is no longer true in the larger matroid geometry Aff V. Affine lines have geometric dimension 2 in the affine geometry, even though affine lines that belong to Pro$V$ have geometric dimension 1 in the projective geometry. When speaking about geometric dimension, it is important to specify in which geometry it is understood. When speaking of a nonempty flat $F \in \text{Aff } V$, its dimension $\dim F$ as defined in Chapter VI is sometimes emphatically called *algebraic dimension*. No one will then mistake an affine flat of algebraic dimension 1 for a singleton.

EXERCISES

1. Write a program that, given any matroid on a finite set M and an independent set I, finds a basis of M containing I. Use a list of circuits to specify the matroid.

2. Show that if (M,σ) is any matroid, then there is a (unique) matroid (M,σ^*) on the same set of elements such that the bases of (M,σ^*) are the various sets $M\backslash B$ where B is a basis of (M,σ). Describe (E,σ^*) if (E,σ) is the cycle matroid of a graph.

3. Describe those graphs (V,E) where the graphically convex sets form a matroid geometry on V. Are there lattices where the order-convex sublattices form a matroid geometry?

4. Let $E:F$ be a field extension. Call a subset K of E closed if no polynomial over the subfield of E generated by $F \cup K$ has a root in $E\backslash K$. Verify that this indeed defines a closure system on E. What is the smallest possible closed set? Describe situations when this closure system is a matroid, and situations when it is not.

5. Let V be the set of bases of a matroid. Define a graph on vertex set V by letting a pair $\{B,C\}$ of bases form an edge whenever $\text{Card}(B + C) = 2$. What can you say about this graph?

6. A maximal proper flat in any matroid is called a *hyperplane*. Verify that this generalizes the affine hyperplane concept. What results of affine geometry involving hyperplanes can be carried over to matroids?

7. Let σ be a closure operator on a set V and let $M \subseteq V$. Verify that the closed sets of the relative closure opertor σ_M are the sets $K \cap M$, where K is a closed set of σ.

2. MINORS AND TUTTE POLYNOMIALS

Let (M,σ) be a matroid and let $S \subseteq M$, $T = M\backslash S$. Remember that the relative closure operator σ_S was defined on S by

$$\sigma_S(A) = \sigma(A) \cap S \qquad \text{for all} \quad A \subseteq S$$

The matroid (S, σ_S) is referred to as the *restriction* of the matroid structure of M to S, or synonymously, as the matroid obtained by *deletion* of T. Another important derivative matroid on S is the *contraction* (S, τ) where

$$\tau(A) = \sigma(A \cup T) \cap S \quad \text{for all} \quad A \subseteq S$$

This matroid on S is said to be obtained by *contraction to S* or by *contraction of the elements of T*. Omitting the operator σ from the notation, the restriction to S of a matroid M is simply denoted by $M|S$ or $M - T$. The contraction to S is denoted by $M : S$ or $M \cdot T$. (Warning for insiders: this is not a universally accepted notation.) If T is a singleton $\{e\}$, then we write simply $M - e$ or $M \cdot e$. This is made to look on purpose like deletion and contraction of graph edges.

Let S be a set of elements of a matroid M. Observe that for every $A \subseteq S$ the geometric dimension of A in $M|S$ coincides with its geometric dimension in M.

For $S \subseteq T \subseteq M$ the matroid $(M|T) \cdot S$ is called a *minor* of M. Most remarkable is the case where both S and T are closed:

Matroid Interval Theorem. *Let A be a flat of a matroid M. For any matroid N, let $\mathcal{L}(N)$ denote the lattice of flats of N.*

(i) *The lower section $(\leftarrow, A]$ in $\mathcal{L}(M)$ coincides with $\mathcal{L}(M|A)$.*

(ii) *The upper section $[A, \rightarrow)$ in $\mathcal{L}(M)$ is order isomorphic to $\mathcal{L}(M \cdot A)$.*

(iii) *If $S \subseteq T$ are flats of M, then the segment $[S, T]$ in $\mathcal{L}(M)$ is order isomorphic to $\mathcal{L}((M|T) \cdot S)$.*

Proof. (i) Closure in $M|A$ is given by

$$\overline{X} = \sigma(X) \cap A \quad \text{for all} \quad X \subseteq A$$

Since A is closed in M, $X \subseteq A$ is closed in $M|A$ if and only if X is closed in M.

(ii) With B the set complement of A in M, closure in $M \cdot A = M : B$ is given by

$$\overline{X} = \sigma(X \cup A) \cap B \quad \text{for} \quad X \subseteq B$$

Clearly $\overline{X} = X$ if and only if $X \cup A$ is closed in M. The order isomorphism from $[A, \to)$ to $\mathcal{L}(M \cdot A)$ is given by associating with each flat F of M containing A the set $F \cap B$ (the set of elements of F not in A).

(iii) Combine (i) and (ii). □

In the remainder of this section we restrict our attention to matroids with a finite number of elements. We define the *Tutte polynomial* $T(M)$ of a matroid as a certain polynomial in two indeterminates X, Y over $\mathbb{Z}$. If M has geometric dimension $\gd M = d$ then $T(M)$ is

$$\sum_{S \subseteq M} (X-1)^{d - \gd S} (Y-1)^{\Card S - \gd S}$$

Note that $\gd S \leq d$ and $\gd S \leq \Card S$ for every $S \subseteq M$, and remember that $(X-1)^0 = (Y-1)^0 = 1$.

An element e of a matroid M is called an *isthmus* if it belongs to every basis of M, and it is called a *loop* if it belongs to no basis of M. Clearly e is an isthmus if and only if it does not belong to any circuit, and it is a loop if and only if the singleton $\{e\}$ is a circuit. If e is an isthmus or a loop of a matroid M, then $M \cdot e = M - e$. Otherwise

$$\gd(M \cdot e) = \gd M - 1 \quad \text{and} \quad \gd(M - e) = \gd M$$

Example. In the cycle matroid on the edge set of a finite connected graph G, an edge e is an isthmus if and only if $G - e$ is disconnected. There are no loops in graphic cycle matroids, according to our definition of graphs.

Recursion Formulas. *Let e be any element of a matroid M. Then the Tutte polynomial of M is expressed in terms of the Tutte polynomials of $M - e$ and $M \cdot e$ as follows:*

$$T(M) = \begin{cases} X \cdot T(M \cdot e) & \text{if e is an isthmus} \\ Y \cdot T(M - e) & \text{if e is a loop} \\ T(M - e) + T(M \cdot e) & \text{otherwise} \end{cases}$$

Proof. In the definition of $T(M)$ the sum is taken over all $S \in \mathcal{P}(M)$. Let
$$P_1 = \{S \in \mathcal{P}(M) : e \notin S\}, \qquad P_2 = \mathcal{P}(M) \backslash P_1$$

Let us write d for $\operatorname{gd} M$ and for any $S \subseteq M$ let us abbreviate $\operatorname{Card} S$ as $|S|$ and the term $(X-1)^{d-\operatorname{gd} S}(Y-1)^{|S|-\operatorname{gd} S}$ as $t(S)$. We have

$$T(M) = \sum_{S \in P_1} t(S) + \sum_{S \in P_2} t(S) = \sum_{S \in P_1} [t(S) + t(S \cup \{e\})]$$

If e is an isthmus, then $\operatorname{gd}(S \cup \{e\}) = \operatorname{gd} S + 1$ for every $S \in P_1$ and

$$t(S) + t(S \cup \{e\}) = (X-1)^{d-\operatorname{gd} S}(Y-1)^{|S|-\operatorname{gd} S}$$
$$+ (X-1)^{d-\operatorname{gd} S-1}(Y-1)^{|S|-\operatorname{gd} S}$$
$$= X(X-1)^{d-\operatorname{gd} S-1}(Y-1)^{|S|-\operatorname{gd} S}$$

Since e is an isthmus, we have $\operatorname{gd} M \cdot e = d - 1$, and for every $S \in P_1$ the geometric dimension of S in $M \cdot e$ equals $\operatorname{gd} S$ in M. Therefore

$$T(M) = X \sum_{S \in P_1} (X-1)^{(d-1)-\operatorname{gd} S}(Y-1)^{|S|-\operatorname{gd} S} = X \cdot T(M \cdot e)$$

If e is a loop then $\operatorname{gd}(S \cup \{e\}) = \operatorname{gd} S$ for every $S \in P_1$ and

$$t(S) + t(S \cup \{e\}) = (X-1)^{d-\operatorname{gd} S}(Y-1)^{|S|-\operatorname{gd} S}$$
$$+ (X-1)^{d-\operatorname{gd} S}(Y-1)^{|S|+1-\operatorname{gd} S}$$
$$= Y(X-1)^{d-\operatorname{gd} S}(Y-1)^{|S|-\operatorname{gd} S}$$

Since e is a loop, we have $\operatorname{gd}(M-e) = d$. Also, for every $S \in P_1$ the geometric dimension of S in $M - e$ coincides with $\operatorname{gd} S$ in M. Therefore

$$T(M) = Y \sum_{S \in P_1} (X-1)^{d-\operatorname{gd} S}(Y-1)^{|S|-\operatorname{gd} S} = Y \cdot T(M-e)$$

If e is neither a loop nor an isthmus, then $\operatorname{gd}(M-e) = d$ and for every $S \in P_1$ the geometric dimension of S in $M - e$ coincides with gd S in M. Therefore

$$\sum_{S \in P_1} t(S) = T(M - e)$$

We also have $\gd(M \cdot e) = d - 1$, and for every $S \in P_1$ the geometric dimension $\gamma\delta S$ of S in $M \cdot e$ is 1 less than the geometric dimension $\gd(S \cup \{e\})$ of $S \cup \{e\}$ in M. For every $S \in P_1$ we have

$$d - \gd(S \cup \{e\}) = \gd M \cdot e + 1 - (\gamma\delta S + 1) = \gd M \cdot e - \gamma\delta S$$
$$|S \cup \{e\}| - \gd(S \cup \{e\}) = |S| + 1 - (\gamma\delta S + 1) = |S| - \gamma\delta S$$

and therefore

$$\sum_{S \in P_1} t(S \cup \{e\}) = T(M \cdot e)$$

We conclude that $T(M) = T(M - e) + T(M \cdot e)$. □

The values of the Tutte polynomial of a matroid provide much combinatorial information. Here we discuss evaluations in the ring $\mathbb{Z}[X,Y]$ itself. Consider any polynomial $P \in \mathbb{Z}[X,Y]$,

$$P = \sum_{i,j} c_{ij} X^i Y^j$$

If $a, b \in \mathbb{Z}[X,Y]$, then the *value* $P(a,b)$ of P at a and b is the element

$$\sum_{i,j} c_{ij} a^i b^j$$

of $\mathbb{Z}[X,Y]$. We also say that $P(a,b)$ is the value of P at $X = a$ and $Y = b$. Of course $P(X,Y) = P$ for all $P \in \mathbb{Z}[X,Y]$. For another example, the value of $XY + 3$ at $X - 1$ and $2X + 2$ is $2X^2 + 1$. The value of XY at 3 and -2 is -6. The computation of values is greatly facilitated by the observation that the function mapping P to $P(a,b)$ is a ring homomorphism from $\mathbb{Z}[X,Y]$ to itself. This is all that we need to know about $\mathbb{Z}[X,Y]$ for the present purposes.

Basis Counting Theorem. *The number of bases of a finite matroid is the value of the Tutte polynomial at $X = 1, Y = 1$.*

Proof. $0^0 = 1$. □

If M is a finite matroid, L the set of its loops, and $G = (V, E)$ a finite graph, then a surjective function $g : M \setminus L \to E$ is called a *graphical representation* of M in G provided the closed sets of M are precisely those of the form $L \cup g^{-1}[K]$ where K is closed in the cycle matroid of G. Note that g is not required to be injective. Indeed, for $e' \neq e$ we have $g(e') = g(e)$ if any only if $\{e', e\}$ is a circuit. In this case e' and e are said to be *parallel*. The element e' is then a loop in $M \cdot e$.

Examples. (1) Let M be the matroid on $\{1, 2, 3, 4, 5, 6\}$ with closed sets

$$\{5,6\}, \{1,5,6\}, \{2,5,6\}, \{3,4,5,6\}, \{1,2,3,4,5,6\}$$

Then a graphical representation of M in K_3 is depicted in Figure 9.1 (top left). (Each edge is labeled by the matroid elements mapped to it.) (2) The two-uniform matroid on a five-element set has no graphical representation.

For the cycle matroid of any finite graph $G = (V, E)$ the identity mapping on E is a graphical representation in G, called the *canonical representation*. Any matroid that has a graphical representation is called a *graphic matroid*. Obviously the cycle matroid of any finite graph is graphic. The proof of the following is left as an exercise.

Proposition 3 *In a graphic matroid M with loop set L and representation g in a graph G, let $e \in M \setminus L$.*

(i) $f = g|(M \setminus L) \setminus \{e\}$ *is a graphical representation of the matroid $M - e$, either in G or in the edge-deleted graph $G - g(e)$, according to whether M does or does not have an element parallel to e.*

(ii) *The matroid $M \cdot e$ has a graphical representation h in $G \cdot g(e)$ given for every nonloop element a of $M \cdot e$ by*

$$h(a) = g(a) \quad \text{for} \quad g(a) \cap g(e) = \emptyset,$$

$$h(a) = \{v, \bar{e}\} \quad \text{for} \quad g(a) = \{v, x\} \text{ with } x \in g(e),$$

where $\bar{e}$ is the vertex of $G \cdot e$ not in $V \setminus \{e\}$.

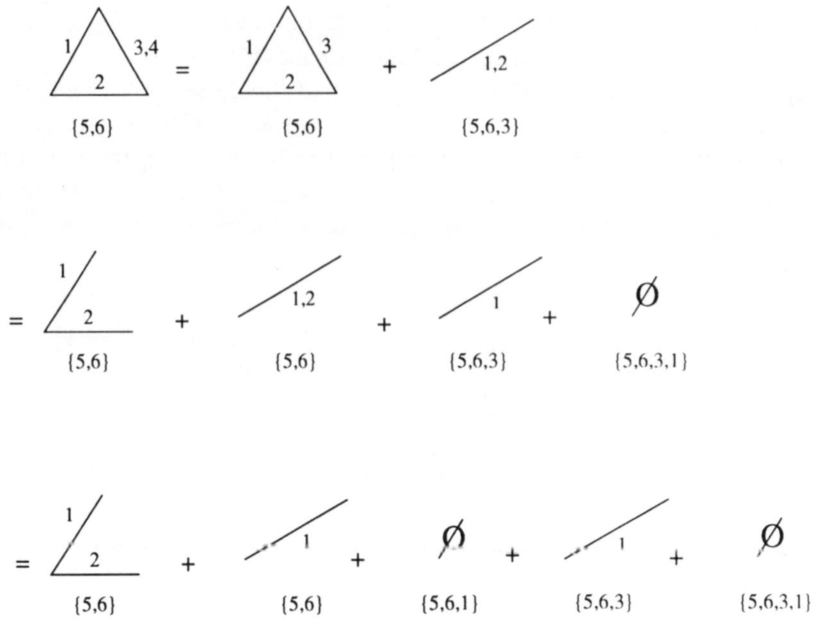

FIGURE 9.1 Computing the Tutte polynomial.

(diagram showing step-by-step Tutte polynomial computation)

$$= X^2Y^2 + XY^2 + Y^3 + XY^3 + Y^4$$

Readers are urged to visualize graphic matroids as if they were "graphs with multiple edges."

The Recursion Formulas, together with the observation that the Tutte polynomial of the empty matroid ($M = \emptyset$) is 1, allow the actual computation of $T(M)$ for any matroid M. If M consists of m isthmuses and n loops and no other elements, then $T(M) = X^m Y^n$. Otherwise we can start the computation with any element e that is neither a loop nor an isthmus. Let us illustrate this for a graphic matroid on $\{1,\ldots,6\}$. Proposition 3 should be kept in mind. In Figure 9.1 graphical representations are specified pictorially by writing each nonloop matroid element next to the corresponding graph edge and listing the loop set underneath. In manual calculations, this lines-and-numbers hieroglyph is used as a notation for the Tutte polynomial itself.

Let g be a graphical representation of a matroid M in a graph G with k connected components. Since the chromatic polynomial $p(G)$ is divisible by X^k, we have $p(G) = X^k Q$ for a certain $Q \in \mathbb{Z}[X]$. The *chromatic polynomial* $Q(M,g)$ of the representation g of M is defined by letting

$$Q(M,g) = Q \quad \text{if } M \text{ is loopless}$$

and letting $Q(M,g)$ be the zero polynomial if M has any loops. A matroid can have not only different representations in a given graph but also representations in different, nonisomorphic graphs. For example, an m-uniform matroid on m elements has representations in every forest with m edges. However, in view of the form the chromatic polynomial of a forest takes, it is obvious that all these graphical representations have the same chromatic polynomial. We shall see below that indeed any two representations of any matroid M have the same chromatic polynomial depending on M only.

Let us show that if g is any graphical representation of a matroid M in a graph G, then for the chromatic polynomial $Q(M,g)$ we have

$$Q(M,g) = (-1)^{\operatorname{gd} M} T(M; 1-X, 0)$$

where $T(M; 1-X, 0)$ denotes the value of the Tutte polynomial of M at $1-X$ and 0. We shall use induction on the number of nonisthmus elements of M.

The case of M having any loop e is quickly disposed of, because in this case $Q(M,g)$ is by definition the zero polynomial, and using loop recursion,

$$T(M; 1-X, 0) = 0 \cdot T(M-e; 1-X, 0) = 0$$

If every element of M is an isthmus, then G is a forest, and $\operatorname{gd} S = \operatorname{Card} S$ for every $S \subseteq M$. If $\operatorname{Card} M = \operatorname{gd} M = m$ and G has k components, then m is necessarily the number of edges of G,

$$p(G) = X^k (X-1)^m = X^k Q(M,g)$$
$$Q(M,g) = (X-1)^m$$

and

$$T(M; 1-X, 0) = \sum_{S \subseteq M} (-X)^{m - \operatorname{Card} S} (-1)^0 = \sum_{S \subseteq M} (-X)^{m - \operatorname{Card} S}$$

Since for every finite set M we have the polynomial identity

$$(1-X)^{\operatorname{Card} M} = \sum_{S \subseteq M} (-X)^{\operatorname{Card} M - \operatorname{Card} S}$$

(easily verifiable by induction on Card M), the desired equality

$$Q(M,g) = (-1)^m T(M; 1-X, 0)$$

follows in the case where every element of M is an isthmus.

Now let us proceed with the inductive step. Assume that

$$Q(N,h) = (-1)^{\operatorname{gd} N} T(N; 1-X, 0)$$

for every graphical representation h of any matroid N having less nonisthmus elements than M. We can assume that M is loopless. Let e be a matroid element that is not an isthmus. We have

$$\operatorname{gd}(M - e) = \operatorname{gd} M = \operatorname{gd}(M \cdot e) + 1$$

because e is neither an isthmus nor a loop. By Proposition 3, $M - e$ has a representation f in G or $G - g(e)$, and $M \cdot e$ has a representation h in $G \cdot g(e)$. We distinguish two cases.

Case 1. There is a parallel matroid element $e' \neq e$ such that $g(e') = g(e)$. In this case $Q(M,g)$ coincides with the chromatic polynomial of the representation $f = g|M\backslash\{e\}$ of $M - e$ in G. Also e' is a loop in $M \cdot e$. Since $M - e$ has less nonisthmus elements than M,

$$Q(M,g) = Q(M-e, f) = (-1)^{\operatorname{gd}(M-e)} T(M-e; 1-X, 0)$$
$$= (-1)^{\operatorname{gd} M} T(M-e; 1-X, 0) + 0$$
$$= (-1)^{\operatorname{gd} M} T(M-e; 1-X, 0) + (-1)^{\operatorname{gd} M} T(M \cdot e; 1-X, 0)$$
$$= (-1)^{\operatorname{gd} M} T(M; 1-X, 0)$$

Case 2. There is no parallel $e' \neq e$ with $g(e') = g(e)$. Then $M \cdot e$ is loopless as well. Both graphs $G - g(e)$ and $G \cdot g(e)$ have the same number k of connected components as G. We have

$$p(G) = p(G - g(e)) - p(G \cdot g(e))$$
$$X^k Q(M,g) = X^k Q(M-e, f) - X^k Q(M \cdot e, h)$$
$$Q(M,g) = Q(M-e, f) - Q(M \cdot e, h)$$

Since both $M - e$ and $M \cdot e$ have less nonisthmus elements than M, the inductive hypothesis implies

$$Q(M,g) = (-1)^{\text{gd}(M-e)} T(M - e; 1 - X, 0)$$

$$- (-1)^{\text{gd}(M \cdot e)} T(M \cdot e; 1 - X, 0)$$

$$= (-1)^{\text{gd} M} T(M - e; 1 - X, 0) + (-1)^{\text{gd} M} T(M \cdot e; 1 - X, 0)$$

$$= (-1)^{\text{gd} M} T(M; 1 - X, 0)$$

By applying this to the canonical representation of the cycle matroid of a graph G, we obtain the following result.

Proposition 4 *Let T be the Tutte polynomial of the cycle matroid of a finite graph G with n vertices and k connected components. Then the chromatic polynomial of G is*

$$(-1)^{n-k} X^k T(1 - X, 0)$$

Thus the Tutte polynomial of a graph's cycle matroid encodes the essence of the graph's chromatic polynomial. (In the literature, the reader may encounter this encoding via other related polynomials.) Also, for any connected graph, the Basis Counting Theorem, applied to the cycle matroid, implies, via the canonical representation that maps bases to edge sets of spanning trees, that the number of spanning trees is given by the value $T(1,1)$ of the Tutte polynomial.

EXERCISES

1. Let σ be a closure operator on a set M, $S \subseteq M$, and $T = M \setminus S$. Verify that a closure operator τ is defined on S by

$$\tau(A) = \sigma(A \cup T) \cap S \quad \text{for} \quad A \subseteq S$$

Verify that the closed sets of τ are the sets $K \cap S$ where K is a closed set of σ containing T.

2. Let e be any element of a matroid M and let $C \subseteq M \setminus \{e\}$. Show that C is a circuit of $M - e$ if and only if it is a circuit of M.

3. Let S be any set of elements of a matroid M and let $T = M\setminus S$. Show that a set $D \subseteq S$ is dependent in $M \cdot T$ if and only if there is a set $I \subseteq T$, independent in M, such that $D \cup I$ is dependent in M.

4. Verify that the rank of any minor of a matroid M is at most $\gd M$.

5. Verify that an element e of a matroid M is an isthmus if and only if $\gd(M - e) = \gd M - 1$, and it is a loop if and only if $\gd(M \cdot e) = \gd M$.

6. Verify that two elements e, d of a matroid M form a circuit if and only if neither one is a loop but d is a loop in $M \cdot e$.

7. Let g be a graphical representation of a matroid M. Show that $I \subseteq M$ is independent if and only if I contains no loops and no parallel elements and $g[I]$ contains no cycle.

8. Show that all minors of a graphic matroid are graphic.

9. For a finite dimensional vector space V, show that every minor of $\Aff V$ is similar to $\Pro W$ or to $\Aff W$ for some subspace W. Conversely, verify that the affine geometry of each subspace W is a minor of $\Aff V$. Is $\Pro W$ similar to a minor of $\Aff V$?

10. Let T be the Tutte polynomial of a finite matroid M. What is the significance of the values $T(0,0), T(1,2), T(2,1), T(2,2)$?

11. Write a program to compute the Tutte polynomial of any finite matroid.

12. Write a program to decide if a given finite matroid is graphic.

3. GREEDY OPTIMIZATION PROCEDURES

For a finite set S, let $\mathcal{I} \subseteq \mathcal{P}(S)$. A *greedy procedure in S with respect to $\mathcal{I}$* is a finite sequence $(a_i : i < n) = (a_0, \ldots, a_{n-1})$ of distinct elements of S such that

(i) $\{a_i : i < m\}$ is in $\mathcal{I}$ for every $m \leq n$,
(ii) $\{a_i : i < n\} \cup \{b\}$ is not in $\mathcal{I}$ for any $b \in S \setminus \{a_i : i < n\}$.

Note that (i) applied to $m = 0$ implies $\emptyset \in \mathcal{I}$. The set
$$\{a_i : i < n\} = \{a_0, \ldots, a_{n-1}\}$$
is of cardinality n; it is called the *output* or *result* of the procedure. Obviously
$$n \leq \max\{\operatorname{Card} X : X \in \mathcal{I}\}.$$
If equality holds, then the output is called *optimal in S*.

The term "greedy" refers to the simple-mindedness of the step-by-step enlargement process followed to find a maximal "interesting" subset of S. Only those subsets of S that belong to $\mathcal{I}$ are interesting.

Example. Consider a graph whose edges form a path of length 5. Let S be the set of those five edges. Let $\mathcal{I}$ consist of those edge subsets whose members are pairwise disjoint. There are $3 + 2 + 2 + 2 + 3$ greedy procedures in S with respect to $\mathcal{I}$. Of these only 6 result in a set of $3 = \max\{\operatorname{Card} X : X \in \mathcal{I}\}$ edges.

A set $\mathcal{H}$ of sets is called *hereditary* if every subset of every member of $\mathcal{H}$ belongs to $\mathcal{H}$, i.e., if $\mathcal{H}$ is a lower section of $(\mathcal{P}(\cup \mathcal{H}), \subseteq)$.

Examples. Every power set is hereditary. The set of independent sets of a matroid is hereditary. For any graph $G = (V, E)$ the set of those subsets of V that induce in G a complete subgraph is hereditary. The set of those subsets of V that induce in G a cycle is not hereditary, unless it is void.

For any set S, the set of all hereditary subsets of $\mathcal{P}(S)$ is a closure system on $\mathcal{P}(S)$. In particular, if $\mathcal{I} \subseteq \mathcal{P}(S)$ is hereditary and $A \subseteq S$, then $\mathcal{I} \cap \mathcal{P}(A)$ is hereditary as well.

Proposition 5 *For any hereditary set $\mathcal{I}$ of subsets of a finite set S the following are equivalent:*

(i) *For every $A \subseteq S$, every greedy procedure in A with respect to $\mathcal{I} \cap \mathcal{P}(A)$ yields a result optimal in A.*

(ii) *For every $A \subseteq S$, all maximal members of the inclusion-ordered set $\mathcal{I} \cap \mathcal{P}(A)$ have the same cardinality.*

Proof. Assume (i). If M is a maximal member of $\mathcal{I} \cap \mathcal{P}(A)$, let

$$M = \{a_0, \ldots, a_{n-1}\}$$

The sequence $(a_i : i < n)$ is a greedy procedure in A with the optimal result M, and therefore

$$\operatorname{Card} M = \max\{\operatorname{Card} X : X \in \mathcal{I} \cap \mathcal{P}(A)\}$$

Conversely, assume (ii). Take an $M \in \mathcal{I} \cap \mathcal{P}(A)$ with largest possible cardinality m, $M = \{a_0, \ldots, a_{m-1}\}$. Certainly $(a_i : i < m)$ is a greedy procedure with optimal output in A. If $(b_i : i < n)$ is another greedy procedure in A, then the output set $B = \{b_i : i < n\}$ is a maximal member of $\mathcal{I} \cap \mathcal{P}(A)$. By (ii) $n = m$ and therefore the output B is optimal as well. □

A hereditary subset $\mathcal{I}$ of $\mathcal{P}(S)$ satisfying the conditions of the above proposition is called *greedy optimizable* in S. It follows from the Basis Equipotence Theorem for matroids that the set $\mathcal{I}$ of independent sets of a matroid M is greedy optimizable in M. This accounts for much of the attention matroid theory has received in operational research. However, the remarkable combinatorial properties of linear and other independence structures were identified by van der Waerden and Hassler Whitney in the 1930s, and it was only later that preoccupations with military and industrial optimization evolved into a paramathematical discipline of management science.

Matroids are fully determined by their independent sets. In a matroid M, a set C is closed if and only if for every independent set $I \subseteq C$ and every $x \in M \setminus C$, the set $I \cup \{x\}$ is independent.

Greedy Characterization of Matroids. *A nonempty set $\mathcal{I}$ of subsets of a finite set S is the set of independent sets of some matroid on S if and only if $\mathcal{I}$ is hereditary and greedy optimizable in S.*

Proof. Only the sufficiency of the conditions remains to be proved. Let $\mathcal{I} \subseteq \mathcal{P}(S)$ be nonempty, hereditary and greedy optimizable in S. Consider the set $\mathcal{C}$ of subsets $C \subseteq S$ with the property that for every $I \in \mathcal{I}$, $I \subseteq C$, $x \in S \setminus C$, the set $I \cup \{x\}$ again belongs to $\mathcal{I}$. Clearly $\mathcal{C}$ is a closure system on S. For every $A \subseteq S$ let

$$d(A) = \max\{\operatorname{Card} I : I \in \mathcal{I}, I \subseteq A\}$$

Obviously $A \in \mathcal{I}$ if and only if $d(A) = \text{Card } A$. Using condition (ii) of Proposition 5, verify that the closure of any $A \subseteq S$ in the system $\mathcal{C}$ is given by

$$\overline{A} = \{x \in S : d(A \cup \{x\}) = d(A)\}$$

Note that $d(A) = d(\overline{A})$ for all $A \subseteq S$. Do we then have a matroid closure? Let $x \in \overline{A \cup \{y\}}$, $x \notin \overline{A}$.

Clearly

$$d(\overline{A \cup \{x\}}) = d(\overline{A}) + 1 = d(A) + 1$$

If y did not belong to $\overline{A \cup \{x\}}$, then

$$d(\overline{A \cup \{x\}} \cup \{y\}) = d(\overline{A \cup \{x\}}) + 1 = d(\overline{A} \cup \{x\}) + 1 = d(A) + 2$$

Since $x \in \overline{A \cup \{y\}}$, the sets $\overline{A \cup \{x\} \cup \{y\}}$ and $A \cup \{y\}$ have the same closure. This implies

$$d(\overline{A \cup \{x\}} \cup \{y\}) = d(A \cup \{y\}) = d(A) + 1$$

which is contradictory. Thus y must belong to $\overline{A \cup \{x\}}$ and $\mathcal{C}$ is a matroid closure system. The independent sets of this matroid are precisely the members of $\mathcal{I}$. $\square$

Matroids arising from graphs have application potential in operations research to the extent that graphs can be used to model operational situations. Trees are highly regarded as modeling tools, and thanks to the cycle matroid structure, spanning trees in any graph can be constructed by a greedy procedure. Another useful graph-theoretical concept is that of a *matching M* in a graph $G = (V, E)$. By definition this is a set of edges no two of which have a common vertex, such as in Figure 9.2.

FIGURE 9.2 A matching with three edges.

The example given at the beginning of this section shows that the set of matchings in a graph (V,E), while being a hereditary subset of $\mathcal{P}(E)$, is generally not greedy optimizable in E. However, let us consider those subsets S of V that are contained in the edges of some matching M, $S \subseteq \cup M$. Call such a set S *matchable* and say that M is a *matching for* S. Clearly the set of matchable subsets of V is hereditary. The result that it is actually greedy optimizable in V is due to Edmonds and Fulkerson (1965).

Matching Matroid Theorem. *In any finite graph $G = (V,E)$ the matchable subsets of V are the independent sets of a matroid on V.*

Proof. Preliminary observation: in a nonempty connected graph, if no vertex is in more than two edges, then the edges altogether must form either a path or a cycle.

Let us show now that in any finite graph $G = (V,E)$, if $T \subseteq V$, then any two maximal matchable subsets S_1 and S_2 of T contain the same number of vertices. Let M_1 and M_2 be matchings for S_1 and S_2, respectively. Let

$$V_1 = \cup M_1, \qquad V_2 = \cup M_2$$

We have $S_1 = V_1 \cap T$, $S_2 = V_2 \cap T$. Consider the symmetric difference

$$M = (M_1 \setminus M_2) \cup (M_2 \setminus M_1)$$

Applying the preliminary observation to the components of the graph $D = (\cup M, M)$, we see that the edge sets of the components of D are paths and cycles. If they are cycles only, that means $V_1 = V_2$ (because every vertex on every such cycle belongs to an edge in M_1 as well as to an edge in M_2). Consequently we have

$$S_1 = V_1 \cap T = V_2 \cap T = S_2$$

if M is a union of cycles. Otherwise let $P_1,\ldots,P_n$ be those maximal paths of $D = (\cup M, M)$ that are not contained in a cycle. If P_i is a path from vertex a_i to b_i, write $E_i = \{a_i, b_i\}$. Clearly neither a_i nor b_i can belong to $V_1 \cap V_2$. All other vertices in $\cup M$ belong to $V_1 \cap V_2$. Could we have $\operatorname{Card} S_1 < \operatorname{Card} S_2$? This would only be possible if for some $1 \leq i \leq n$,

$$\operatorname{Card}(E_i \cap S_1) < \operatorname{Card}(E_i \cap S_2)$$

Then one of the two vertices in E_i, say a_i, would belong to $S_2 \setminus S_1$, without the other belonging to $S_1 \setminus S_2$:

$$a_i \in S_2 \setminus S_1, \quad b_i \notin S_1 \setminus S_2$$

(In fact $b_i \in S_2$ or $b_i \in V \setminus T$.) Certainly $a_i \in T$. Then

$$M' = (M_1 \setminus P_i) \cup (P_i \cap M_2)$$

is a matching with $S_1 \cup \{a_i\} \subseteq \cup M'$, contradicting the maximality of the matchable set S_1. This forces $\text{Card } S_1 = \text{Card } S_2$ for any two maximal matchable subsets S_1, S_2 of T. □

EXERCISES

1. Let $\mathcal{I}$ be a hereditary set of subsets of a finite set S. Are the following equivalent?
   (a) Every greedy procedure in S with respect to $\mathcal{I}$ yields a result optimal in S.
   (b) All maximal members of the inclusion-ordered set $\mathcal{I}$ have the same cardinality.

2. If M is a matching in a finite graph G, then a path P between vertices a, b is called an *alternating path* for M if $a, b \notin \cup M$ and the symmetric difference $M + P$ is a matching of higher cardinality than M. Show that for every matching of less than maximum size there is an alternating path in G.

3. Write a computer program to find a maximum size matching in any finite graph. How would you sell this program to a marriage broker who complains about too much paperwork?

4. (a) Explain how you can find a largest possible number of linearly independent rows in a given matrix, say over the field $\mathbb{Q}$.
   (b) Given a finite set of points in $\mathbb{R}^n$ (specified by rational coordinates if you wish), how can you find the dimension of the affine flat that they generate? Down to earth, describe explicitly the case $n = 3$.
   (c) In a finite connected graph, how can you find the largest possible number of edges the removal of which does not disconnect the graph?

BIBLIOGRAPHY

Anders BJÖRNER and Günter M. ZIEGLER, Introduction to greedoids. In *Matroid Applications*, ed. Neil White, Cambridge University Press, 1992. Greedoids generalize matroids and provide the proper structural framework for greedy algorithms.

J. A. BONDY and U. S. R. MURTY, *Graph Theory with Applications*. North-Holland 1979. In the 20 pages that the authors of this introductory book devote to matchings (Chapter 5), the reader will find a full and rigorous presentation of selected nontrivial structural and algorithmic results of basic importance to matching theory.

Thomas BRYLAWSKI and James OXLEY, The Tutte polynomial and its applications. In *Martroid Applications*, ed. Neil White, Cambridge University Press, 1992. Over 100 pages of up-to-date material and references. For the researcher.

Eugene L. LAWLER, *Combinatorial Optimization: Networks and Matroids*. Holt Rinehart and Winston 1976. Linear programming, graphs, matroid structures, and algorithms.

L. LOVÁSZ and M. D. PLUMMER, *Matching Theory*. North-Holland 1986. Despite its advanced level, this many-faceted monograph is accessible to the motivated novice. It contains an extensive exploration of connections between matchings and linear programming.

Satoru FUJISHIGE, *Submodular Functions and Optimization*. North-Holland 1991. A research monograph accessible to graduate students. Submodular functions include and generalize the rank function of a matroid and clarify the role that matroids play in optimization.

W. T. TUTTE, *Introduction to the Theory of Matroids*. American Elsevier 1971. Concise, even limited in scope, when seen in light of the accumulation of research since its writing and personal in approach. Its rigor and the essential nature of the concepts discussed make this text a highly effective introduction to matroids for the algebraically minded.

D. J. A. WELSH, *Matroid Theory*. Academic Press 1976. A classic reference volume on matroids. Sufficiently self-contained for the autodidact at the graduate level.

CHAPTER X

TOPOLOGICAL SPACES

1. FILTERS

An upper section F of a lattice L that is closed under the meet operation,

$$x \wedge y \in F \quad \text{for} \quad x, y \in F$$

is called a *filter in the lattice* L. The set of filters in L constitutes an algebraic closure system on L. Thus, ordered by inclusion, the set of filters is a lattice. A *proper filter* is a filter that is distinct from L itself. For any $x \in L$, the section $[x, \rightarrow)$ is a filter, called the *principal filter* generated by x. It is proper if and only if x is not a minimum of L. In a finite lattice, every nonempty filter is principal. On the other hand, consider the power set lattice $(\mathcal{P}(S), \subseteq)$ of an infinite set S. Those subsets of S that have a finite complement in S form a nonprincipal filter.

A filter's closure property under meet implies, by induction, that if B is a finite nonempty subset of a filter F, then $\text{glb}\, B \in F$.

Let L be a lattice with a minimum u. A filter F is proper if and only if $u \notin F$. An *ultrafilter* is a maximal member of the inclusion-ordered set of proper filters. Zorn's Lemma applies to this ordered set: every proper filter is included in some ultrafilter. A principal filter $[x, \to)$ is an ultrafilter if and only if x is an atom.

Let A be any nonempty set of elements in a lattice L with minimum u. If A is contained in some ultrafilter, then obviously $\text{glb } B \neq u$ for each finite nonempty subset B of A. Conversely, if $\text{glb } B \neq u$ for every finite nonempty $B \subseteq A$, then

$$\{x \in L : x \geq \text{glb } B \text{ for some finite nonempty } B \subseteq A\}$$

is a proper filter containing A and contained in some ultrafilter.

If L is a nontrivial Boolean lattice with minimum u, and $x \in L$, $x \neq u$, then clearly no ultrafilter can have as members both x and its complement x'. On the other hand, suppose x' is not a member of some ultrafilter F. Consider the set $A = F \cup \{x\}$. Could it be that for some finite subset B of A, $\text{glb } B = u$? Only if $x \in B$. We would have

$$x' = x' \vee \text{glb } B = \text{glb}\{x' \vee b : b \in B\}$$

$$= \text{glb}\{x' \vee b : b \in B, b \neq x\} = x' \vee \text{glb}(B\setminus\{x\})$$

and therefore $\text{glb}(B\setminus\{x\}) \leq x'$. Since $B\setminus\{x\} \subseteq F$, we have

$$\text{glb}(B\setminus\{x\}) \in F$$

and x' would have to belong to F as well, which is contrary to assumption. Thus $\text{glb } B \neq u$ for all finite subsets B of $A = F \cup \{x\}$ and therefore $A \subseteq F'$ for some ultrafilter F'. Since $F \subseteq A$ is already an ultrafilter, we must have $F' = F$ and $x \in F$. We have shown that if F is an ultrafilter in a nontrivial Boolean lattice L, and $x \in L$, then one and only one of x or its complement x' belongs to F.

If F is a proper filter in a Boolean lattice L, but not an ultrafilter, then it is properly contained in some ultrafilter F', $F \subset F'$. For x in $F'\setminus F$, neither x nor x' belongs to F. Thus a filter F in a nontrivial Boolean lattice L is an ultrafilter if and only if for each $x \in L$, one and only one of x or x' belongs to F.

Let L be a Boolean lattice and let L^* be its dual. Of course L^* too is a Boolean lattice. Let R^* be the ring associated with the Boolean lattice L^*. The ideals of R^* are precisely the nonempty filters in L. The principal ideals are the principal filters. The maximal ideals are the nonempty ultrafilters.

In the sequel we shall only be interested in filters in a power set lattice $(\mathcal{P}(S), \subseteq)$. Such a filter F is called a *filter on the set* S. If S' is another set and $g : S \to S'$ any function, then

$$\{g[A] : A \in F\}$$

generates a filter on S' that we call the *image filter* of F and denote by $g(F)$. The reader should verify that if F is a proper filter, then so is $g(F)$, and if F is an ultrafilter on S, then $g(F)$ is an ultrafilter on S'.

If F and F' are filters on a set S and $F \subseteq F'$, then F' is called a *refinement* of F. For $S \neq \emptyset$, the ultrafilters F on S are characterized among all proper filters by the property that if S is expressed as a finite union of subsets

$$A_1 \cup \cdots \cup A_n = S$$

then some of the A_i must belong to F. The principal filters on S are the sets

$$F_X = \{Y \in \mathcal{P}(S) : X \subseteq Y\}$$

for $X \subseteq S$. The principal ultrafilters are the sets

$$U_x = \{Y \in \mathcal{P}(S) : x \in Y\}$$

for $x \in S$. There are nonprincipal ultrafilters if and only if S is infinite. In this case all ultrafilter refinements of the *cofinite filter*

$$\{X \in \mathcal{P}(S) : S \backslash X \text{ is finite}\}$$

are nonprincipal. On the other hand, if an ultrafilter F on S has a finite member M, we can take one with least possible cardinality, and M must then be a singleton. This is so because otherwise

$$\bigcap_{x \in M} (S \backslash \{x\}) = S \backslash M$$

would belong to F, which is absurd. It follows that a nonprincipal ultrafilter has only infinite members. In conclusion, on any infinite set S, the nonprincipal ultrafilters are precisely the ultrafilter refinements of the cofinite filter.

Some discrete mathematics is definitely not finite mathematics.

EXERCISES

1. Prove the "Stone Representation Theorem": every Boolean lattice, finite or infinite, is isomorphic to a sublattice of a power set lattice. (Hint: consider the set of ultrafilters.)

2. Verify that ultrafilters are characterized among all nonempty proper filters F in a Boolean lattice L by the property that for all $x, y \in L$, $x \vee y \in L$ if and only if x or y belongs to L.

3. Verify that a nonempty proper filter F in a Boolean lattice L is an ultrafilter if and only if $L \setminus F$ is a filter in the dual lattice L^*.

4. Let L be any lattice. Is the lattice of filters in L a sublattice of the lattice of all sublattices of L?

5. Describe the ultrafilter refinements of a given principal filter on a set S.

6. Let F be a filter on a set S and let $g : S \to S'$ be any map. Verify that the image filter $g(F)$ on S' coincides with the upper section of $(\mathcal{P}(S'), \subseteq)$ generated by the sets $g[A]$, $A \in F$.

2. CLOSURE, CONVERGENCE, AND CONTINUITY

Algebra concerns itself, initially at least, with the question of how to obtain a result from certain "components" by applying certain operations: we get 5 from 1 and 4 by applying addition, we get 6 from 2 and 3 by taking an lcm. Topology's basic question is how to get to a result from certain "approximants" by applying some concept of closeness: we get to 0 from the approximants $1/n$, $n \in \omega$; and we get to a circle from a set of approximating polygons. The concept of closeness is clear when we approximate 0 with $1/n$, $n \in \omega$; it is less obvious for the circle and the polygons.

Let $\mathcal{C}$ be a closure system on a set T such that

(i) the empty set is closed,
(ii) the union of any two closed sets is closed.

Then $(T, \mathcal{C})$ is called a *topological space*. Abusively, we often refer to the "topological space T" when $(T, \mathcal{C})$ is understood. The closure system $\mathcal{C}$ is also called a *topology* or *topological closure system* on T. The elements of T are often called *points*. A point x of T is *close to*, or *approximated by*, a set $A \subseteq T$ if x belongs to the closure $\overline{A}$ of A in the closure system $\mathcal{C}$. We also say in this case that x is a *closure point*, or *adherence point*, of A.

A subset of T is called *open* if its complement in T is closed.

Examples. Let T be any totally ordered set. Call a set $C \subseteq T$ closed if no nonempty subset of C has a lub or glb outside C. The closure system thus defined is a topology on T, called *chain topology*. If $T = \mathbb{R}$, with the standard order, then we have the *standard real topology*. Here, indeed, 0 is approximated by the set $\{1/n : n \in \omega\}$. The interval $[0,1]$ is closed in this topology, while the negative real numbers form an open set. The set $\mathbb{Q}$ of rationals is neither closed nor open, and the entire set $\mathbb{R}$ is of course both closed and open. The set $\mathbb{Q}$ approximates every real number.

On any set T, there is at least the *discrete topology* in which all subsets of T are closed and the *trivial topology* having only $\emptyset$ and T as closed sets.

If $\mathcal{C} \subseteq \mathcal{C}'$ for two topologies on the same set T, then $\mathcal{C}'$ is said to be *finer*, or to be a *refinement* of $\mathcal{C}$. This defines the *refinement order* on the set τ of all topologies on T. The trivial and discrete topologies are the min and max of τ, respectively. Also, the set τ of all topologies on T is a closure system on the power set $\mathcal{P}(T)$, and thus it is a complete lattice under the refinement order.

For any set S of points of a topological space $(T, \mathcal{C})$ the *inherited topology* on S is by definition $\{S \cap C : C \in \mathcal{C}\}$. When we refer to the topological properties of such a subset S of T, it is understood that we have the inherited topology in mind.

Example. The set of integers is discrete in the standard topology of the real numbers.

Algebraic homomorphisms preserve operations. In topology, we are concerned with functions that preserve approximation. A function $h : T \to T'$ between topological spaces T and T' is said to be *continuous* when for every $x \in T$ and $A \subseteq T$, if A approximates x, then $h[A]$ approximates $h(x)$.

Proposition 1 *The following conditions are equivalent for any function $h : T \to T'$ between topological spaces:*

(i) *h is continuous,*
(ii) *whenever $B \subseteq T'$ is closed, the inverse image $h^{-1}[B]$ is closed in T,*
(iii) *whenever $B \subseteq T'$ is open, $h^{-1}[B]$ is open in T.*

Proof. Assume (i). If $h^{-1}[B]$ is not closed for some closed $B \subseteq T'$, then $h^{-1}[B]$ approximates some x that is not in $h^{-1}[B]$. By continuity $h[h^{-1}[B]] = B$ approximates $h(x)$ and of course $h(x) \notin B$. But this is impossible since B is closed: we have proved (ii).

Next, (ii) implies (iii) because

$$h^{-1}[T' \setminus B] = T \setminus h^{-1}[B] \quad \text{for every} \quad B \subseteq T'$$

Finally, assume (iii). Let x be a closure point of some $A \subseteq T$. If $h(x)$ were not a closure point of $h[A]$, then $h(x)$ would belong to the open set $B = T' \setminus \overline{h[A]}$. The set

$$h^{-1}[B] = T \setminus h^{-1}[\overline{h[A]}]$$

would be open and its complement $h^{-1}[\overline{h[A]}]$ would be closed. As a closed superset of A not containing x, it could not approximate x: a contradiction. Thus $h(x) \in \overline{h[A]}$, proving (i). $\square$

The following is easy to verify:

Proposition 2 *The composition of continuous functions $h : T \to R$ and $g : R \to S$ is a continuous function from T to S. The identity function on each topological space is continuous. All constant functions are continuous.*

In contrast with algebraic homomorphisms, the inverse of a bijective continuous function is not necessarily continuous: take any bijection from a discrete topological space to a trivial one. Continuous bijections with continuous inverse are called *homeomorphisms*. The domain and codomain spaces are then said to be *homeomorphic*.

Example. Every order isomorphism between totally ordered sets is a chain topology homeomorphism. In particular, any two real segments $[a,b]$, $a \neq b$, and $[c,d]$, $c \neq d$, are homeomorphic in the standard topology: map $x \in [a,b]$ to $c + (d-c) \cdot (x-a)/(b-a)$.

Topology is often defined as the study of continuous transformations. Which functions are continuous depends of course on the topological structure selected. Consider, for example, the set T of real numbers whose absolute value is not less than 1, with the topological closure system generated by sets of the form

$$[-x,-y] \cup [x,y], \quad 1 \leq x \leq y$$

Then only one of the three functions illustrated in Figure 10.1 is continuous from T to the standard real space.

Let $\mathcal{F}$ be a filter on a topological space T [i.e., a filter in the lattice $(\mathcal{P}(T), \subseteq)$]. We say that $\mathcal{F}$ *approximates* $x \in T$ if every member of $\mathcal{F}$ approximates x. Obviously, a set $A \subseteq T$ approximates a point x if and only if the principal filter generated by A approximates x. A proper filter $\mathcal{F}$ *converges* to x if every proper filter refinement of $\mathcal{F}$ approximates x. Then x is said to be a *limit point* of $\mathcal{F}$. The filter $\mathcal{F}$ is called *convergent* if it converges to some $x \in T$. An ultrafilter, of course, converges to a point x if and only if it approximates x. Every principal ultrafilter, generated by a singleton $\{x\}$, converges to x.

Examples. In the standard real topology, the filter generated by the sets

$$[0, 1/n] \setminus \{0\}, \quad n \in \omega$$

converges to 0. The principal filter generated by any nonempty nonsingleton subset of $\mathbb{R}$ is not convergent. Neither is the filter generated by the sets $[n, \rightarrow)$, $n \in \omega$. This latter filter does not even approximate any point whatsoever. For a different example, consider a trivial topology. Here every proper filter converges to every point.

Let $(x_n : n \in \omega)$ be a sequence of points in a topological space T. The sets

$$\{x_m : n \leq m\}, \quad n \in \omega$$

generate a filter $\mathcal{F}$ on T. By *convergence* and *limit points* of the sequence we mean those of the filter $\mathcal{F}$. Thus sequence convergence is a particular case of filter convergence.

Example. The sequence $(1/n : n \in \omega)$ converges to the limit point 0 in the standard real topology.

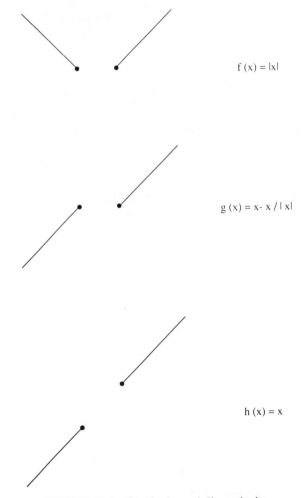

FIGURE 10.1 Continuity and discontinuity.

For each element x of a topological space T the open sets containing x generate a filter $\mathcal{N}_x$ on T. Any member of $\mathcal{N}_x$ is called a *neighborhood* of x. Observe that

$$\mathcal{N}_x = \{N \in \mathcal{P}(T) : x \in O \text{ and } O \subseteq N \text{ for some open } O\}$$

$$= \{N \in \mathcal{P}(T) : x \notin \overline{T \setminus N}\}$$

By a *neighborhood* of a set $A \subseteq T$ we mean any set N that is a neighborhood of every point in A.

Example. In standard real topology, the closed set $[0,3]$ is a neighborhood of 1. It is not a neighborhood of 0.

Proposition 3 *Let $\mathcal{F}$ be a proper filter on a topological space T and let $x \in T$.*

  (i) *$\mathcal{F}$ approximates x if and only if in the lattice of all filters on T, $\mathcal{F} \vee \mathcal{N}_x$ is a proper filter.*
  (ii) *$\mathcal{F}$ converges to x if and only if $\mathcal{N}_x \subseteq \mathcal{F}$.*

Proof. For the first assertion, note that the join of two filters is a proper filter if and only if no two filter members taken from the respective filters are disjoint.

For the second assertion, note that if $\mathcal{N}_x \not\subseteq \mathcal{F}$, then for some neighborhood N of x the filter generated by $\mathcal{F} \cup \{T \setminus N\}$ is proper and it does not approximate x. □

Proposition 4 *Let $h : T \to T'$ be a function between topological spaces. The following conditions are equivalent:*

  (i) *h is continuous,*
  (ii) *for every $x \in T$, $\mathcal{N}_{h(x)}$ is a subset of the image filter $h(\mathcal{N}_x)$,*
  (iii) *if a filter $\mathcal{F}$ on T approximates a point x, then the image filter $h(\mathcal{F})$ approximates $h(x)$,*
  (iv) *if a filter $\mathcal{F}$ on T converges to a point x, then $h(\mathcal{F})$ converges to $h(x)$.*

Proof. Use Propositions 1 and 3, and remark that (ii) can be restated as follows: for every $x \in T$ and every neighborhood N of $h(x)$, there is a neighborhood M of x such that $h[M] \subseteq N$. □

An *open cover* of a topological space T is a set $\mathcal{O}$ of open sets the union of which is T. A *subcover* is any subset $\mathcal{S}$ of $\mathcal{O}$ the union of which is still the whole of T. For example, in standard real topology, the set of open intervals

$$(a,b) = [a,b] \setminus \{a,b\}, \qquad a,b \in \mathbb{R}, \quad a < b$$

is an open cover of $\mathbb{R}$, and those (a,b) for which both a and b are rational form a subcover.

Proposition 5 *For any nonempty topological space T the following conditions are equivalent:*

  (i) *Each proper filter on T approximates some point of T.*
  (ii) *Each ultrafilter on T is convergent.*
  (iii) *Each nonempty set of closed sets in T having empty intersection has a nonempty finite subset with empty intersection.*
  (iv) *Each open cover of T has a finite subcover.*

Proof. (i) clearly implies (ii), and (iii) is obviously equivalent to (iv).

To show that (ii) implies (iii) assume that some nonempty set S of closed sets has no finite nonempty subset with empty intersection. Thus S generates a proper filter $\mathcal{F}$. Take any ultrafilter refinement $\mathcal{U}$ of $\mathcal{F}$ and apply (ii) to show that $\cap S \neq \emptyset$.

Finally, we derive (i) from (iv). Suppose that $\mathcal{F}$ is a proper filter that does not approximate any point of T. This implies that every $x \in T$ has an open neighborhood $N(x)$ disjoint from some member $F(x)$ of $\mathcal{F}$. The various $N(x)$, $x \in T$, form an open cover of T. According to (iv) we can take a finite subcover $N(x_1), \ldots, N(x_n)$. But then

$$F = F(x_1) \cap \cdots \cap F(x_n)$$

is disjoint from each of the sets $N(x_1), \ldots, N(x_n)$, which contradicts the fact that $F \in \mathcal{F}$ cannot be empty. $\square$

A topological space satisfying the conditions of Proposition 5 is called *compact*.

Examples. Every nonempty finite topological space is compact. The standard topology on $\mathbb{R}$ is not compact. (Why?)

Let $(T_i, C_i)_{i \in I}$ be a family of topological spaces. The *product topology* is defined on the product set $T = \prod_{i \in I} T_i$ as the smallest topology on T generated by the set of all product sets $\prod_{i \in I} C_i$, where

$C_i \in \mathcal{C}_i$. Indeed the closed sets of the product topology are the intersections of sets of the form

$$\left(\prod_{i \in I} C_{1i}\right) \cup \cdots \cup \left(\prod_{i \in I} C_{ni}\right)$$

where $n \in \omega$ and $C_{1i}, \ldots, C_{ni}$ are in $\mathcal{C}_i$. Clearly the projection maps $\text{pr}_j : (\prod_{i \in I} T_i) \to T_j$ are continuous. Indeed the product topology is the smallest topology on the set-theoretical product set $\prod T_i$ for which all projections pr_j are continuous. Moreover, the following is easy to verify:

Proposition 6 *A function h from a topological space Q to $\prod T_i$ is continuous if and only if the compositions $\text{pr}_j \circ h$ with the projection maps are continuous.*

Recall that in Chapter I, the Axiom of Choice was needed to establish that the product of nonempty sets is not empty (Proposition 10 of Chapter I). There is a conceptual kinship between nonemptiness and compactness. Compactness means the nonemptiness of the set of filter-approximated points. Compare:

Tychonoff Product Theorem. *The product of any family of compact topological spaces is compact.*

Proof. Let $\mathcal{F}$ be any ultrafilter on $\prod T_i$. Let A_i be the set of points approximated by the ultrafilter $\text{pr}_i(\mathcal{F})$ on T_i. By Proposition 10 of Chapter I, the set $\prod_{i \in I} A_i$ has at least one element. This element is approximated by $\mathcal{F}$. □

Infinite products of finite spaces now naturally present themselves as examples of infinite compact spaces. The product topology $\mathbb{R}^n = \prod_{i \in I} \mathbb{R}_i$, where $n \in \omega$ and each $\mathbb{R}_i$ coincides with $\mathbb{R}$ (standard real topology), has of course no reason to be compact.

Using condition (iii) or (iv) of Proposition 5, one can easily deduce the following:

Proposition 7 *The image of a continuous function defined on a compact domain is always compact.*

Whence the well-known theorem of analysis that every real-valued continuous function defined on a compact domain "attains its bounds," i.e., the image set has a smallest and a largest element. Indeed, the reader should verify that the existence of such min and max is a necessary condition of compactness for any subset of R. More of this in the sequel.

EXERCISES

1. Show that every polynomial of odd degree in $R[X]$ has a real root.

2. (a) Show that the binary operations of sum and product on the real numbers are continuous functions from R^2 to R. Can you make analogous statements for integers or rationals? What can you say about polynomial functions?
   (b) Show that every linear map $R^n \to R^m$ is continuous.

3. (a) Verify that the chain topology on a discrete chain is always discrete. Can the chain topology be discrete without the chain being discrete?
   (b) Let C be a chain, and let $K \subseteq C$. Verify that the chain topology on K coincides with the chain topology of C inherited by K.

4. Verify that a set is open if and only if it is a neighborhood of itself.

5. Verify that a set A approximates a point x if and only if A intersects every neighborhood of x.

6. Let $(T, \mathcal{C})$ be a topological space. Verify that if a filter converges to a point x in some refinement topology, then it converges to x in $(T, \mathcal{C})$.

7. Verify that if a filter converges to a point x, then every refinement filter converges to x as well.

8. Is the set of all topologies on a set T an algebraic closure system on $\mathcal{P}(T)$?

3. DISTANCES AND ENTOURAGES

Consider a filter $\mathcal{U}$ on the set V^2 of all couples of elements of a set V. Each member of $\mathcal{U}$ is a binary relation on V. We wish to think of these relations as expressing some measure of proximity, not necessarily in a numerical sense. Therefore, assume that

(i) for every $x \in V$ and $E \in \mathcal{U}$, $(x,x) \in E$,
(ii) for every $E \in \mathcal{U}$ the dual relation E^* belongs to $\mathcal{U}$ as well,
(iii) for every $E \in \mathcal{U}$ there is some $D \in \mathcal{U}$ such that for every $x, y, z \in V$, $(x,y) \in D$ and $(y,z) \in D$ together imply that $(x,z) \in E$.

The couple $(V, \mathcal{U})$ is called a *uniform space*, or *uniformity*, on V. (Abusively, we can refer to the "uniform space V" if it is clearly understood which filter $\mathcal{U}$ we have in mind.) The members of $\mathcal{U}$ are called *entourages*, or *uniform entourages* for emphasis. Condition (iii) should be reminiscent of epsilon–delta (E–D) argumentation used in analysis, although it involves no numerical reference whatsoever. For an entourage D, it will be convenient to denote by $D \circ D$ the relation

$$\{(x,z) \in V^2 : (x,y) \in D \text{ and } (y,z) \in D \text{ for some } y \in V\}$$

Condition (iii) then says that for every E there is some D such that $D \circ D \subseteq E$.

The most natural example is the *standard uniformity* on $V = \mathbb{R}$, where $\mathcal{U}$ is the filter generated by relations of the form

$$\{(x,y) \in \mathbb{R}^2 : |x - y| < \varepsilon\}$$

for positive real ε. Less natural but nonetheless obvious examples are the *discrete uniformity* on any set V, consisting of all reflexive relations, and the *trivial uniformity*, where $\mathcal{U} = \{V^2\}$.

The standard uniformity on $\mathbb{R}$ belongs to a remarkable class of uniformities, arising from a generalization of the "absolute value of difference" concept. Precisely, let d be a function from V^2 to $\mathbb{R}$ for any set V. Assume that for every $x, y, z \in V$

(i) $d(x,y) \geq 0$,
(ii) $d(x,y) = 0$ if and only if $x = y$,
(iii) $d(x,y) = d(y,x)$,
(iv) $d(x,y) + d(y,z) \geq d(x,z)$.

The map d is called a *distance* function on V, and (V,d) is called a *metric space*. Condition (iv) is usually referred to as the *triangle inequality*.

Examples. For any connected graph (V,E) the graph-theoretical distance function on V^2 makes the vertex set into a metric space. The function $d(x,y) = |x-y|$ makes $\mathbb{R}$ into a metric space.

Every metric space (V,d) gives rise to a uniform space $(V,\mathcal{U}_d)$ where $\mathcal{U}_d$ is the filter on V^2 generated by the set of relations of the form
$$\{(x,y) \in V^2 : d(x,y) < \varepsilon\}$$
for real positive ε. The metric space defined on a connected graph by its graphic distance function gives rise to a discrete uniformity. The distance $|x-y|$ on $\mathbb{R}$ gives rise to the standard uniformity on $\mathbb{R}$. Not all uniformities arise from metric spaces: consider the trivial uniformity on any set with at least two elements. Those uniformities that do arise from metric spaces are called *metric uniformities*.

The inner (dot) product of the vector space $\mathbb{R}^n$ allows us to define a metric space structure on $\mathbb{R}^n$ as follows. The *Euclidean distance* $d(x,y)$ is given, for $x,y \in \mathbb{R}^n$, by
$$d(x,y) = \sqrt{(x-y) \bullet (x-y)}$$
The reader who verifies that this is indeed a distance will have proved a form of the *"Schwarz inequality."* The resulting uniformity on $\mathbb{R}^n$ is called *Euclidean uniformity*. By considering the case $n = 1$, one can see that we truly have a generalization to n dimensions of the distance $|x-y|$ and of the corresponding standard uniformity.

Let $(T,\mathcal{U})$ be any uniform space. For $E \in \mathcal{U}$ and $x \in T$ define
$$N_E(x) = \{y : (x,y) \in E\}$$
and for $A \subseteq T$ define
$$N_E(A) = \bigcup_{x \in A} N_E(x)$$

Define the closure $\overline{A}$ of each subset A of T by

$$\overline{A} = \bigcap_{E \in \mathcal{U}} N_E(A)$$

The reader should verify the closure operator properties. Further, the set

$$\mathcal{C} = \{\overline{A} : A \subseteq T\}$$

of closed sets is easily seen to be a topological closure system on T. The topological space $(T, \mathcal{C})$ is called the *uniform topology* corresponding to, or determined by, the uniformity $\mathcal{U}$. If the uniformity $\mathcal{U}$ is metric, then the corresponding uniform topology is called a *metric topology*. In a uniform topology, an element $x \in T$ is approximated by a set $A \subseteq T$ if there are elements of A "related to x" in every entourage relation. The uniform topology on $\mathbb{R}^n$ arising from the Euclidean uniformity is called *Euclidean topology* on $\mathbb{R}^n$. It is a metric topology. By considering the case $n = 1$, it is immediately seen to be a generalization to n dimensions of the standard real topology.

Note that discrete uniformities determine discrete topologies and trivial uniformities determine trivial topologies.

Clearly, in a topological space T corresponding to a uniformity, for every $x \in T$ and entourage E, $N_E(x)$ is a topological neighborhood of x. Also, for every subset $A \subseteq T$, $N_E(A)$ is a topological neighborhood of A. Assume now that A is closed and $x \notin A$. Then there is a uniform entourage E such that $N_E(x) \cap A = \emptyset$. Let D be an entourage with $D \circ D \subseteq E$. For the uniform entourage $G = D \cap D^*$ we must have

$$N_G(A) \cap N_G(x) = \emptyset$$

This motivates the following definition: a topological space in which every closed set and any point not belonging to it are contained in disjoint respective neighborhoods is called *regular*. We also say that the point is *"separated"* from the closed set. It is a major concern of topology to ascertain which pairs of sets or points can be *separated*, meaning "enclosed in disjoint neighborhoods." Thus regular topological spaces are those where all closed sets can be separated from all points not belonging to them. We can state:

Regular Separation Theorem. *All uniform topologies are regular.*

It is now easy to see that not every topology is uniform. On any infinite set T, consider the topology whose closed sets are all the finite subsets of T plus T itself. All nonempty neighborhoods have finite complements in T, and no two can be disjoint. In this topology, traditionally called the *cofinite topology* on T, in reference to the open sets, not even pairs of distinct points can be separated. Topological spaces with pairwise separable points are called *Hausdorff spaces*.

Hausdorff Separation Theorem. *All metric topologies are Hausdorff.*

Proof. For two distinct points x, y, the set of points whose distance from x is less than $d(x,y)/3$ is a neighborhood of x. Define a neighborhood of y similarly, and separation is achieved. □

Uniform and regular topologies need not be Hausdorff: the simplest counterexamples arise from trivial uniformities. Thus regularity does not imply Hausdorff separation. Conversely, are Hausdorff spaces necessarily regular? A counterexample would of course have to be nonmetric. Among the monumental constructions of mathematical theories, topology is not unlike a gothic cathedral with many surprising representations of a countergeometric netherworld. Here is a beautiful gargoyle:

Counterexample. On the set $\mathbb{R}$ of real numbers, consider the smallest topology generated by the standard closed sets plus $\mathbb{Q}$ (the rationals). Its closed sets are the sets of the form

$$C \cup (K \cap \mathbb{Q})$$

where C and K are standard closed. Equivalently, these are also the sets of the form $C \cap (K \cup \mathbb{Q})$. By De Morgan's rule, the open sets are the sets of the form

$$O \cup (P \cap \mathbb{I})$$

or equivalently $O \cap (P \cup \mathbb{I})$, with O, P standard open in $\mathbb{R}$, and $\mathbb{I} = \mathbb{R} \setminus \mathbb{Q}$, the set of irrational numbers. Now try to separate $\mathbb{Q}$ from an irrational!

In a topological space T arising from a uniformity, let E be a uniform entourage and let $x \in T$. For some entourage D we have

$D \circ D \subseteq E$. It follows that for $G = D \cap D^*$ and the topological neighborhood $N_G(x)$ of x the inclusion

$$N_G(x)^2 \subseteq E$$

holds. Also, by Proposition 3, $N_G(x)$ belongs to every filter on T that converges to x. We conclude that every convergent filter $\mathcal{F}$ on T has the following property: for every entourage E there is a member N of $\mathcal{F}$ such that $N^2 \subseteq E$. In any uniform space, a proper filter $\mathcal{F}$ having this property is called a *Cauchy filter*. Thus, in any uniform space, every convergent filter is Cauchy. A uniform space where every Cauchy filter is convergent is called *complete*.

The classic examples of incomplete and complete spaces are $\mathbb{Q}$ and $\mathbb{R}$, with the metric uniformities of the distance $|x - y|$. The arising topologies are also precisely the standard chain topologies. The uniform space $\mathbb{Q}$ is not complete: for any fixed irrational number α consider the filter on $\mathbb{Q}$ generated by the sets

$$F_q = \{x \in \mathbb{Q} : q \leq x < \alpha\}$$

for rational $q < \alpha$. The space $\mathbb{R}$ is complete: for a given Cauchy filter $\mathcal{F}$ and real positive ε, let F_ε be a filter member such that $|x - y| < \varepsilon$ for all $x, y \in F_\varepsilon$. Then $\mathcal{F}$ converges to

$$\text{lub}\{\text{glb}\, F_\varepsilon : \varepsilon > 0\}$$

The concept of completeness enables us to provide yet another characterization of compactness in uniform topologies. Call a uniformity on a set T *bounded* if for each uniform entourage E there is some finite subset A of T such that $N_E(A) = T$. Obviously every finite uniform space is bounded, and it is easy to see that the Euclidean uniformity on $\mathbb{R}^n$ is not bounded.

Compact Uniformity Theorem. *A nonempty topological space arising from a uniformity is compact if and only if the uniformity is both complete and bounded.*

Proof. Assume compactness. First, let $\mathcal{F}$ be a Cauchy filter. If $\mathcal{F}$ did not converge, then for some proper filter $\mathcal{G} \supseteq \mathcal{F}$ the intersection of all the closures $\overline{A}$, $A \in \mathcal{G}$, would be empty, and thus for some finite number of members $A_1, \ldots, A_n$ of $\mathcal{G}$ we would have

$$\overline{A_1} \cap \cdots \cap \overline{A_n} = \emptyset$$

and a fortiori
$$A_1 \cap \cdots \cap A_n = \emptyset$$

But this is impossible since $A_1 \cap \cdots \cap A_n \in \mathcal{G}$. Second, take any entourage E. For every element x of the space, let $O(x)$ be an open neighborhood of x contained in $N_E(x)$. [Any doubt that such an $O(x)$ exists?] The various $O(x)$ form an open cover of the entire topological space. Take any finite subcover
$$\{O(x_1), \ldots, O(x_n)\}$$
Clearly $N_E(\{x_1, \ldots, x_n\})$ is the entire space.

Conversely, assume completeness and boundedness. Let $\mathcal{F}$ be any ultrafilter: we shall show that it is convergent. Let E be any uniform entourage. Let D be an entourage such that $D \circ D \subseteq E$ and let $G = D \cap D^*$. Let A be a finite set of points such that $N_G(A)$ includes the entire space. If A consists of $x_1, \ldots, x_n$, then
$$N_G(A) = N_G(x_1) \cup \cdots \cup N_G(x_n)$$
For some i, $N_G(x_i) \in \mathcal{F}$ and $N_G(x_i)^2 \subseteq E$. This shows that $\mathcal{F}$ is Cauchy. By completeness, $\mathcal{F}$ is convergent. $\square$

We conclude this section by applying the concepts and results to the Euclidean uniform topologies $\mathbb{R}^n$. These were defined with the Euclidean distance
$$d(x, y) = \sqrt{(x - y) \bullet (x - y)}$$
for vectors $x, y \in \mathbb{R}^n$, but the reader may wish to verify that they actually coincide with the topological product
$$\prod_{i \in [1, n]} \mathbb{R}_i$$
where each $\mathbb{R}_i = \mathbb{R}$. For any subset S of $\mathbb{R}^n$, the Euclidean distance, defined between elements of S, makes S into a metric space and therefore into a uniform space that may or may not be bounded. The reader is urged to verify that this Euclidean uniformity on S is bounded if and only if the image set of the distance function $d : S^2 \to \mathbb{R}$ possesses an upper bound in $\mathbb{R}$. The uniform topology

of S derived from its metric structure coincides with the Euclidean topology of $\mathbb{R}^n$ inherited by the subset S: it may or may not be compact.

Borel–Heine–Lebesgue Theorem. *A nonempty subset S of $\mathbb{R}^n$ is compact if and only if S is closed in $\mathbb{R}^n$ and bounded.*

Proof. We show that being closed in $\mathbb{R}^n$ is equivalent to completeness under metric uniformity. The argument is an extension of the one used earlier to show that $\mathbb{R}$ is complete but $\mathbb{Q}$ is not.

If S is not closed, $S \subset \overline{S}$, let $x \in \overline{S}\setminus S$. Let $\mathcal{N}_x$ be the neighborhood filter of x in $\mathbb{R}^n$. Let

$$\mathcal{F} = \{N \cap S : N \in \mathcal{N}_x\}$$

Then $\mathcal{F}$ is a nonconvergent Cauchy filter on S.

Conversely, assume that S is closed. Let $\mathcal{F}$ be a Cauchy filter on S. Let

$$pr_i : \mathbb{R}^n \to \mathbb{R}$$

be the ith projection. For each i, $1 \leq i \leq n$, the image filter

$$\mathcal{F}_i = pr_i(\mathcal{F})$$

on $\mathbb{R}$ is a Cauchy filter under standard uniformity. Since $\mathbb{R}$ is complete, $\mathcal{F}_i$ converges to some real number x_i. The exercise of verifying that $\mathcal{F}$ converges to $x = (x_1,\ldots,x_i,\ldots,x_n)$ is left to the reader. □

Examples. Every segment $[a,b]$ of $\mathbb{R}$ is compact. If $a,b \in \mathbb{Q}$, $a < b$, then the segment $[a,b]$ of $\mathbb{Q}$ is not compact. The set

$$\{1/n : n \in \omega\} \cup \{0\}$$

is compact.

Historical Reference. We have presented a view of topological spaces in the spirit of Hausdorff's *Foundations of Set Theory* (1914). This corresponds to our preoccupation with the algebra of nonnumbers rather than with the geometry of nondistances. Nonnumerical algebra and nonmensurational geometry are components of the same Leibnizian research program.

EXERCISES

1. Show that a nonempty subset of $\mathbb{R}$ is compact if and only if it is a complete lattice.

2. Show that there is a metric structure $(\mathbb{C}, d)$ on the field of complex numbers such that
   (a) for $x, y \in \mathbb{R}$, $d(x,y) = |x - y|$,
   (b) the metric topology of $\mathbb{C}$ is homeomorphic to the Euclidean topology on $\mathbb{R}^2$,
   (c) the function mapping $x \in \mathbb{C}^*$ to $d(0, x)$ is a continuous surjective group homomorphism $(\mathbb{C}^*, \cdot) \to (\mathbb{R}^*, \cdot)$,
   (d) the function mapping $x \in \mathbb{C}^*$ to $x/d(0, x)$ is a continuous group endomorphism of $(\mathbb{C}^*, \cdot)$ whose image set contains all roots of unity in $\mathbb{C}$.
   What can you say about the algebraic and topological structure of the set $\{x/d(0, x) : x \in \mathbb{C}^*\}$?

3. Let L be an affine line in $\mathbb{R}^n$. Verify that the chain topology defined on L by any of the two natural orders coincides with the Euclidean topology inherited by L.

4. Verify that the Euclidean topology inherited by any nonempty affine flat in $\mathbb{R}^n$ is homeomorphic to the Euclidean topology of some $\mathbb{R}^m$.

5. Let S be any set of reflexive and symmetric relations on a set V such that for every $E \in S$ there is some $D \in S$ with $D \circ D \subseteq E$. Let $\mathcal{U}$ be the filter on V^2 generated by S. Show that $(V, \mathcal{U})$ is a uniformity.

BIBLIOGRAPHY

Claude BERGE, *Topological Spaces, Including a Treatment of Multi-Valued Functions, Vector Spaces and Convexity*. Oliver & Boyd 1963. A natural progression of ideas from sets through topology to Banach spaces.

James DUGUNDJI, *Topology*. Allyn & Bacon 1978. Graduate and undergraduate students alike can use this text to gain a well-rounded general knowledge of topology. The only prerequisite is some acquaintance with elementary analysis. The breadth of coverage and depth of treatment indeed allow the working mathematician to use this volume as a reference to classical results. There are two introductory chapters on set theory, and a fair amount of material on homotopy links topology to group theory.

Riszard ENGELKING, *General Topology*. Heldermann Verlag 1989. An up-to-date and thorough exposition of topology, including uniform spaces. For graduate students and researchers.

J. R. ISBELL, *Uniform Spaces*. American Mathematical Society 1964. A classical exposition of the theory by one of the early contributors. Attention: the term "mapping" is used in a sense more restrictive than "function."

I. M. JAMES, *Topological and Uniform Spaces*. Springer 1987. A concise and limpid textbook. For undergraduates and above.

Kazimierz KURATOWSKI, *Introduction to Set Theory and Topology*. Pergamon Press 1961. Written in the spirit of Hausdorff, this relatively short book starts with considerations of elementary logic, reviews a fair amount of material on sets and ordered sets, and develops topology to the point of proving several advanced theorems. The last one of these is the Jordan curve theorem.

Warren PAGE, *Topological Uniform Structures*. John Wiley & Sons 1978. Groups, vector spaces, and other algebraic structures endowed with a compatible topology are presented in an investigative spirit. Accessible at the advanced undergraduate level.

Horst SCHUBERT, *Topology*. Allyn & Bacon 1968. The first half of this volume is a standard presentation of topological and uniform spaces. The second half is devoted to homotopy and homology, introducing the reader to algebraic topology. Requires only a minimal knowledge of elementary analysis.

L. A. STEEN and J. A. SEEBACH, *Counterexamples in Topology*. Holt, Rinehart and Winston 1970. Mathematical teratology par excellence.

Wolfgang J. THRON, *Topological Structures*. Holt, Rinehart and Winston 1966. This accessible textbook has dedicated chapters on metric and pseudometric spaces, uniform spaces, and proximity spaces. The reader must cope with some distracting typos. Numerous interesting historical references are included.

Steven VICKERS, *Topology Via Logic*. Cambridge University Press 1989. An unusual approach to fundamental concepts of topology. Addressed to the graduate student in computer science. Ordered sets, lattices, universal algebras, and categories are referred to throughout and play a significant role.

Gordon WHYBURN and Edwin DUDA, *Dynamic Topology*. Springer 1979. A prove-it-yourself presentation of topological spaces and continuous functions, for undergraduates. The material is organized as a collection of definitions, examples, exercises, and solutions. Attention: "mapping" means "continuous function."

CHAPTER XI

UNIVERSAL ALGEBRAS

1. HOMOMORPHISMS AND CONGRUENCES

The Romans were accomplished engineers of roads, bridges, and plumbing systems. They used letters to denote numbers, but this was notation only. They could add and multiply numbers, but not letters. If compared either with the earlier Euclid or the later al-Khwarizmi, the Romans may be said to have been algebraically illiterate. There was progress but no revolution in this respect until Galois's group theory. His was a century of structural inventions, rich in nonnumerical algebra. Whitehead's *Treatise on Universal Algebra*, published in 1898, was a first attempt at unification. (The second attempt is twentieth-century category theory.)

A *universal algebra*, or simply *algebra*, is a couple $(U, (f_i : i \in I))$ where U is any set and $(f_i : i \in I)$ is a family of operations,

$$f_i : U^{n(i)} \to U, \qquad n(i) \in \omega$$

for $i \in I$. The family $(n(i) : i \in I)$ is said to be the *type* of the algebra. For notational convenience we suppose that I is an ordinal. If $n(i) = 0$, then the nullary operation f_i is necessarily constant (since U^0 is a singleton) and it is completely determined by the unique element $c_i \in U$ in the image of f_i. We shall refer to f_i as the "constant operation c_i." Another notational simplification consists in writing

$(U, f_i : i \in I)$ instead of $(U, (f_i : i \in I))$. Often we refer to the "algebra U" when the operations f_i are understood from the context.

If a subset S of U is closed under all the operations f_i, $i \in I$, then

$$(S, f_i | S^{n(i)} : i \in I)$$

is also an algebra of type $(n(i) : i \in I)$, called a *subalgebra* of U. We generally denote this subalgebra simply by $(S, f_i : i \in I)$. By a slight abuse of terminology, the subset S itself may also be called a "subalgebra."

Let $(U, f_i : i \in I)$ and $(V, f_i' : i \in I)$ be algebras of the same type. A *homomorphism* from the algebra U to the algebra V is a map $h : U \to V$ such that for all $i \in I$ and $(x_0, \ldots, x_{n-1}) \in U^{n(i)}$,

$$h(f_i(x_0, \ldots, x_{n-1})) = f_i'(h(x_0), \ldots, h(x_{n-1}))$$

We shall write $h : (U, f_i : i \in I) \to (V, f_i' : i \in I)$ in this case. A bijective homomorphism is called an *isomorphism*, and if such an isomorphism exists, then the algebras concerned are said to be *isomorphic*.

Examples. A number of classical algebraic structures seen in previous chapters correspond to some universal algebra $(U, f_i : i \in I)$. For a *groupoid* $I = 1 = \{0\}$ and f_0 is binary, $n(0) = 2$. A *monoid* with neutral element e corresponds to an algebra (U, f_0, f_1) where f_0 is the binary product and f_1 is the nullary constant operation e. A *group* with neutral element e corresponds to an algebra (U, f_0, f_1, f_2) of type $(2, 0, 1)$ where f_0 and f_1 are as for monoids and f_2 is the unary operation giving the inverse of each group element. A *ring* $(U, +, \cdot)$ corresponds to the algebra $(U, +, 0_U, f_2, \cdot)$ of type $(2, 0, 1, 2)$ where 0_U is the nullary constant corresponding to the ring's zero element and f_2 is the unary additive inverse ("negative") operation. It is better to write simply $(U, +, 0, -, \cdot)$ for this algebra. A *lattice* corresponds to an algebra $(U, \vee, \wedge)$ with the two binary operations of join and meet. Every *median graph* $G = (V, E)$ gives rise to an algebra on V with a single ternary operation m that associates to every $(x, y, z) \in V^3$ the median vertex $m(x, y, z)$. *Subgroupoids, submonoids, subgroups, subrings, and sublattices* are precisely the subalgebras of the respective universal algebras. Among the subalgebras of the ternary algebra of a median graph G we find all convex sets of vertices of G, and more. Universal algebra *homomorphisms* and *isomorphisms* for groupoids, monoids, groups, and rings are the same as the structure-specific homomorphisms and isomorphisms defined in Chapters III and IV.

For any universal algebra $(U, f_i : i \in I)$ the set of subalgebras constitutes an algebraic closure system on U. We are now able to justify the term "algebraic."

Algebraic Closure Theorem. *A closure system $\mathcal{C}$ on a set U is algebraic if and only if $\mathcal{C}$ coincides with the subalgebra closure system of some universal algebra $(U, f_i : i \in I)$.*

Proof. In view of the preceding observation, we need only to find the operations f_i for a given algebraic closure system $\mathcal{C}$ on U. As usual, we shall write $\overline{A}$ for the closure of any set $A \subseteq U$. Let F be the set of all possible choice functions c associating with each nonempty closed set $K \in \mathcal{C}$ an element $c(K) \in K$. Let I be an ordinal equipotent to $F \times \omega^+$ and let

$$g : F \times \omega^+ \to I$$

be a bijection. For $c \in F$ and $n \geq 1$, let $f_{g(c,n)}$ be the n-ary operation on U given by

$$f_{g(c,n)}(x_0, \ldots, x_{n-1}) = c(\overline{\{x_0, \ldots, x_{n-1}\}})$$

Then refer to the definition of algebraic closure. □

Many properties of homomorphisms between groupoids, monoids, groups, or rings remain valid in the universal algebra context.

Proposition 1 *The composition of two universal algebra homomorphisms*

$$h : (U, f_i : i \in I) \to (V, f'_i : i \in I)$$

and

$$g : (V, f'_i : i \in I) \to (W, f''_i : i \in I)$$

is a homomorphism from U to W. The identity mapping on any universal algebra is a homomorphism from that algebra to itself. The inverse of a bijective universal algebra homomorphism is again a homomorphism.

Proposition 2 *Let $h : U \to V$ be a homomorphism between universal algebras. The image by h of any subalgebra of U is a subalgebra of V. The inverse image of any subalgebra of V is a subalgebra of U.*

A *congruence relation* on a universal algebra $(U, f_i : i \in I)$ is an equivalence relation $\equiv$ on U such that for every $i \in I$, with $n(i) = n$, if

$$x_0 \equiv x'_0, \ldots, x_{n-1} \equiv x'_{n-1}$$

then

$$f_i(x_0, \ldots, x_{n-1}) \equiv f_i(x'_0, \ldots, x'_{n-1})$$

For groupoids, groups, and rings, viewed as universal algebras, the congruences in the sense of universal algebra are the same as the congruences defined earlier in the respective theory of each structure. For a totally ordered set, viewed as an algebra with the operations $\vee$ (max) and $\wedge$ (min), an equivalence relation is an algebra congruence if and only if each equivalence class is order convex. The reader can find a counterexample showing that this is not true in arbitrary lattices.

An equivalence relation E on a universal algebra U with operations f_i, $i \in I$, is a congruence if and only if for every $i \in I$ and $n = n(i)$ equivalence classes $C_0, \ldots, C_{n-1}$ the set

$$\{f_i(x_0, \ldots, x_{n-1}) : x_j \in C_j \text{ for all } j \in n\}$$

is included in some class D. As such a class is unique, this defines an n-ary operation $\overline{f_i}$ on the quotient set U/E of all equivalence classes by

$$\overline{f_i}(C_0, \ldots, C_{n-1}) = D$$

The algebra

$$(U/E, \overline{f_i} : i \in I)$$

is called the *quotient algebra* of $(U, f_i : i \in I)$ by the congruence E. Often we use the same symbols f_i to denote the quotient operations $\overline{f_i}$ as well. Note that the canonical surjection $U \to U/E$ is a universal algebra homomorphism. Conversely, if h is any universal algebra homomorphism, then the induced equivalence defined on its domain is a congruence of the domain algebra. The corresponding quotient is isomorphic to the algebra $\text{Im} \, h$.

Examples. Groupoids, groups, rings.

EXERCISES

1. (a) Let $(U, \leq)$ be any linear order. Verify that for $a_1, a_2, a_3 \in U$, the intersection of the three segments generated by the pairs $\{a_i, a_j\}$, $i \neq j$, is a singleton and indeed it is one of the $\{a_k\}$. Write $m(a_1, a_2, a_3)$ for this a_k. Viewing m as a ternary operation on U, describe the algebra (U, m). In particular, what are the subalgebras, congruences, and quotients? What connections do you see with median graphs?
   (b) Define a ternary operation μ on $\mathbb{R}^n$ by
   $$\mu(x, y, z) = ((m(x_1, y_1, z_1), \ldots, m(x_n, y_n, z_n)))$$
   Describe the subalgebras, congruences, and quotients of $(\mathbb{R}^n, \mu)$.

2. Show that the algebraic closure systems on a set U constitute a closure system $\mathcal{A}$ on $\mathcal{P}(U)$. Is this closure system $\mathcal{A}$ algebraic?

3. Review the chapters on groups, rings, and lattices and see what you can generalize to universal algebras. Take any book on universal algebra and see what you can particularize to groups.

2. ALGEBRA OF SYNTAX

We now develop a rudimentary theory of formal expressions. Take any set $\mathcal{A}$, call it the *alphabet*, and call its elements *letters*. For $n \in \omega$, any n-tuple

$$(a_0, \ldots, a_{n-1})$$

in $\mathcal{A}^n$ is called a *word* of *length* n over the alphabet $\mathcal{A}$. We usually denote words by simple juxtaposition of the letters, as in

$$a_0 \cdots a_{n-1}$$

For $a \in \mathcal{A}$, the word of length 1 whose only letter is a is called the *word corresponding to a*, and it is generally denoted, somewhat abusively, by the same symbol a.

Example. Take the set $2 = \{0,1\}$ as alphabet. Then 1, 10, 101, 1101, and 0011 are words. Not numbers, just plain, meaningless words over $\{0,1\}$. Of course, $\emptyset$ is a word as well, of length zero.

Let $A = a_0 \cdots a_{n-1}$ and $B = b_0 \cdots b_{m-1}$ be two words of length n and m. The *concatenation* AB is defined as the word $c_0 \cdots c_{n+m-1}$ of length $n + m$ where

$$c_i = \begin{cases} a_i & \text{for } 0 \leq i \leq n-1 \\ b_{i-n} & \text{for } n \leq i \leq n+m-1 \end{cases}$$

Example. The concatenation of 101 and 1110 is 1011110. The concatenation of 01101 and $\emptyset$ is 01101.

Observe that the set of all nonempty words over a given alphabet forms a semigroup under concatenation.

Consider a given algebra type $(n(i) : i \in I)$. Let $(\sigma_i : i \in I)$ be an injective family indexed by I, that is, $\sigma_i \neq \sigma_j$ for $i \neq j$. What the various σ_i are does not matter. Call them *operation symbols*. Write $\mathcal{O}$ for $\{\sigma_i : i \in I\}$. Let $\mathcal{V}$ be any set disjoint from $\mathcal{O}$: its members shall be called *variables*. Consider the semigroup S of all nonempty words over the alphabet $\mathcal{A} = \mathcal{O} \cup \mathcal{V}$. For every $i \in I$, with $n(i) = n$, an n-ary *formative operation* s_i is defined on the word set S by concatenation,

$$s_i(A_0, \ldots, A_{n-1}) = \sigma_i A_0 \cdots A_{n-1}$$

If $n(i) = 0$, then s_i is nullary too, and its value is the one-letter word σ_i. If A_j has length $l(j)$, then $s_i(A_0, \ldots, A_{n-1})$ has length

$$1 + l(0) + \cdots + l(n-1)$$

Consider the universal algebra $(S, s_i : i \in I)$. It is of the given type $(n(i) : i \in I)$ as well. Let V be the set of all length 1 words with their single letter in $\mathcal{V}$. The members of the subalgebra of $(S, s_i : i \in I)$ generated by V are called *terms* for the algebra type under consideration and with variable set $\mathcal{V}$. Terms constitute an algebra under the formative operations applied to terms, and this *algebra of terms* is naturally of the type originally given.

Examples of Ring Terms. Consider the algebra type $(2,0,1,2)$ of rings. Denote the operation symbols $\sigma_0, \sigma_1, \sigma_2, \sigma_3$ more suggestively by $+, 0, -, \bullet$. Let $\mathcal{V}$ consist of two variables, x and y. Here are some terms:

$$\mathbf{0} \quad x \quad y \quad -x \quad -0 \quad +xy \quad \bullet +xy-y$$

The reader may see in the last term an artificial way of saying "multiply the sum of x and y with the negative of y."

Proposition 3 *If A is a term and B is a nonempty word, then the concatenation AB is not a term.*

Proof. Let us first establish that every term is either a single variable or of the form $\sigma_i A_0 \cdots A_{n-1}$, where σ_i is an n-ary operation symbol and $A_0, \ldots, A_{n-1}$ are terms. Indeed, the set U of terms subject to this apparent constraint of form is already closed under all the formative operations s_i. Since U includes all terms corresponding to variables, which generate the algebra of terms, U must be the entire algebra of terms.

Assuming the proposition false, let A be a term of minimal length for which it fails. Obviously A is not reduced to a single variable. Let

$$A = \sigma_i A_0 \cdots A_{n-1}$$

If $AB = \sigma_i A_0 \cdots A_{n-1} B$ is a term, then

$$A_0 \cdots A_{n-1} B = C_0 \cdots C_{n-1}$$

where the C_j are terms as well. The term C_0 cannot be longer than A_0, for then $C_0 = A_0 D$ would be a term for some nonempty word D. Similarly, A_0 cannot be longer than C_0, and therefore $A_0 = C_0$. Inductively, $A_j = C_j$ for $j = 0, \ldots, n-1$. But then B must be empty. $\square$

Corollary. *Every term that is not reduced to a single variable can be written, in a unique fashion, as $\sigma_i A_0 \cdots A_{n-1}$ with some n-ary operation symbol σ_i and terms A_j.*

Let $A = a_0 \cdots a_{n-1}$ be any word of length n over any given alphabet. Take any segment $[j, k]$ of the natural order on $n = \{0, \ldots, n-1\}$. Consider the word $B = a_j \cdots a_k$ of length $k - j + 1$. The segment $[j, k]$ is called an *occurrence* of the word B in A. The word B is said to *occur* in A. If A is a universal algebra term, the word B

may or may not be a term. If $j = k$, B consisting of a single letter b, then we also say that the letter b *occurs at position j* in the word A.

Example. In the word *singing*, the word *ing* has two occurrences, namely $\{1,2,3\}$ and $\{4,5,6\}$. The letter s occurs at position 0 only; other letters occur at two positions. Also occurring are *gin* and *sin*, but *gig* does not occur.

Let A be a term. Let $T(A)$ be the set of all occurrences of whatever terms may occur in A. Consider the dually inclusion-ordered set $(T(A), \supseteq)$. Proposition 3 implies that incomparable occurrences of terms are disjoint.

Term Structure Theorem. *Let A be a term of length n. Let $(T(A), \supseteq)$ be the dually inclusion-ordered set of all term occurrences in A. Then the covering relation of $(T(A), \supseteq)$ is a directed tree with basepoint $[0, n-1] = n$.*

Proof. We proceed by induction on the length of A. The inductive step uses the observation that if $A = \sigma_i A_0 \cdots A_{n-1}$, where the length of A_j is l_j, then the term occurrences covering $[0, n-1]$ are precisely the segments

$$\left[1 + \sum_{j<i} l(j), \sum_{j \leq i} l(j) \right], \quad i \in n \qquad \square$$

Example. Consider the ring term $A = + \bullet xy - + x - y$ The term structure tree is displayed in Figure 11.1.

Free Algebra Theorem. *For a given universal algebra type and a variable set V, let $(T, s_i : i \in I)$ be the algebra of terms, and let $V \subseteq T$ be the set of one-letter words reduced to a variable. Then every mapping g from V to any algebra $(U, f_i : i \in I)$ of the given type can be extended to a unique algebra homomorphism h from T to U.*

Proof. $h(x) = g(x)$ for every $x \in V$ and

$$h(\sigma_i A_0 \cdots A_{n-1}) = f_i(h(A_0), \ldots, h(A_{n-1}))$$

where $n = n(i)$. $\qquad \square$

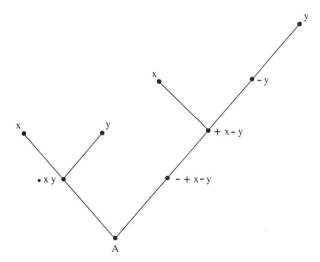

FIGURE 11.1 A term structure tree.

EXERCISES

1. Take the positive integers as variables. As operation symbols σ_i, take couples of the form $(n(i), i)$, $i \in \omega$. For finite types of algebras, write programs to
   (a) decide if a given word, composed of variables and operation symbols, is a term,
   (b) list all the terms occurring within a given term and determine the inclusion order of the occurrences,
   (c) decide whether a given term occurs in another given term.

2. Show that the image of the homomorphism $h : T \to U$ described in the Free Algebra Theorem coincides with the subalgebra of U generated by $g[V]$.

3. Observe that the set of all words over a given alphabet is a semigroup under concatenation. What can you say about this semigroup?

4. Why do we use brackets in mathematical expressions? Why do we use different kinds of brackets?

3. TRUTH AND FORMAL PROOF

Over a given alphabet, consider words A and B of length n and m and a letter b. Let J be the set of positions $j \in n$ at which b occurs in A. Let $\operatorname{Card} J = k$. Then

$$A = A_0 b \cdots A_{k-1} b A_k$$

for $k+1$ words $A_0, \ldots, A_{k-1}, A_k$ in which b does not occur any more. The word

$$A_0 B \cdots A_{k-1} B A_k$$

is called the *substitution* of B for b in A, and it is denoted by $A(b:B)$. If b does not occur in A, $k = 0$, then $A(b:B) = A$. Note also that $A(b:b) = A$.

Examples and Comment. If A is the ring word $+x \bullet xy$ and B is $+xz$, then $A(x:B)$ is $++xz\bullet+xzy$. The reader is correct in thinking of this as the substitution of $x + z$ for x in $x + x \cdot y$ or $x + (x \cdot y)$, but our "Polish notation" has the advantage of being free of brackets and precedence rules. It is mathematically simpler.

Proposition 4 *If A and B are terms and b is a variable, then $A(b:B)$ is a term.*

Proof. We can use induction on the length of A. If A is not of length 1, then write

$$A = \sigma_i A_0 \cdots A_{n-1}$$

and observe that

$$A(b:B) = \sigma_i A_0(b:B) \cdots A_{n-1}(b:B) \qquad \square$$

Let $\cong$ be a letter not in $\mathcal{O} \cup \mathcal{V}$, i.e., neither an operation symbol nor a variable: we shall call it the *equality symbol*. If A and B are terms, then the concatenated word $A \cong B$ is called an *equational sentence*. Let S be any set of such sentences. We say that a sentence $D \cong E$ is an *immediate consequence* of S if any one of the following holds:

(i) $D \cong E$ is in S,

(ii) D and E are identical words,
(iii) $E \cong D$ is in $\mathcal{S}$,
(iv) for some term F, both $D \cong F$ and $F \cong E$ are in $\mathcal{S}$,
(v) for some $A \cong B$ and $F \cong G$ in $\mathcal{S}$ and a variable x, D is $A(x : F)$ and E is $B(x : G)$.

A *formal proof* of a sentence $A \cong B$ from $\mathcal{S}$ is a sequence of $n \geq 1$ sentences

$$(S_0, \ldots, S_{n-1})$$

such that S_{n-1} is $A \cong B$ and for each $0 \leq j \leq n-1$, S_j is an immediate consequence of

$$\mathcal{S} \cup \{S_k : k < j\}$$

Obviously every member of $\mathcal{S}$ has a formal proof from $\mathcal{S}$. If only members of $\mathcal{S}$ have a formal proof from $\mathcal{S}$, then $\mathcal{S}$ is called an *equational theory*, or *formal theory*. The set of all equational theories is a closure system on the set of equational sentences. It is an excellent exercise in logic to verify that this closure system is algebraic: if a sentence has a formal proof from a set $\mathcal{S}$ of sentences, then it has a formal proof from a finite subset of $\mathcal{S}$.

A few general technical comments will help to understand the formal theory concept. Suppose that a sentence $A \cong B$ belongs to a theory T, that $x_1, \ldots, x_n$ are all the distinct variables occurring in $A \cong B$, and that $y_1, \ldots, y_n$ are some other variables,

$$\{x_1, \ldots, x_n\} \cap \{y_1, \ldots, y_n\} = \emptyset$$

Define terms $A_0, A_1, \ldots, A_n$ by

$$A_0 = A \quad \text{and} \quad A_i = A_{i-1}(x_i : y_i) \quad \text{for} \quad i = 1, \ldots, n$$

and terms $B_0, B_1, \ldots, B_n$ by

$$B_0 = B \quad \text{and} \quad B_i = B_{i-1}(x_i : y_i)$$

Then $A_n \cong B_n$ belongs to T. Informally speaking, each variable x_i has been exchanged for y_i. Let now $F_i \cong G_i$ be a sentence in T for each $i = 1, \ldots, n$. If the set of variables is infinite, then $y_1, \ldots, y_n$ can be assumed to have been chosen so as not to occur in any of the F_i or G_i. Define terms $C_0, C_1, \ldots, C_n$ by

$$C_0 = A_n \quad \text{and} \quad C_i = C_{i-1}(y_i : F_i) \quad \text{for} \quad i = 1, \ldots, n$$

and terms $D_0, D_1, \ldots, D_n$ by

$$D_0 = B_n \quad \text{and} \quad D_i = D_{i-1}(y_i : G_i)$$

Then the sentence $C_n \cong D_n$ belongs to T. Informally speaking, to get $C_n \cong D_n$ from $A \cong B$, each variable x_i has been replaced by the term F_i in A and by the term G_i in B. Note how convenient it is to have an infinite supply of variables. We assume henceforth the infinity of the variable set.

The theory T generated by any set S of sentences consists of those sentences that have a formal proof from S. The generating set S is referred to as a set of *formal axioms* of the theory T. In all of the following examples, we assume that the alphabet contains an infinity of distinct variables, five of which are denoted by x, y, z, v, w.

(1) Let $\mathcal{O}$ consist of a single binary operation symbol $\bullet$. Let S consist of the single formal axiom $\bullet x \bullet yz \cong \bullet\bullet xyz$. The theory generated is called the *formal theory of semigroups*.

(2) For the algebra type $(2,0,1)$ let $+, 0, -$ be binary, nullary, and unary operation symbols. Take five formal axioms:

$$+x+yz \cong ++xyz$$

$$+x\mathbf{0} \cong x$$

$$+\mathbf{0}x \cong x$$

$$+x-x \cong \mathbf{0}$$

$$+-xx \cong \mathbf{0}$$

The theory generated is called the *formal theory of groups*. If we add as sixth axiom the sentence

$$+xy \cong +yx$$

then the theory generated is called the *formal theory of commutative groups*. (At the end of this section the reader should be able to demonstrate rigorously that these two theories are distinct, i.e., that the sixth axiom is not a sentence belonging to the formal theory of groups.)

(3) For type $(2,0,1,2)$ let the corresponding four operation symbols be $+$, $\mathbf{0}$, $-$, as in the formal theory of groups, plus the binary symbol $\bullet$. As axioms, take all the six formal axioms listed above for the formal theory of commutative groups plus the additional three axioms

$$\bullet x \bullet yz \cong \bullet \bullet xyz$$

$$\bullet xy \cong \bullet yx$$

$$\bullet x+yz \cong +\bullet xy \bullet xz$$

The theory generated by these nine axioms is called the *formal theory of rings*. The following sequence of sentences constitutes a formal proof of $\bullet + yzx \cong + \bullet yx \bullet zx$ from these nine axioms:

$$+yz \cong +yz$$
$$\bullet x+yz \cong \bullet +yzx$$
$$+vw \cong +vw$$
$$+\bullet xyw \cong +\bullet yxw$$
$$+\bullet yxw \cong +\bullet xyw$$
$$z \cong z$$
$$\bullet xz \cong \bullet zx$$
$$\bullet zx \cong \bullet xz$$
$$+\bullet yx \bullet zx \cong +\bullet xy \bullet xz$$
$$+\bullet xy \bullet xz \cong \bullet x+yz$$
$$+\bullet yx \bullet zx \cong \bullet x+yz$$
$$+\bullet yx \bullet zx \cong \bullet +yzx$$
$$\bullet +yzx \cong +\bullet yx \bullet zx$$

Fortunately it is not necessary to exhibit a formal proof in order to establish that one exists. The reader is challenged to show, beyond reasonable doubt, that

$$\bullet + xy + x - y \cong + \bullet xx - \bullet yy$$

belongs to the formal theory of rings. In the course of this exercise such creative notations as "$(x+y)(x-y)$" and "$x^2 - y^2$" may be used.

(4) For type (2,2) let the operation symbols be $\vee$ and $\wedge$. The *formal theory of lattices* is generated by the following axioms:

$$\vee x \vee yz \cong \vee xyz$$
$$\wedge x \wedge yz \cong \wedge \wedge xyz$$
$$\vee xy \cong \vee yx$$
$$\wedge xy \cong \wedge yx$$
$$\vee x \wedge xy \cong x$$
$$\wedge x \vee xy \cong x$$

(5) Consider the type (2,2,1,0,0). Let the binary operation symbols be those of the formal theory of lattices, $\vee$ and $\wedge$. Let $-$ be the unary operation symbol and let **1** and **0** be the nullary symbols. Take as axioms all the six axioms of the formal theory of lattices given above plus the following:

$$\vee x \wedge yz \cong \wedge \vee xy \vee xz$$
$$\wedge x \vee yz \cong \vee \wedge xy \wedge xz$$
$$\vee x - x \cong \mathbf{1}$$
$$\wedge x - x \cong \mathbf{0}$$

The theory generated by all these axioms is called the *formal theory of Boolean lattices*. The terms are called *Boolean lattice terms*.

Do all these formal theories mean anything? For a given type $(n(i) : i \in I)$, operation symbols σ_i, $i \in I$, and variable set $\mathcal{V}$, an *interpretation* of the variables in a universal algebra $(U, f_i : i \in I)$ is a function $g : \mathcal{V} \to U$. According to the Free Algebra Theorem, g defines a unique homomorphism h from the algebra of terms to U. For a term A, $h(A)$ is called the *semantic value* of A under g. A sentence $A \cong B$ is said to be *true* under the interpretation g if $h(A) = h(B)$. It is said to be *valid* in $(U, f_i : i \in I)$ if it is true under every interpretation of the variables in U. A *model* for a set S of sentences is an algebra $(U, f_i : i \in I)$ in which every member of S is valid.

Examples. Groups, when viewed as type (2,0,1) algebras, are precisely the models of the set of five defining axioms of the formal theory of groups. Rings are the models of the nine axioms of the formal theory of rings.

If U is a model for a set S of sentences, then it is easy to verify that every immediate consequence of S is also valid in U. It follows by induction that every equational sentence that has a proof from S is valid in every model of S. An equational sentence that is valid in every model of S is called a *semantic consequence* of S.

A priori certain semantic consequences of S could fail to have a formal proof from S, and this would in some sense indicate the logical inadequacy of the "formal proof" concept. But semantic consequences turn out to be formally provable. The disturbing possibility of "truth without formal proof" was indeed ruled out for a much broader class of situations than those considered here, by Gödel and Herbrand, independently, in 1930. The following theorem is more limited and mathematically less complex than the Gödel–Herbrand result, but it does capture some of its philosophical essence.

Remark. Let us point out that the result that follows, or the general Gödel–Herbrand theorem, does not say that the validity of any given sentence can be decided mechanically. Unjustified hopes in that sense were addressed by Gödel's "incompleteness theorem" of 1931.

Completeness Theorem for Equational Calculus. *Every semantic consequence of a set S of equational sentences has a formal proof from S.*

Proof. We show that every formal theory T has a model in which no sentence outside T is valid. Applying this to the set $\overline{S}$ of those sentences that admit a formal proof from S yields the intended result.

Consider the algebra of terms $(T, s_i : i \in I)$. Define a congruence relation $\equiv$ on T by

$$A \equiv B \quad \text{if and only if} \quad A \cong B \quad \text{belongs to the theory } T$$

(Make sure to dispel any doubt that we have a congruence relation on the algebra of terms.) Take the quotient algebra of T by this congruence. □

Consequently, the formal theory of semigroups consists of all equational sentences valid in every semigroup, i.e., true under every interpretation in any semigroup. The formal theory of groups consists of all sentences valid in every group. The formal theory of rings is precisely the set of sentences valid in rings. And the formal theory of lattices is the set of sentences valid in lattices.

The formal theory of Boolean lattices is most remarkable.

Let us view every Boolean lattice L as a type $(2,2,1,0,0)$ algebra. The binary operations are join and meet, the unary operation is complementation, and the nullary constant operations have as values the maximum 1_L and the minimum 0_L of the lattice. Letting the operation symbols $\vee, \wedge, -, \mathbf{1}, \mathbf{0}$ correspond to these operations, in this order, we can say that the formal theory of Boolean lattices consists precisely of those equational sentences that are valid in every Boolean lattice. Such a sentence is called a *tautology*.

Examples. The sentences $x \cong x$, $--x \cong x$, $\vee xx \cong x$, and $\wedge xx \cong x$ as well as the sentence $\vee xy \cong -\wedge -x-y$ are tautologies.

Proposition 5 *A sentence $A \cong B$ is a tautology if and only if*

$$\vee \wedge AB \wedge -A-B \cong \mathbf{1}$$

is a tautology.

Proof. Under a given interpretation of variables in a Boolean lattice L, let $h(A)$ and $h(B)$ denote the semantic values of A and B. If $A \cong B$ is a tautology, then $h(A) = h(B)$ for every interpretation, and the semantic value of $\vee \wedge AB \wedge -A-B$ is

$$[h(A) \wedge h(B)] \vee [h(A)' \wedge h(B)'] = h(A) \vee h(A)' = 1_L$$

under every interpretation.

Conversely, if $\vee \wedge AB \wedge -A-B \cong \mathbf{1}$ is a tautology, then under every interpretation

$$[h(A) \wedge h(B)] \vee [h(A)' \wedge h(B)'] = 1_L$$

$$[h(A) \vee h(A)'] \wedge [h(A) \vee h(B)'] \wedge [h(B) \vee h(A)'] \wedge [h(B) \vee h(B)'] = 1_L$$

which means

$$h(A) \vee h(B)' = 1_L$$
$$h(B) \vee h(A)' = 1_L$$

Therefore

$$h(A) = h(A) \wedge 1_L = h(A) \wedge [h(B) \vee h(A)'] = h(A) \wedge h(B)$$

$$h(B) = h(B) \wedge 1_L = h(B) \wedge [h(A) \vee h(B)'] = h(B) \wedge h(A)$$

and consequently $h(A) = h(B)$. □

Proposition 6 *A sentence of the form* $A \cong 1$ *is a tautology if and only if under every interpretation of the variables in the two-element Boolean lattice* $2 = \{0, 1\}$, *the semantic value of A is 1.*

Proof. The "if" part alone requires proof. Suppose we have a term A whose only possible semantic value in $2 = \{0, 1\}$ is 1. If $A \cong 1$ is not a tautology, then under some interpretation g in some Boolean lattice L, the semantic value $h(A)$ of A is not the lattice maximum 1_L, $h(A) < 1_L$. (Keep in mind that h is the one homomorphism, from the algebra of Boolean lattice terms to L, that extends g.) Let F be an ultrafilter in L that contains the complement $h(A)'$. Convince yourself that the map $k : L \to 2$ given by

$$k(x) = \begin{cases} 0 & \text{for } x \in L \setminus F \\ 1 & \text{for } x \in F \end{cases}$$

is a homomorphism from L to 2.

Look at the semantic value of A under the interpretation $k \circ g$ in the Boolean lattice 2. □

It follows from the last two propositions that $A \cong B$ is a tautology if and only if the terms A and B have the same semantic value under every interpretation in the Boolean lattice 2. Since there are only a finite number of such interpretations to worry about, for any given candidate sentence $A \cong B$, the task of deciding whether $A \cong B$ is a tautology becomes a matter of simple mechanical verification.

A further consequence is that tautologies can be characterized in terms of Boolean functions. Let us fix a finite number of variables, say n variables $x_0, \ldots, x_{n-1} \in \mathcal{V}$, and consider those Boolean lattice terms whose variables are among the chosen x_i. Every such term A defines a Boolean function $f_A : 2^n \to 2$, where $f_A(a_0, \ldots, a_{n-1})$ is the semantic value of A under any interpretation $g : \mathcal{V} \to 2$ of the variables such that $g(x_i) = a_i$. Using the representation of a Boolean function as a disjunction of simple conjunctions, it is easily seen that every Boolean function $f : 2^n \to 2$ is obtained as an f_A. Two terms A, B give rise to the same Boolean function, $f_A = f_B$, if and only if $A \cong B$ is a tautology.

The equational sentences valid in the Boolean lattice 2 are often called "theorems of propositional calculus." Since they are precisely the sentences that make up the formal theory of Boolean lattices, each of them admits a formal proof from the simple axioms given earlier. This fact is customarily referred to as the completeness of propositional calculus.

EXERCISES

1. Write a program to verify the correctness of formal proofs from a given finite set of formal axioms.

2. Write a program to verify tautologies.

3. Write a program to exhibit a formal proof, from the ten axioms of the formal theory of Boolean lattices, for any given tautology.

4. Can you develop analogues or generalizations of Propositions 5 and 6 in the context of some other equational theories?

BIBLIOGRAPHY

Hans-J. BANDELT and Jarmila HEDLÍKOVÁ, Median algebras. *Discrete Math.* 45, 1983, pp. 1–30. A very readable research paper on a class of algebras with a ternary operation, closely related to median graphs and certain semilattices.

J. L. BELL and A. B. SLOMSON, *Models and Ultraproducts: An Introduction*. North-Holland 1969. A clear and self-contained introduction to model theory centered around a construction based on ultrafilters. Indeed the book can serve as an introduction to mathematical logic. Unlike in our Chapter XI, the syntax is not limited to equational sentences but includes the full predicate calculus with existential and universal quantifiers.

Paul M. COHN, *Universal Algebra*. Reidel 1981. Connections with logic are quite apparent in this classical text on universal algebra. Besides n-ary operations, n-ary relations are considered as well, and Gödel's completeness theorem is proved in its full generality.

George GRÄTZER, *Universal Algebra*. Springer 1979. This classic volume devotes an entire chapter to the intriguing concept of partial algebra.

Hans LAUSCH and Wilfried NÖBAUER, *Algebra of Polynomials*. North-Holland 1973. Polynomials over a universal algebra are a generalization of polynomials over a ring. They may also be viewed as an algebra of terms in the sense of our text. Most books on universal algebra have some material on polynomials; this one is devoted to them entirely. More precisely, about half of the volume deals with polynomials over universal algebras, and the other half with polynomials over particular species of structures, mostly rings, fields, and groups.

Roger C. LYNDON, *Notes on Logic*. Van Nostrand 1966. The reader who wishes to gain a general insight into mathematical logic can hardly do it more economically than by perusing these 90 pages. The topics addressed include the syntax of terms and quantified sentences, models, decidability, axiomatizability, and the unprovability of consistency established by Gödel.

Ralph N. MCKENZIE, George F. MCNULTY and Walter F. TAYLOR, *Algebras, Lattices, Varieties*, Volume I. Wadsworth & Brooks/Cole 1987. The researcher will find this an excellent initial reference to current topics.

G. SZÁSZ, *Introduction to Lattice Theory*. Academic Press 1963. This book contains selected material on universal algebras that can serve as an introduction. Universal algebras are defined at the beginning of Chapter II, preceding the definition of lattices themselves. A section in Chapter III deals with subalgebra lattices, Chapter IX is devoted to universal algebra congruences, and Chapter X treats of direct and subdirect products.

CHAPTER XII

CATEGORIES

Inside numbers we saw numbers. Sets were made up of sets. Order relations were compared in the framework of larger meta-orders. We have outlined the beginning of a theory of theories. The mathematician, like Archimedes, plays with mirrors and catapults. Structures are invented to hold and manipulate objects. At the end the structures themselves become objects of algebraic manipulation. The theory of categories is general enough to serve as a foundation for mathematics. Here it will be a looking-glass for a cursory retrospection on algebra.

We saw in Chapter I that multiplication of numbers reposes on the set product concept. The language of categories will now be used to transfer the product concept to proper algebraic structures. Products will then be used in classifying algebraic structures according to formal equational theories. We begin with a technical lemma and some definitions.

Countable Universe Lemma. *There exists some set U of sets with the following properties:*

(i) $\emptyset \in U$,
(ii) *if* $A \in U$ *and* $B \subseteq A$, *then* $B \in U$,

(iii) if $A \in U$, then $\mathcal{P}(A) \in U$,
(iv) if $A, B \in U$, then $\{A, B\} \in U$,
(v) if $(A_i : i \in I)$ is a family of members of U indexed by a set $I \in U$, then the union $\cup \{A_i : i \in I\}$ belongs to U.

Proof. Using induction on natural numbers, it is easy to verify that there is a family of sets $(U_n : n \in \omega)$ such that

$$U_0 = \emptyset$$

$$U_n = \bigcup_{m \in n} \mathcal{P}(U_m) \quad \text{for all} \quad n \in \omega$$

Let $U = \bigcup_{n \in \omega} U_n$.

The verification of properties (i) to (v) is left to the reader as an exercise in elementary set theory. A key fact to notice at the outset is that for $m \in n$, U_m is both a member and a subset of U_n. □

Any set U satisfying conditions (i) to (v) of this lemma is called a *universe*.

Properties (i) to (v) are reminiscent of axioms (A1) to (A5) postulated in Section 1 of Chapter I. We shall not deal here with the significance of this resemblance. Note that every member of the particular universe constructed in the proof of the lemma is finite, even though the universe itself is infinite. For our purposes it is a poor universe. We shall once again postulate what we cannot prove, the existence of a much larger universe. The ninth and last axiom to be introduced in this volume would deserve to be called the "second axiom of existential infinity." We only refrain from this anabaptism in deference to the extensive literature on the subject.

(A9) Inaccessibility Axiom. *There is a universe having among its elements the set ω of natural numbers.*

Once and for all, we choose one such large universe and denote it by **U**. The reader should have no difficulty verifying the following facts.

(1) If $A, B \in \mathbf{U}$, then every function $f : A \to B$ belongs to **U** as well. Also B^A, the set of all such functions, belongs to **U**.

(2) Every natural number n is a member of **U**. Every n-tuple of members of **U** belongs to **U**. Every n-ary operation on any mem-

ber of **U** belongs to **U**. Every relation on any member of **U** belongs to **U**.

A *(partial) associative structure* on a set C is a map from a subset $\text{Comp} \subseteq C^2$ to C, associating with every $(g,f) \in \text{Comp}$ the *composition* $g \circ f$ and such that

$$(h,g) \in \text{Comp} \quad \text{and} \quad (h \circ g, f) \in \text{Comp}$$

if and only if

$$(g,f) \in \text{Comp} \quad \text{and} \quad (h, g \circ f) \in \text{Comp}$$

and in this case $(h \circ g) \circ f = h \circ (g \circ f)$. An *identity* is any element u of C such that

$$u \circ f = f \quad \text{for all } f \text{ such that } (u,f) \in \text{Comp}$$

and

$$g \circ u = g \quad \text{if } (g,u) \in \text{Comp}$$

The set C, together with its associative structure, is called a *category* if for every $f \in C$ there are identities u, w such that

$$(w,f) \in \text{Comp}, \quad (f,u) \in \text{Comp}$$

The elements of a category C are called *morphisms*. For $(g,f) \in \text{Comp}$ we say that g and f are *composable*, or that "$g \circ f$ is defined." Of course this does not mean that "$f \circ g$ is defined" as well. The order of composition matters.

Elementary Example. Every monoid, with its binary operation as an associative structure, is a category. This category has only one identity, namely the neutral element of the monoid.

Category of Sets. Let C be the set of triples (B, f, A) where A, B are sets belonging to the universe **U** and $f : A \to B$ is a function. Let (E, g, D) and (B, f, A) be composable if $D = B$ and in that case let their composition be defined by

$$(E, g, B) \circ (B, f, A) = (E, g \circ f, A)$$

where the same symbol $\circ$ is used, somewhat abusively but according to custom, to denote the usual composition of functions. The identities are the triples (A, id_A, A) with the familiar identity maps $id_A, A \in U$. This category is called the *category of sets*. Note that any statement like "the category of sets comprises all sets" is nonsense. The liar's paradox of Crete must be kept on a short leash.

Consider universal algebras of a given type $(n(i) : i \in I)$ with I in U. Let C be the set of all triples (B, h, A) where A, B are algebras of this type with underlying sets in U and $h : A \to B$ is an algebra homomorphism. Define the composition of (E, g, D) and (B, h, A) as $(E, g \circ h, A)$ if $D = B$; otherwise the composition is not defined. This is the *category of algebras of type* $(n(i) : i \in I)$.

Given a morphism h in a category C, the identity u with $h \circ u = h$ is unique. Indeed, if u' is another such identity, $h \circ u' = h$, then
$$h = (h \circ u) \circ u'$$
Thus $u \circ u'$ is defined, and since both u and u' are identities,
$$u = u \circ u' = u'$$
The unique identity u with $h \circ u = h$ is called the *domain identity* of h. Similarly, the identity w with $w \circ h = h$ is unique too, and it is called the *codomain identity* of h.

Example. In the category of sets, the domain identity of a morphism (B, f, A) is (A, id_A, A). The codomain identity is (B, id_B, B).

Let S be a set of morphisms in a category C such that

(i) whenever g and h are composable morphisms belonging to S, we have $g \circ h \in S$,
(ii) for every $h \in S$, both the domain and codomain identities of h belong to S.

Then S, together with the composition of morphisms restricted to $Comp \cap S^2$, is a category. It is called a *subcategory* of C. Obviously, the set of all subcategories of C is an algebraic closure system on C.

A subcategory S of C is said to be *full* if it contains every morphism whose domain and codomain identities belong to S.

Examples. (1) If a monoid is viewed as a single-identity category, then the subcategories are precisely the submonoids. (2) In the category of sets, those morphisms (B, f, A) for which h is injective constitute a subcategory. This subcategory is not full. Given a cardinal number n, those (B, f, A) for which

$$\operatorname{Card} A = \operatorname{Card} B = n$$

form a full subcategory.

Consider the category C of universal algebras of a given type $(n(i) : i \in I)$. Let T be any formal equational theory with the corresponding operation symbols. Then

$$V = \{(B, h, A) \in C : A, B \text{ are models of } T\}$$

is a full subcategory of C. It is called the *variety*, or *equational class*, determined by the equational sentences in T. Take the groupoid type, for example (a single binary operation). The variety determined by the formal theory of semigroups (the theory generated by the formal axiom $\bullet x \bullet yz \cong \bullet \bullet xyz$) is called the *category of semigroups*. We hope the reader finds this terminology justified. For type $(2, 0, 1)$, the *category of groups* is the equational class determined by the formal theory of groups. The formal theory of commutative groups determines a subcategory of this, called the *category of commutative groups*. For type $(2, 0, 1, 2)$ the formal theory of rings gives rise to the equational *category of rings*. For type $(2, 2)$ we get the *category of lattices* from the formal theory of lattices.

Proposition 1 *Let C be the category of universal algebras of a given type $(n(i) : i \in I)$. Then the equational classes within C form a closure system on C.*

Proof. Let S be any set of equational classes. Each $V \in S$ is defined by some formal theory $T(V)$. Let T be the set of the various theories $T(V)$. Consider the theory generated by $\cup T$. Verify that the equational class defined by this latter theory coincides with $\cap S$.
□

The category of sets, the category of universal algebras of a given type, and any of their subcategories are examples of what we call a *category of structures*. Also called *concrete categories*, these are by

definition categories in which
  (i) all morphisms are triples (B, h, A) where h is a function,
  (ii) (E, g, D) is composable with (B, h, A) if and only if $D = B$, and their composition is then $(E, g \circ h, A)$ where $g \circ h$ is the usual composition of functions,
  (iii) the identities are the triples (A, h, A) where h is an identity function.

In a category of structures a morphism (B, h, A) is said to be a morphism *from A to B*; A is called the *domain structure* and B is called the *codomain structure* of the morphism. Whatever appears as a domain or codomain structure is then called simply a *structure* of the concrete category under consideration.

Example. As noted above, the category of sets is a category of structures. For a morphism (B, h, A) the sets A and B are the domain and codomain structures, respectively. Observe that if (E, h, D) is another morphism with the same function h, then $A = D$ but B need not coincide with E. (Functions specify their images, but not their codomains. The author has adopted this view to facilitate the set-theoretical handling of functions.)

Given a field F, whose underlying set belongs to the universe U, the *category of vector spaces* over F is the concrete category consisting of triples (B, h, A) where A, B are vector spaces over F with underlying sets belonging to U and $h : A \to B$ is linear.

The concrete *category of relational structures* consists of the triples (B, h, A) where A, B are relational structures on sets belonging to the universe U and $h : A \to B$ is relation preserving. Those (B, h, A) for which A and B are ordered sets form a subcategory, called the *category of ordered sets*. On the other hand, *to every preordered set* $(S, \lesssim)$ *corresponds a concrete category*: the set of morphisms is

$$\{(B, h, A) : A, B \in S,\ A \lesssim B \text{ and } h = \{\langle A, B \rangle\}\}$$

The *category of graphs* is the concrete category consisting of the triples (H, h, G) where G, H are graphs on vertex sets belonging to the universe U and $h : G \to H$ is a graph homomorphism.

The triples (P, h, T) where T, P are topological spaces with underlying sets in U and $h : T \to P$ is a continuous function form the concrete *category of topological spaces*.

Most of these categories, as well as the category of sets, are large enough to allow certain constructions to be carried out. This is owing to the richness of the universe U. Among such constructions we shall focus our attention on products, to be defined in a moment. Categories that are not so large may of course still be interesting from some algebraic point of view. The term *small category* will be reserved for those categories whose morphism set is a member of the universe U. For example, every monoid with underlying set in U is a small category. For any $S \in U$, and preorder $\lesssim$ on S, the concrete category corresponding to $(S, \lesssim)$ as defined above is small.

A *functor* from a category C to a category K is a function $F: C \to K$ such that

(i) if g and h are composable morphisms in C, then $F(g)$ and $F(h)$ are composable in K and $F(g) \circ F(h) = F(g \circ h)$,
(ii) if u is an identity in C, then $F(u)$ is an identity in K.

For example, if C and K are single-identity categories, i.e., monoids, then the functors are precisely the monoid homomorphisms. Functors are a natural generalization of the homomorphism concept, from monoids to partial associative structures.

Examples. (1) The *power set functor* is defined from the category of sets to itself by associating to each morphism (B, f, A) the morphism

$$(\mathcal{P}(B), \overline{f}, \mathcal{P}(A))$$

where
$$\overline{f}(S) = f[S] \quad \text{for every} \quad S \subseteq A$$

(2) A *stripping functor* is defined from the category of universal algebras of any given type $(n(i) : i \in I)$ to the category of sets by associating to every algebra morphism

$$((S, f_i : i \in I), h, (T, g_i : i \in I))$$

the set morphism (S, h, T). This functor strips the domain and codomain structures of their algebraic operations. (3) Let us associate to every ring morphism

$$((S, +, \cdot), h, (T, +, \cdot))$$

the group morphism
$$((S, +), h, (T, +))$$

Such a functor is called a *forgetful functor*, from the category of rings to the category of groups, since it "forgets" the rings' multiplicative operation.

Proposition 2 *If $F : C \to K$ and $G : K \to L$ are functors, then the composition $G \circ F$ is a functor from C to L. For any category C, the identity map on C is a functor from C to itself.*

In view of this proposition, we can consider the concrete category whose morphisms are the triples (K, F, C) where K and C are small categories and $F : C \to K$ is a functor. It is called the *category of categories*.

Let C be a category of structures. A structure B is called *terminal* if for every structure A there is a unique morphism from A to B, and B is *initial* if for every structure A there is a unique morphism from B to A.

Examples. In the category of sets, $\emptyset$ is initial. Every singleton is a terminal structure in this category. In the category of groups, every one-element group is both initial and terminal.

From any given category C, new categories can be constructed in many ways. In what follows we confine the exposition to categories of structures, but the development can be carried out for general categories without mathematical difficulty, and it may be even simpler, depending on notation. Let $(A_i : i \in I)$ be a family of structures in C. Throughout the rest of this chapter, any such indexing set I for a family of structures is assumed to belong to the universe **U**. A *projection cone* to this family is a family $(p_i : i \in I)$ of morphisms that have the same domain structure

$$D(p_i : i \in I)$$

and such that the codomain structure of each p_i is A_i.

Example. In the category of sets, if $(A_i : i \in I)$ is a family of sets in the universe **U**, then with the projections $\text{pr}_i : \prod_{j \in I} A_j \to A_i$, the family

$$\left(\left(A_i, \text{pr}_i, \prod_{j \in I} A_j \right) : i \in I \right)$$

is a projection cone. An even simpler cone is

$$((A_i,\emptyset,\emptyset) : i \in I)$$

The reader can similarly describe some projection cones in the category of topological spaces.

Given a family of structures $(A_i : i \in I)$ in a concrete category C, the *category of projection cones* to $(A_i : i \in I)$ is the concrete category whose morphisms are the triples (ρ, h, τ) such that

(i) $\tau = (t_i : i \in I)$ and $\rho = (r_i : i \in I)$ are projection cones to the family $(A_i : i \in I)$,
(ii) $(D(\rho), h, D(\tau))$ is in C,
(iii) the composition $r_i \circ (D(\rho), h, D(\tau))$ is defined in C and coincides with t_i, for each $i \in I$.

Any terminal structure $(p_i : i \in I)$ in the category of projection cones to $(A_i : i \in I)$ is called a *product* of the family $(A_i : i \in I)$. The common domain structure $D(p_i : i \in I)$ is called a *product structure* or *product object* of the A_i, $i \in I$.

Examples. (1) In the category of sets, the cone

$$\left(\left(A_i, \mathrm{pr}_i, \prod_{j \in I} A_j\right) : i \in I\right)$$

is a product of the family of sets $(A_i : i \in I)$ and $\prod_{j \in I} A_j$ is a product object. Indeed any set in $\mathbf{U}$ that is equipotent to $\prod A_j$ is a product object, and there are no other product objects for a given family $(A_i : i \in I)$ of sets. (2) In the category of topological spaces, the product space $\prod T_i$ is a product structure for a given family of spaces $(T_i : i \in I)$. (3) In the concrete category corresponding to a preordered set $(S, \leqslant)$ the structures are the elements of S. A projection cone to a family $(a_i : i \in I)$ of elements is any family of the form

$$((a_i, \{\langle b, a_i \rangle\}, b) : i \in I)$$

such that $b \leqslant a_i$ for each i. A product object, if there is one, is any greatest lower bound of the a_i, $i \in I$.

Let C be the category of universal algebras of a specified type $(n(j) : j \in J)$. Let $(A_i : i \in I)$ be a family of algebras (structures)

in C. For $i \in I$ we have
$$A_i = (S_i, f_{ij} : j \in J)$$
where the f_{ij}'s are operations on the set S_i. Let the product set $\prod S_i$ be denoted in short by S. For every $j \in J$ define an operation $f_j : S^{n(j)} \to S$ by
$$[f_j(x_0, \ldots, x_{n(j)-1})](i) = f_{ij}(x_0(i), \ldots, x_{n(j)-1}(i))$$
Then the algebra $A = (S, f_j : j \in J)$ is a structure in C, and it is a product object of the family $(A_i : i \in I)$ with the cone
$$((A_i, \mathrm{pr}_i, A) : i \in I)$$
as a product. The other product objects of $(A_i : i \in I)$ are the different algebras isomorphic to A, with underlying set in the universe $\mathbf{U}$. We shall write $\prod_{i \in I} A_i$ for the algebra A. Observe that every projection map
$$\mathrm{pr}_k : \prod_{i \in I} A_i \to A_k$$
is an algebra homomorphism. If B is any algebra of the same type, and
$$h : B \to \prod_{i \in I} A_i$$
any function, then h is an algebra homomorphism if and only if every composition $\mathrm{pr}_k \circ h$ is a homomorphism. (Compare this with Proposition 6 of Chapter X.)

Let now $V \subseteq C$ be an equational class. If every A_i, $i \in I$, is a structure in V, then the product algebra $\prod A_i$ is a structure in V as well: we say that every equational class is *closed under the formation of products*.

Two further properties of equational classes are perhaps even more evident. First, equational classes are *closed under the formation of subalgebras*: every subalgebra of any structure of a variety V is again a structure in V. Second, equational classes are *closed under the formation of homomorphic images*: if A is a structure in V and $h : A \to B$ is a surjective algebra homomorphism whose underlying set is in the universe $\mathbf{U}$, then B is a structure in V as well. In particular, all quotient algebras of a structure A in V are in V, and all algebras B with underlying set in $\mathbf{U}$ that are isomorphic to A are in V.

The following theorem states that the above three properties, closure under the formation of homomorphic images (**H**), subalgebras (**S**), and products (**P**), are characteristic of equational classes.

HSP Theorem. *Consider the category of universal algebras of a given type. A full subcategory V is an equational class if and only if it is closed under the formation of homomorphic images, subalgebras, and products.*

Proof. Only the "if" part needs to be proved. Suppose that V is closed under the formation of homomorphic images, subalgebras, and products. Consider the equational class closure $\overline{V}$ of V (Proposition 1). We need to show that $\overline{V} \subseteq V$. Let M be a structure in $\overline{V}$. Keep in mind that the underlying set S of M belongs to the universe U. We shall prove that M is in V. Since V is full, this will suffice.

The case $\operatorname{Card} S \leq 1$ is easily taken care of, so assume that S has at least two elements. We can actually assume that S is infinite, for otherwise we could replace M, in the argument that follows, by the product of an infinite number of factors identical with M.

Take S as a set of variables and consider the algebra T of terms. Make sure the set of operation symbols and the symbols themselves are in the universe U. Let $\mathcal{T}$ be the set of equational sentences valid in every structure of V. Obviously $\mathcal{T}$ is a formal theory. Let $\equiv$ be the congruence on T defined by

$$A \equiv B \quad \text{if and only if} \quad A \cong B \quad \text{is a sentence in } \mathcal{T}$$

(Any doubt that we have a congruence?) Let $\mathcal{N}$ be the set of sentences (with terms in T) that are not in the theory $\mathcal{T}$. The reader should verify that $\mathcal{N}$ belongs to the universe U. For every $\sigma \in \mathcal{N}$ there is an interpretation of the variables $g_\sigma : S \to G_\sigma$ in some structure G_σ of V such that σ is not true under g_σ. We may assume that the algebra G_σ is generated by $g_\sigma[S]$. The product algebra $\prod_{\sigma \in \mathcal{N}} G_\sigma$ is a structure in V. Consider the interpretation $g : S \to \prod G_\sigma$ given by

$$g(x) = (g_\sigma(x) : \sigma \in \mathcal{N})$$

Take the subalgebra P of $\prod G_\sigma$ generated by the set $\{g(x) : x \in S\}$. This too is a structure in V. We claim that it is isomorphic to the quotient algebra $T/\equiv$. By the Free Algebra Theorem of Chapter XI we have a homomorphism $h: T \to P$ such that

$$h(x) = g(x) \quad \text{for all} \quad x \in S$$

Obviously h is surjective onto P. We need to show only that

$$h(A) = h(B) \quad \text{if and only if} \quad A \equiv B$$

If $A \equiv B$, then $A \cong B$ is a sentence in T; therefore it is valid in P and $h(A) = h(B)$. If $A \equiv B$ does not hold, then $A \cong B$ coincides with a sentence ν in $\mathcal{N}$. Thus $A \cong B$ is not true under g_ν in G_ν and therefore it cannot be true under g in $\prod G_\sigma$, $h(A) \neq h(B)$. We have shown that P is isomorphic to $T/\equiv$. Therefore $T/\equiv$ is a structure in V as well.

Since the algebra M on the set S belongs to the equational class closure $\overline{V}$, it is a model of the theory T. Let h be the homomorphism $T \to M$ such that

$$h(x) = x \quad \text{for all} \quad x \in S$$

Obviously h is surjective onto M and $A \equiv B$ implies $h(A) = h(B)$. Therefore a surjective homomorphism $\bar{h}: T/\equiv \to M$ is defined by mapping each class of $\equiv$ to the element $h(A)$ where A is any term in that congruence class. As a homomorphic image of $T/\equiv$, M must be a structure in V. □

Products were defined as terminal structures in the category of projection cones of a given family $(A_i : i \in I)$ of structures of a basic concrete category C. Dually, an *injection cone* is a family $(p_i : i \in I)$ of morphisms with common codomain structure $T(p_i : i \in I)$ and such that the domain structure of each p_i is A_i. The *category of injection cones* for $(A_i : i \in I)$ is the concrete category whose morphisms are the triples (ρ, h, τ) such that

(i) $\tau = (t_i : i \in I)$ and $\rho = (r_i : i \in I)$ are injection cones for $(A_i : i \in I)$,
(ii) $(T(\rho), h, T(\tau))$ is in C,
(iii) the composition $(T(\rho), h, T(\tau)) \circ t_i$ is defined in C and coincides with r_i, for each $i \in I$.

An initial structure $(p_i : i \in I)$ in the category of injection cones is called a *coproduct* of $(A_i : i \in I)$, and the common codomain struc-

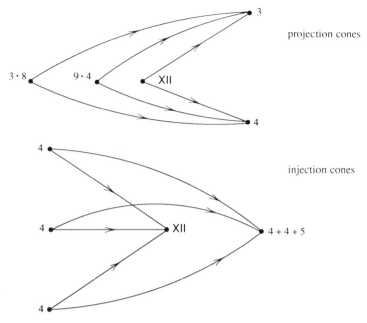

FIGURE 12.1 Product and sum.

ture $T(p_i : i \in I)$ is called a *coproduct object*. We confine ourselves to illustrate the simplicity of this concept in the category of sets. Figure 12.1 shows a product and a coproduct construction for the cardinal number commonly known as *twelve*.

The reader should be able to carry out the proof of the following theorem without difficulty.

Product–Sum Duality Theorem. *For any cardinal numbers A and B, the cardinal product $A \cdot B$ is their product object in the category of sets. The cardinal sum $A + B$ is their coproduct object.*

This duality is all the more remarkable in that the traditional genesis of product from sum, as in $3 \cdot 4 = 4 + 4 + 4$, places these arithmetic operations in a definitely asymmetrical relationship.

EXERCISES

1. Clarify the meaning of injection cones in a category corresponding to a given preordered set $(S, \lesssim)$, in particular when $\lesssim$ is an

order relation. Examine the question of existence of products and coproducts.

2. Develop an elementary theory of products in the category of ordered sets. What is the order dimension of a product of chains? What is the order dimension of a finite Boolean lattice?

3. Develop an elementary theory of products and coproducts for vector spaces. What can you say about dimension?

4. Characterize products in the category of graphs. Describe the product when all factors coincide with K_2.

5. Given a universal algebra type, what connections can you find between the inclusion-ordered set of equational theories and the inclusion-ordered set of equational classes?

6. What similarities and dissimilarities can you identify when you compare the categories of (a) sets, (b) ordered sets, (c) groups, (d) vector spaces, (e) graphs, and (f) topological spaces?

BIBLIOGRAPHY

Jiří ADÁMEK, Horst HERRLICH, and George E. STRECKER, *Abstract and Concrete Categories: The Joy of Cats*. John Wiley & Sons 1990. An essentially self-contained exposition of basic category theory oriented towards the idea of "structure." The book discusses how algebraic structures differ from topological and other structures.

Jiří ADÁMEK and Věra TRNKOVÁ, *Automata and Algebras in Categories*. Kluwer 1990. The relationship between automata and universal algebras is set in a categorical framework. An advanced monograph for researchers.

Garrett BIRKHOFF and Saunders MAC LANE, *Algebra*. Macmillan 1979. The reader who seeks an undergraduate-level refresher on categories and functors will find a number of basic concepts assembled here, in a general algebra context, in Chapter XV.

T. S. BLYTH, *Categories*. Longman 1986. A concise introduction accessible at the advanced undergraduate level.

Peter J. FREYD and Andre SCEDROV, *Categories, Allegories*. North-Holland 1990. An up-to-date graduate text with a metamathematical agenda. Allegories are categories enriched with an inverse operation (reciprocation) and a largest common restriction operation (intersection). Think about partially defined multivalued functions.

Robert GOLDBLATT, *Topoi: The Categorical Analysis of Logic*. North-Holland 1984. Starting from elementary set theory, a clear exposition of basic category theory leads to metamathematical considerations.

Nathan JACOBSON, *Basic Algebra II*. Freeman 1989. The first two chapters of this volume are devoted to categories and universal algebras. The concepts presented include natural transformations, subdirect products, and limits.

Saunders MAC LANE, *Categories for the Working Mathematician*. Springer 1971. A classical reference volume for the generalist.

INDEX OF DEFINITIONS

Absolute value, 135
Absorption, 19, 22, 213
Abstract group, 70
Action of:
  group on set, 70
  permutation group, 71
Activity, 52
Acyclic relation, 40
Addition:
  cardinal, 23
  ideal, 97
  integer, 76
  modulo an integer, 79
  in ring, 93
  of vectors, 157
Adherence point, 273
Adjacency, 197
Affine geometry:
  of flat, 179
  on vector space, 176
Algebra, 291
  of terms, 296
Algebraic:
  closure, 45
  dimension, 251
  field extension, 142

Algebras, category of, 314
Alphabet, 295
Alternating group, 85
Antichain, 33
Antisymmetry, 13, 33
Approximation, 273, 275
Ascending chain condition, 39
Associative structure, 313
Associativity, 4, 9, 19, 22, 23, 59, 158
Atom, 233
Automorphisms of:
  fields, 95
  groupoids, 65
  relations, 32
  rings, 95
  vector spaces, 167
Axiom, 302

Basepoint and basepoint order, 200
Basis:
  in matroid, 247
  in vector space, 161
Bijection, 7
Binary operation, 25
Binary relation, 31
Bipartite graph, 208

328 INDEX OF DEFINITIONS

Boolean:
  function, 231
  sum, 22
Boolean lattices, 230
  formal theory of, 304
Boolean lattice terms, 304
Boundary flat, 184
Bounded:
  lattice and sublattice, 214
  uniform space, 285
Bounds, upper and lower, 36

Canonical:
  field injection, 140
  representation of cycle matroid, 257
  surjection, 35
Cardinal, 16, 17
Cartesian product, 5
Category, 313
  equationally defined, 315
Category of:
  categories, 318
  injection cones, 322
  projection cones, 319
  sets, 314
  various structures, 314–316
Cauchy filter, 285
Causality, 187
Chain, 33
Chain topology, 273
Characteristic set, 231
Choice function, 15
Chromatic:
  function, 208
  number, 207
Chromatic polynomial of:
  graph, 209
  matroid representation, 259
Circuit:
  in matroid, 246
  in vector space, 159
Class, 34
Closed sets, 24
  under function, 25
  under operation, 25
  in a topology, 272–273

Closeness, 273
Closure point, 273
Closure systems and operators, 24
  algebraic, 45
  matroid, 245
  Noetherian, 48
  relative, 246
  topological, 272–273
Closure under HSP, 320–321
Codomain, 6
  identity, 314
  structure, 316
Coefficient, 105
Cofinite:
  filter, 271
  topology, 284
Coloring of graph, 207
Column, 171, 172
Comaximal ideals, 97
Commutative groups:
  category of, 315
  formal theory of, 302
Commutativity, 4, 19, 22, 23, 59, 170
Commuting monoid elements, 63
Compactness, 278
Compatibility of:
  order with group, 128
  scenario with strategy, 203–204
Complement:
  in lattice, 214
  of subset, 22
Complete:
  lattice, 43
  graph, 197
  uniform space, 285
Completion date, 52
Complex number, 141
Composition of:
  functions, 8
  morphisms, 313
Concatenation, 296
Concrete category, 315
Cone:
  injection, 322
  projection, 319

Congruence:
  of groupoid, 61
  modulo a subgroup, 67
  of universal algebra, 294
Conjugation, 65, 71
Conjunction of literals, 235
Connected graph, 198
  components, 199
Consequence:
  immediate, 300
  semantic, 305
Constant, 7
Constraint:
  function, 183
  in linear programming, 184
  of precedence, 52
Containment concept, 26
Continuous:
  function, 273
  order, 130
Contraction:
  of edge in graph, 208
  matroid, 253
Contradictory constraints, 52
Convergence, 275
Convex hull, 163
Convexity in:
  graphs, 198
  ordered sets, 44
  vector spaces, 163
Coordinate function, 168
Coprime:
  ideals, 97
  ring elements, 114
Coproduct, 322–323
Coset, 73, 95
Couple, 18
Covering conditions, 220
Covering:
  graph, 200
  relation, 49
Critical activity, 52
Critical path, 53
Cut, 130
Cycle:
  in a binary relation, 40

  factor, 83
  in a graph, 198
  matroid, 246
  permutation, 82
  structure, 83
Cyclic group, 78

Deletion:
  of edge in graph, 208
  matroid, 253
Dependence:
  linear, 159
  in matroid, 246
Descending chain condition, 39
Difference of:
  numbers, 77
  sets, 15
Dilatation, 191
Dimension:
  algebraic, 251
  geometric, 249
Dimension of:
  affine flat, 179
  order, 41
  vector space, 163
Directed tree, 200
Discrete:
  order, 51
  topology, 273
  uniformity, 281
Disjoint:
  permutation cycles, 82
  sets, 23
Disjunction, 235
Distance:
  of integers, 76
  in metric space, 282
  between vertices, 198
Distinctness relation, 32
Distributive lattice, 218
Distributivity, 22, 23, 78, 93, 158, 170
Divisibility, 111
Domain, 6
  identity, 314
  structure, 316

Dot product, 170
Dual relation, 35
Duration, 52

Early start schedule, 52
Edge, 197
   deletion and contraction, 208
   recursion, 208
Eight, 12
Element, 3
Elevation, 222
Embedding of geometries, 177
Empty graph, 197
Empty set, 3
Endomorphism of:
   groupoid, 63
   vector space, 168
Entire ring, 116
Entourage, 281
Entry of matrix, 171
Equality symbol, 300
Equational:
   class, 315
   sentence, 300
   theory, 301
Equilibrium strategies, 204
Equipotence, 9
Equivalence, 33, 35
   associated with a partition, 34
   associated with a preorder, 35
   induced by a function, 34
Euclidean:
   algorithm, 121
   distance, 282
   norm, 115
   ring, 115
   topology, 283
   uniformity, 282
Even:
   integer, 80
   permutation, 84
Event, 187
Exponent of radical, 141
Exponentiation of sets, 6
Extension of:
   function, 6
   relation, 39
Extension of fields, 141
   radical, 146
   simple, 141
   simple algebraic, 142
Extensive law, 24

Factor, 19, 83
Factorial, 69
Family, 18
Fano plane, 181
Feasibility in LP, 184
Field, 94
   extensions, 141, 142, 146
Filter:
   in a lattice, 269
   on a set, 271
Finer topology, 273
Finite:
   graph, 197
   ordinal, 14
   set, 17
First element, 12, 14
First kind ordinal, 14
Five, 12
Fixed field, 142
Fixed point, 72
Fixing group, 142
Flat, 176
   boundary, 184
   at infinity, 180
   lying in, 179
Forest, 199
Forgetful functor, 318
Formal:
   axiom, 302
   proof, 301
   theory, 301
Formal theory of:
   lattices, 304
   rings, 303
   semigroups and groups, 302
Formative operation, 296
Four, 12
Fraction field, 128

Full:
　subcategory, 314
　subgraph, 199
Full slack schedule, 53
Function, 5
Functor, 317

Galois group, 143
Game, 203
Gaussian elimination, 174
gcd, 113
Generating an order, 49
Generators, 25
Geodesics, 198
　high and low, 222
Geometric:
　dimension, 249
　lattice, 240
Geometry, 176
　matroid, 249
glb, 36
Graphical representation, 257
Graphic matroid, 257
Graphs, 197
　category of, 316
Greater element, 35
Greater ordinal, 13
Greatest:
　common divisor, 113
　lower bound, 36
Greedy optimization, 264
Greedy procedure, 262
Grid lattice, 228
Groupoid, 59
　as universal algebra, 292
Groups, 64
　category of, 315
　formal theory of, 302

Hausdorff space, 284
Height:
　in geometric lattice, 241
　relative, 220
Helly number, 55
Hereditary set of sets, 263
High geodesic, 222

Homeomorphism, 274
Homomorphism of:
　graphs, 207
　groupoids, 60
　groups, 65
　monoids, 62
　rings, 94
　semigroups, 62
　universal algebras, 292
Hyperplane, 177
　constraint, 184

Ideal decomposition property, 101
Ideals, 95
　addition of, 97
　cosets of, 95, 100
　product of, 98
Idempotence, 4, 22, 24
Idempotent ring, 232
Identity:
　function, 6
　morphism, 313, 314
　relation, 32
　ring element, 93
Image, 6, 7
　filter, 271
Immediate consequence, 300
Inclusion concept, 26
Inclusion order, 33
Incomparable elements, 35
Independence:
　in closure system, 245
　linear, 159
　in matroid, 245–246
Indeterminates, 104–106
Index set, 18
Index of subgroup, 73
Induced equivalence, 34
Induced subgraph, 199
Inertial line, 189
Infinite:
　ordinal, 14
　set, 17
Inherited topology, 273
Initial:
　position in game, 203

332 INDEX OF DEFINITIONS

Initial *(Continued)*
  structure, 318
Injection, 7
  cone, 322
Inner automorphism, 65
Inner product, 170
Inputs, 121
Integer, 75
Integral domain, 116
Interpretation of variables, 304
Intersection, 21
Interval:
  in graph, 198
  of ordered set, 44
Inverse:
  element, 64
  function, 8
  image, 7
Inverted pairs, 84
Involution, 22
Irreducible polynomial, 138, 142
Irreflexive relation, 33
Isomorphism of:
  graphs, 207
  groupoids, 61
  relations, 32
  rings, 95
  universal algebras, 292
  vector spaces, 167
Isotone law, 24
Isthmus, 254

Join, 43
  semigroup, 213
Jordan–Dedekind chain condition, 221

k-cycle, 82
Kernel, 67, 96

Larger element, 35
Largest element, 36
Lattices, 43
  associated with idempotent rings, 233
  bounded, 214
  category of, 315
  formal theory of, 304
  as universal algebras, 292
lcm, 113

Leading coefficient, 105
Least common multiple, 113
Least upper bound, 36
Length of:
  cycle permutation, 82
  path or cycle in binary relation, 40
  path or cycle in graph, 198
  word, 295
Lesser element, 35
Lesser ordinal, 13
Letter, 295
Lightspeed, 187
Limit:
  ordinal, 14
  point, 275
Line, 176
Linear:
  combination, 159
  form, 169
  map, 167
  map defined by matrix, 172–173
  order, 33
  polynomial, 104
  programming problem, 183
Literals, 235
Locally high and low vertices, 222
Location, 188
Loop, 254
Lorentz group, 190
Lower:
  bound, 36
  covering condition, 220
  section, 44
Low geodesic, 222
LP problem, 183
lub, 36

Mapping, 5, 6
Matchable set of vertices, 266
Matching, 265, 266
Material causality, 187
Matrix, 171
  of linear map, 172–173
Matroid, 245–246
  graphical representation of, 257
Matroid geometry, 249

Maximal:
  chain, 39
  element, 35
  ideal, 97
Maximum, 35–36
Median equality, 216
Median graph, 201
  as universal algebra, 292
Meet, 43
  semigroup, 213
Member, 3
Metric:
  topology, 283
  uniform space, 282
Minimal element, 35
Minimum, 35
Minor of a matroid, 253
Model, 304
Modular:
  identity, 216, 237
  lattice, 237
Modulo m addition, 79
Monoid, 62
  as universal algebra, 292
Monomial, 105
Morphisms, 313
Motion:
  inertial, 189
  material, 190
Moves in games, 203
Multiple:
  in a ring, 111
  vector by scalar, 158
Multiplication:
  of ideals, 98–99
  of integers, 77
  in a ring, 93

n-ary operation, 25
Natural number, 14
Natural order:
  on line, 184
  of ordinals, 33
Negative, 75, 129, 132
Neighborhood of:
  point, 276
  set, 277
Neutral element, 62
Neutrality, 9, 19, 23
Nine, 12
Noetherian:
  closure, 48
  ring, 97
Normal subgroup, 65
nth root of unity, 141
n-tuple, 18
Null:
  flat, 176
  linear form, 169
  set, 3
  vector, 157
Nullary operation, 25
Null product ring, 94
Number:
  cardinal, 16, 17
  complex, 141
  of elements, 17
  integer, 75
  natural, 14
  ordinal, 11
  rational, 128
  real, 132

Object:
  coproduct, 324
  product, 319
Objective function, 183
Occurrence in words, 297
Occurrence of event, 188
Odd:
  integer, 80
  permutation, 84
One, 12
Onto map, 7
Open cover, 277
Open set, 273
Operation, 25
Operation symbol, 296
Optical causality, 187
Optimal:
  result of greedy procedure, 263

334 INDEX OF DEFINITIONS

Optimal *(Continued)*
   schedule, 52
   solution, 184
Orbit, 72
Order, in group theory, 119
Ordered pair, 5
Ordered sets, 33
   category of, 316
Order relation, 33
   associated with preorder, 35
Ordinal, 11
Orientation of graph, 200
Orthogonality, 170
Outcome of game, 203
Output of:
   Euclidean algorithm, 121
   greedy procedure, 263

Pair, 4
Pairwise disjoint sets, 34
Parallelism:
   of affine flats, 178
   in matroid, 257
Parity function, 85
Partial associative structure, 313
Partition, 34
Path:
   critical, 53
   in a graph, 198
   in relation, 40
Payoff function, 203
Permutation, 68
   cycle, 82
   group, 26
Photon, 187
Plane, 176
Players in games, 203
Point:
   geometric, 176
   in topological space, 273
Polynomial, 104
   in several indeterminates, 105–106
Polynomial function, 107
Position:
   in game, 203
   of letter occurrence, 298

   in matrix, 171
Positive, 75, 129, 132
Power ideals, 100
Power set, 3
   functor, 317
Precedence, 52
Predecessor ordinal, 14
Preorder, 33
   closure, 40
Preordered set as a category, 316
Prime element, 113
Primitive element, 141
Principal:
   filter, 269
   ideal, 97
   ring, 97
Product:
   cardinal, 19
   closure under, 320
   in concrete category, 319
   of groupoid subsets, 61
   of ideals, 98
   inner or dot, 170
   of integers, 77
   in monoids, 63–64
   in rings, 93
   of sets, 18
   topology, 278
   vector space, 157
Projection, 18
   cone, 318
Projections in vector space, 179–180
Projective geometry, 176
Project schedule, 52
Proof, 301
Proper:
   filter, 269
   ideal, 95
   subgraph, 199
   subgroup, 65
   subset, 3

Quotient:
   group, 67
   groupoid, 61
   ring, 96

INDEX OF DEFINITIONS **335**

set, 35
  of universal algebra, 294
  vector space, 167

Radicals, 141
  solvability by, 150
Range, 7
Rank, 249
Rational fractions, 128
Rational numbers, 128
Real:
  numbers, 132
  topology, 273
  uniformity, 281
Reference system, 188
  optical 189
Refinement of:
  filter, 271
  topology, 273
Refinement order, 273
Reflexivity, 10, 13, 33
Regular graph, 235
Regular topology, 283
Relation, 31
Relational structures, 31
  category of, 316
Relation-preserving map, 32
Relative:
  closure operator, 246
  height, 220
  splitting field, 145
Remainder, 121
Representation of matroids, 257
Restriction of:
  function, 6
  matroid, 253
  relation, 31
Result of:
  game strategies, 204
  greedy procedure, 263
Rings, 93
  associated with Boolean lattices, 232
  category of, 315
  formal theory of, 303
  of sets, 26
  as universal algebras, 292

Root, 110
  of unity, 141
Row, 171, 172

Scalar, 157
Scenario, 203
  compatible with strategy, 203–204
Schedule, 52
Second kind ordinal, 14
Segment, 45
Semantic:
  consequence, 305
  value, 304
Semigroups, 62
  category of, 315
  formal theory of, 302
Semilattices, 43
Sentence, 300
Separation in topological space, 283
Sequence, 18
  limit point of, 275
Sets, 2
  category of, 314
Seven, 12
Similar:
  cycle structures, 83
  geometries, 177
Simple:
  group, 87
  field extension, 141
Singleton, 3
Six, 12
Slope, 189
Small category, 317
Smaller element, 35
Smallest element, 35
Solution of:
  linear equations, 173
  LP problem, 184
Solvability by radicals, 150
Solvable group, 146
Space:
  geometric, 177
  metric, 282
  uniform, 281
  of vectors, 157

Spacetime universe, 187
Spanning subgraph, 199
Speed, 189
Splitting field:
   absolute, 149
   relative, 145
Square matrix, 173
Square root, 135
Stabilizer, 72
Standard:
   rational order, 129
   real order, 132
   real topology, 273
   real uniformity, 281
Start date, 52
Step differential, 222
Strategic couple, 204
Strategy, 203
   in equilibrium, 204
   compatible with scenario, 203–204
Stripping functor, 317
Strongly acyclic relation, 49
Structures, 315–316
   category of, 315
   domain and codomain, 316
   initial and terminal, 318
   product, 319
Subalgebras, 292
   closure under formation of, 320
Subcategory, 314
Subcover, 277
Subfield, 137
Subgraph, 199
Subgroup, 65
Subgroupoid, 60
Sublattice, 214
Submonoid, 62
Subring, 94
Sub-semigroup, 62
Subset, 3
Subspace, 159
Substitution, 300
Successor ordinal, 12
Sum:
   Boolean, 22
   cardinal, 23
   ideal, 97
   integer, 76
   in a ring, 93
   of vectors 157
Superset, 3
Support, 168
Surjection, 7
Symmetric difference, 22
Symmetric group, 68
Symmetry, 10, 33

Tautology, 306
Term, 296
Terminal:
   positions in games, 203
   structure, 318
Ternary operation, 25
Theory, 301
Three, 12
Tight constraint, 184
Time, 188
   shift, 191
Topological spaces, 272–273
   category of, 316
Topology, 273
Total:
   comparability of ordinals, 13
   order, 33
Trace of reference system, 188
Transitive closure, 40
Transitivity, 10, 13, 33
Translate, affine, 178
Translation, 71
Transposition, 82
Tree, 199
   underlying a game, 203
Triangle inequality:
   in graph, 198
   metric, 282
Triple, 18
Trivial:
   group, 65
   ideal, 95
   lattice, 214
   ring, 94
   subspace, 159

topology, 273
uniformity, 281
Truth, 304
Tutte polynomial, 254
Twelve, 323
Two, 12
Type of algebra, 291

Ultrafilter, 269
Unary operation, 25
Uniform:
  matroid, 246
  topology, 283
Uniformity, 281–282
Union, 4, 18
Unit, 112
Unit equivalence, 113
Universal algebras, 291
  category of, 314
Universe, 312
Upper:
  bound, 36
  covering condition, 220
  section, 44

Validity, 304
Value, 6
  of a polynomial, 107
  semantic, 304
  of Tutte polynomial, 256
Variable, 296
Variety of algebras, 315
Vectors, 157
Vector spaces, 157
  category of, 316
Velocity, 189
Vertex, 197
Vertex interval, 198
Void set, 3

Well-ordered set, 37
Word, 295
Worldpoint, 187

Zero, 12
  polynomial, 104
  in a ring, 93

INDEX OF NOTATION

$A \subset B$	Proper subset	3
$\mathcal{P}(A)$	Power set	3
$\cup A$	Union of the members of A	4
$\langle a,b \rangle$	Ordered pair	5
B^A	Function set, exponentiation	6
id_A	Identity map	6
$f^{\text{inv}}[T]$	Inverse image of a set	7
f^*	Inverse map	8
$A \simeq B$	Equipotence of sets	9
α'	Successor ordinal	12
$<$	Less than	13, 35
$\leq$	Less than or equal to	13, 33, 75, 129, 132
ω	Set of natural numbers	15
$\mathcal{P}^*(S)$	Set of nonempty subsets	16
Card S	Cardinality of a set	17
$(u\,v)$	Couple	18
$(u\,v\,w)$	Triple	18
$(u_0, u_1, \ldots, u_i, \ldots)$	n-tuple	18
pr_j	Projection map	18
$\cap E$	Intersection of the members of E	21
$+$	Cardinal sum, Boolean sum	23
$\overline{A}$	Closure of a set	24
(A, R)	Relational structure	31
Aut R	Automorphisms	32, 95
aRb	Binary relation	33
$(A, \leq)$	Ordered set	33
$(A, \subseteq)$	Inclusion-ordered set	33

INDEX OF NOTATION

A/E	Quotient set	35
$\lesssim$	Preorder	35
$a \sim b$	Equivalence in a preorder	35, 111
$a \parallel b$	Incomparability	35
R^*	Dual relation	35
$\gtrsim$	Dual preorder	35
$\geq$	Dual order	35
glb	Greatest lower bound	36
lub	Least upper bound	36
$\mathcal{P}r(A)$	Set of preorders	39
$\mathcal{T}r(A)$	Set of transitive relations	40
$x \wedge y$	Meet in lower semilattice	43, 213
$x \vee y$	Join in upper semilattice	43, 213
$\wedge B$	glb in complete lattice	43
$\vee B$	lub in complete lattice	43
$[x,y]$	Order segment	45
$[x, \rightarrow)$	Upper section	45
$(\leftarrow, x]$	Lower section	45
$x \prec y$	Covering relation	49
$(A/E, \odot)$	Quotient groupoid structure	61
$\operatorname{End} A$	Endomorphism monoid	63
a^n	Product of n identical factors	63
a^*	Inverse of a monoid element	64
$x \equiv y \bmod N$	Congruence modulo a normal subgroup	67
A/N	Quotient by a normal subgroup	67
$\Sigma(A)$	Symmetric group	68
Σ_n	Symmetric group on n elements	69
$\begin{pmatrix} 1 & 2 & 3 & 4 \\ 4 & 2 & 1 & 3 \end{pmatrix}$	Permutation	69
S_x	Stabilizer	72
gH	Coset of subgroup	73
$[G:H]$	Index of subgroup	73
ω^+	Set of positive integers	75
ω^-	Set of negative integers	75
$\mathbb{Z}$	Set of integers	75
$-z$	Negative of an integer	75
$d(x,y)$	Distance	76, 198, 281
$m\mathbb{Z}$	Subgroup of the group of integers	78
$\mathbb{Z}_m$	Quotient of $\mathbb{Z}$	79, 96
$i \oplus j$	Integer addition modulo m	79
g^z	Group element with integer exponent	81
f^{-1}	Inverse map	81
$f^{-1}[T]$	Inverse image of a set	81
$(n_1,\ldots,n_k)$	Permutation	82
A_n	Alternating group	85

INDEX OF NOTATION

$(A, +, \cdot)$	Ring structure	93
0_A	Zero element of ring	93
1_A	Identity element of ring	93
$a + I$	Ideal coset	95
$\mathcal{I}(A)$	Set of ideals	96
(a)	Principal ideal	97
$(a_1, \ldots, a_n)$	Finitely generated ideal	97
$I + J$	Sum of ideals	97
IJ	Product of ideals	98
$P(A)$	Ring of polynomials	104
$A[b]$	Subring generated by A and b	106
$A[X]$	Polynomials in indeterminate X	107
f_p	Polynomial function defined by p	107
$a\|b$	Divisibility	111
gcd	Greatest common divisor	113
lcm	Least common multiple	113
F^*	Multiplicative group of field	125
$\mathbb{Q}$	Field of rational numbers	128
$\mathbb{R}$	Field of real numbers	132
$\sqrt{a}$	Square root	135
$|r|$	Absolute value	135
$\mathbb{C}$	Field of complex numbers	141
i	Root of $X^2 + 1$	141
$E : K$	Field extension	141
$\mathcal{G}(A)$	Fixing group	142
$\mathcal{F}(B)$	Fixed field	142
$\mathcal{G}(E : K)$	Galois group	143
$K(a_1, \ldots, a_j)$	Subfield generated by K and the a_i's	147
$\overline{0}$	Null vector	157
$\dim V$	Dimension	163, 179
c_u	Coordinate function	168
F^n	n-dimensional vector space over F	170
$\overrightarrow{a_i}$	Row vector	172
$a_{[j]}$	Column vector	172
h_a	Linear function defined by matrix a	172
Pro$V$	Projective geometry	177
$v + A$	Affine translate	178
Aff$V$	Affine geometry	179
(f_i, b_i)	Linear programming constraint	184
c	Lightspeed	187
L	Photon of velocity c	187
L^-	Photon of velocity $-c$	187
O	Optical causality relation	187
C	Causality	187
M	Material causality	187
$\mathcal{L}_2$	Lorentz group	190

INDEX OF NOTATION

$G = (V, E)$	Graph	197
$I(a, c)$	Vertex interval	198
$\leq_u$	Basepoint order	200
K_V	Complete graph	207
$G - e$	Edge-deleted graph	208
$G \cdot e$	Edge-contracted graph	208
D_5	Nondistributive lattice	219
M_5	Nonmodular lattice	219
$h(a, b)$	Relative height in discrete order	220
x'	Complement in Boolean lattice	230
B_n	Set of Boolean functions	231
x_i	Literal	235
$h(a)$	Height in geometric lattice	241
σ_M	Relative closure operator	246
(M, σ)	Matroid	246
$A + x$	Union of a set and a singleton	246
$A - x$	Difference of a set and a singleton	246
$\mathrm{gd}\, S$	Geometric dimension	249
$M \| S$	Matroid restriction to set S	253
$M - T$	Matroid obtained by deleting set T	253
$M : S$	Matroid contraction to set S	253
$M \cdot T$	Matroid obtained by contracting set T	253
$M - e$	Deletion of singleton	253
$M \cdot e$	Contraction of singleton	253
$T(M)$	Tutte polynomial	254
$Q(M, g)$	Chromatic polynomial	259
$(T, \mathcal{C})$	Topological space	272
$\mathcal{N}_x$	Neighborhood filter	276
$(V, \mathcal{U})$	Uniform space	281
$N_E(x)$	Neighborhood in specified entourage	282
$(n(i) : i \in I)$	Algebra type	291
$A(b : B)$	Word substitution	300
$\cong$	Equality symbol	300
1_L	Boolean lattice maximum	306
0_L	Boolean lattice minimum	306
$\mathbf{U}$	Inaccessible universe	312
$Comp$	Set of composable morphism couples	313
$g \circ f$	Composition of morphisms	313
(B, h, A)	Morphism in concrete category	316
$D(p_i : i \in I)$	Domain structure of projection cone	318
$T(p_i : i \in I)$	Codomain structure of injection cone	322

INDEX OF THEOREMS

A1 Empty Set Axiom, 3
A2 Subset Axiom, 3
A3 Power Set Axiom, 3
A4 Pair Axiom, 3
A5 Union Axiom, 4
A6 Axiom of Existential Infinity, 14
A7 Axiom of Limited Infinity, 15
A8 Axiom of Choice, 15
A9 Inaccessibility Axiom, 312

Abel–Ruffini, 152
Algebraic Closure, 293
Alternating Groups, Simplicity, 87

Basis, Dual, 169
Basis Characterization, 160
Basis Counting, 256
Basis Equipotence, 162, 249
Basis Existence, 161
Borel–Heine–Lebesgue, 287

Cayley Representation Theorem, 72
Chinese Remainder Theorem, 99
Chromatic Polynomial, 209
Compact Uniformity, 285
Completeness of Equational Calculus, 305
Coordinatization, 168
Countable Universe Lemma, 311

Counting Lemma, 15
Covering Lemma, 220
Cycle Representation, 83
Cyclic Group Structure, 80

De Morgan's Laws, 232
Disjoint Copy Lemma, 26
Dual Basis, 169

Equilibrium Theorem, 205
Euclidean Algorithm, 121
Exchange Theorem:
  Matroid, 248
  Steinitz, 161

Factorization, 116
Forbidden Sublattice Criterion, 218
Forward Pass Scheduling, 52
Fraction Field, 125
Free Algebra, 298

Galois Quotient Theorem, 145
Greedy Characterization of Matroids, 264

Hausdorff Separation, 284
Helly Theorem for:
  Ideal Cosets, 101
  Intervals, 55
  Median Graphs, 202

INDEX OF THEOREMS

Helly Number Two, Lemma, 56
Helly's Theorem, Convex Sets, 165
Hilbert's Transfer Theorem, 109
HSP, 321

Imaginary Root, 141
Injection–Extension Lemma, 26
Integer Factorization, 118
Interval, 224

Jordan–Dedekind Chain Theorem:
   Geometric Lattices, 241
   Modular Lattices, 238

Lagrange's Subgroup Counting
   Theorem, 73
Lemma for Helly Number Two, 56
Linear Conjunction, 41
Linear Extension, 40
Linear Map Dimension Theorem, 168

Matching Matroid, 266
Matroid Exchange, 248
Matroid Interval, 253
Matroid Lattice, 250
Maximal Chain, 39
Maximal Ideal, 98
Median Diagram, 225
Median Equality, 220
Modular Characterization, 237

Nested Union, 46

Orbit Counting, 72

Polynomial Factorization Theorem, 120

Product–Sum Duality, 323
Pythagoras, 171

Radical Extensions, Solvability, 146
Radon, 164
Real Field Structure, 132
Recursion Formulas for:
   Chromatic Polynomials, 209
   Tutte Polynomials, 254
Regular Diagram, 235
Regular Separation, 283
Representation Theorem, 233

Separation Theorem:
   Hausdorff, 284
   Regular, 283
Simplicity of Alternating Groups, 87
Solvability Theorem for Radical
   Extensions, 146
Splitting Field, 149
Square Root, 134
Steinitz Exchange Theorem, 161
Subgroup Counting Theorem,
   Lagrange, 73

Term Structure, 298
Transfer Theorem, Hilbert, 109
Tychonoff Product Theorem, 279

Unique Complementation, 215

Velocity Composition, 193

Zermelo, 16, 37
Zorn's Lemma, 37